HEALTHCARE TECHNOLOGIES SERIES 31

Technologies and Techniques in Gait Analysis

Other volumes in this series:

Volume 1	**Nanobiosensors for Personalized and Onsite Biomedical Diagnosis** P. Chandra (Editor)
Volume 2	**Machine Learning for Healthcare Technologies** D.A. Clifton (Editor)
Volume 3	**Portable Biosensors and Point-of-Care Systems** S.E. Kintzios (Editor)
Volume 4	**Biomedical Nanomaterials: From Design To Implementation** T.J Webster and H. Yazici (Editors)
Volume 6	**Active and Assisted Living: Technologies and Applications** F. Florez-Revuelta and A.A. Chaaraoui (Editors)
Volume 7	**Semiconductor Lasers and Diode-based Light Sources for Biophotonics** P.E. Andersen and P.M. Petersen (Editors)
Volume 9	**Human Monitoring, Smart Health and Assisted Living: Techniques and Technologies** S. Longhi, A. Monteriù and A. Freddi (Editors)
Volume 13	**Handbook of Speckle Filtering and Tracking in Cardiovascular Ultrasound Imaging and Video** C.P. Loizou, C.S. Pattichis and J. D'hooge (Editors)
Volume 14	**Soft Robots for Healthcare Applications: Design, Modelling, and Control** S. Xie, M. Zhang and W. Meng
Volume 16	**EEG Signal Processing: Feature extraction, selection and classification methods** W. Leong
Volume 17	**Patient-Centered Digital Healthcare Technology: Novel applications for next generation healthcare systems** L. Goldschmidt and R. M. Relova (Editors)
Volume 19	**Neurotechnology: Methods, advances and applications** V. de Albuquerque, A. Athanasiou and S. Ribeiro (Editors)
Volume 20	**Security and Privacy of Electronic Healthcare Records: Concepts, paradigms and solutions** S. Tanwar, S. Tyagi and N. Kumar (Editors)
Volume 23	**Advances in Telemedicine for Health Monitoring: Technologies, design and applications** T.A. Rashid, C. Chakraborty and K. Fraser
Volume 24	**Mobile Technologies for Delivering Healthcare in Remote, Rural or Developing Regions** P. Ray, N. Nakashima, A. Ahmed, S. Ro and Y. Soshino (Editors)
Volume 26	**Wireless Medical Sensor Networks for IoT-based eHealth** F. Al-Turjman (Editor)
Volume 29	**Blockchain and Machine Learning for e-Healthcare Systems** B. Balusamy, N. Chilamkurti, L.A. Beena and P. Thangamuthu (Editors)
Volume 33	**Electromagnetic Waves and Antennas for Biomedical Applications** L. Wang
Volume 34	**The Internet of Medical Things: Enabling technologies and emerging applications** S.K. Pani, P. Patra, G. Ferrari, R. Kraleva and H. Wang (Editors)
Volume 38	**Healthcare Monitoring and Data Analysis using IoT: Technologies and applications** V. Jain, J. M. Chatterjee, P. Kumar and U. Kose (Editors)
Volume 39	**Digital Tools and Methods to Support Healthy Ageing** P.K. Ray, S.-T. Liaw and J.A. Serano (Editors)
Volume 42	**Smart Health Technologies for the COVID-19 Pandemic: Internet of medical things perspectives** C. Chakraborty and J. J. P. C. Rodrigues (Editors)

Technologies and Techniques in Gait Analysis

Past, present and future

Edited by
Nachiappan Chockalingam

The Institution of Engineering and Technology

Published by The Institution of Engineering and Technology, London, United Kingdom

The Institution of Engineering and Technology is registered as a Charity in England & Wales (no. 211014) and Scotland (no. SC038698).

First published 2022

The Institution of Engineering and Technology
Michael Faraday House
Six Hills Way, Stevenage
Herts, SG1 2AY, United Kingdom

www.theiet.org

British Library Cataloguing in Publication Data
A catalogue record for this product is available from the British Library

ISBN 978-1-83953-131-6 (hardback)
ISBN 978-1-83953-132-3 (PDF)

Typeset in India by MPS Ltd
Printed in the UK by CPI Group (UK) Ltd, Croydon

Contents

About the editor

Nachiappan Chockalingam is a Professor of Clinical Biomechanics and Director of the Centre for Biomechanics and Rehabilitation Technologies at Staffordshire University, UK. He is a chartered engineer, registered clinical scientist, a Fellow of the Institute of Physics and Engineering in Medicine and a member of numerous professional organisations. He holds visiting professorships in Malta, India and China. He has over 700 research outputs, including over 240 full research manuscripts and he reviews for numerous journals and grant-awarding bodies worldwide. His research has received funding from various international bodies including the European Commission and British Council. He is involved in charitable and non-profit organisations to help the wider global community on healthy ageing and assistive technology which aid mobility. His current activities at Staffordshire University focuses on translational research. He has played a pivotal role in bringing various allied health professionals to the wider biomechanics and medical engineering community.

Chapter 1
Introduction to gait analysis

Kristiaan D'Août[1], Aoife Healy[2], Nicola Eddison[2] and Nachiappan Chockalingam[2]

Gait is the coordinated movement of limbs we use to move from one place to another (locomotion). There are many possible types of gait, as horse riders will testify, but humans only use a few: walking and running (and occasionally skipping). These gaits can be easily distinguished; usually, this is done by determining the timing of footfalls (when our feet touch the ground and when they leave the ground). In this chapter, we will focus on walking, because of its high clinical relevance.

Simple as the classification of gaits seems, there is however much more to it. The variation within gait types is large in healthy people and much more in people with gait disorders. The gait of healthy children and adults has been well documented [1–5]. This chapter will set out how to quantify this variation, which is the subject of gait analysis (see Chapter 2 for a historic introduction). It is worth mentioning that the movements we see as "gait" are the result of complex motor patterns generated at different anatomical levels, i.e., by the Central Pattern Generator in the spine and with additional input from the brain and reflex loops. This book will not focus on control aspects of gait, or on how to measure them directly, but it is important to bear in mind that gait problems, although they can have a mechanical cause (e.g., in amputees) very often have a neurological cause (for a good introduction on motor control see e.g., [6,7]).

1.1 Describing the basic gait pattern: spatiotemporal variables

Human gait is a cyclical movement, in which one complete cycle is called a stride. Conventionally, the gait cycle starts with the initial ground contact of one foot (we will use the left foot as an example) and finishes when the same foot contacts the ground again. At its most basic, the gait cycle contains two phases for each foot: stance, where the foot is in contact with the ground, and swing, where the foot

[1]Department of Musculoskeletal and Ageing Science, Institute of Life Course & Medical Sciences, Faculty of Health & Life Sciences, University of Liverpool, Liverpool, UK
[2]Centre for Biomechanics and Rehabilitation Technologies, Staffordshire University, Stoke on Trent, UK

moves forwards to prepare for the next ground contact. When walking, the contralateral side moves in anti-phase: the right foot has the same footfall pattern as the left foot, except that it happens half a stride duration later (of course in reality there is always a small variation). This is what we call a symmetrical gait. The stance phase of a stride lasts longer than the swing phase and therefore there are two phases in the gait cycle when both feet are on the ground, and there is always at least one foot on the ground. Running is also a symmetrical gait, but stance lasts shorter than swing and therefore there are two phases in the gait cycle where there is no ground contact at all, and during the other parts of the gait cycle there is one foot on the ground (Figure 1.1). Please note that this classification of gait is based on footfall patterns and there are other, more functional, ways to distinguish walking from running [8].

The gait cycle is often further subdivided into sub-phases of stance and swing, for example, initial contact, loading response, pre-swing, initial swing, etc. When focussing on the foot, one might use sub-phases of stance such as heel strike, foot flat and push-off. However, these sub-phases are not always universal (e.g., initial contact might not be done with the heel) and it is important to define them when used.

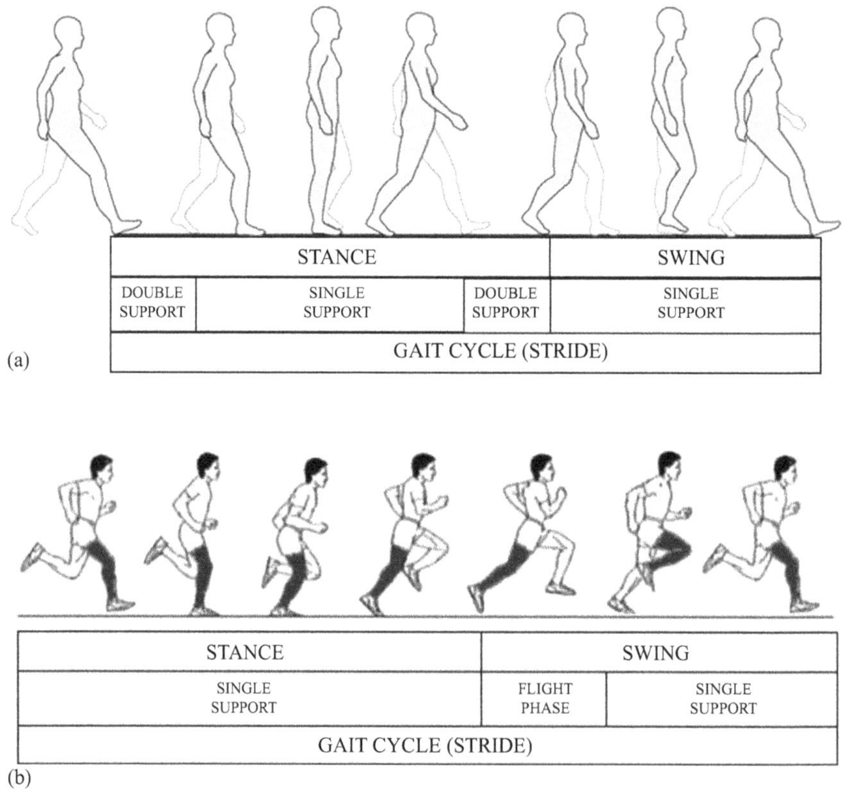

Figure 1.1 The gait cycle: footfall patterns (a) walking and (b) running

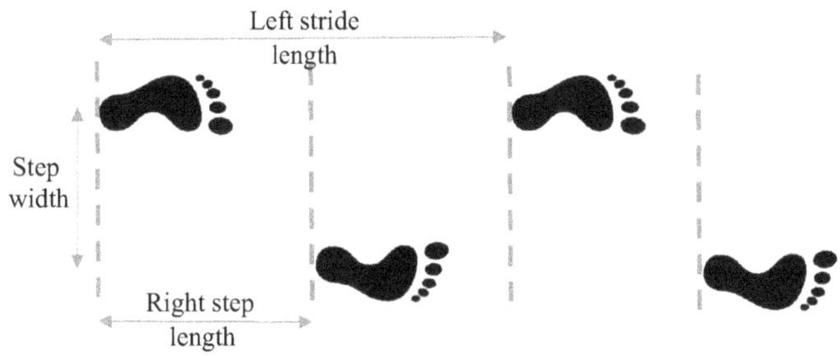

Figure 1.2 Spatiotemporal variables

The most basic description of gait uses spatiotemporal parameters (Figure 1.2). They can be calculated from nothing more than foot touch-down and lift-off times and locations as follows:

- **Stride length** (units: m): the distance travelled during one stride. This can be calculated as the displacement of the body's centre of mass (CoM) during one stride, but also by measuring the distance between the position of a footfall and the position of the next footfall by the same foot. In practice, these two methods will yield very similar results.
- **Stride duration** (units: s): the duration of one stride.
- **Stride frequency** (units: s^{-1}, or Hz): the number of strides taken per unit of time, typically calculated as 1/stride duration.
- **Speed** (units: m/s): speed can be directly measured or calculated as stride length divided by stride duration (or multiplied by stride frequency). Speed can also be calculated over a longer time, but it is ideally done over an integral number of strides, since the instantaneous speed fluctuates over the course of a stride.
- **Stance time** (units: s): the time a foot is on the ground during a stride. Measured from initial contact (in walking this would typically be heel strike) to final contact (in walking this would typically be toe-off).
- **Swing time** (units: s): the time a foot is off the ground during a stride. Measured from final contact to initial contact.
- **Duty factor** (dimensionless, i.e., has no units): the proportion of time a foot is on the ground. Calculated as stance time/stride duration. From the footfalls-based definition of walking and running, it follows that walking has a duty factor greater than 0.5 and running has a duty factor of less than 0.5. The duty factor is often expressed as a percentage.
- A **step** can be defined in two different ways. In the first way, a step is the event starting at initial contact of one foot and ending at initial contact of the contralateral foot (as shown in Figure 1.2). Following this definition, in a symmetrical gait, step length and step duration would be half of stride length and stride duration. A second, alternative definition of a step is by using the distance

travelled by the CoM during stance, and its corresponding duration [9]. In this case, during walking, step length would be more than half of stride length. When publishing step length, it is important to make clear how a step is defined.
- **Step width** (units: m): the medio-lateral distance between the left and right foot.

Spatiotemporal variables form the basis of gait analysis but in most cases, we need more detail than is available from footfall patterns only: we need to know how our segments and joints move.

1.2 Segment and joint kinematics

Kinematics deals with motion (but not the forces causing it) and includes spatio-temporal variables but also includes the movements of segments (e.g., the shank, or the trunk) and joints. These movements can be linear (translations), e.g., the instantaneous speed of the head, or angular (rotations), e.g., the trunk angle, or the knee angle as a function of time. Both linear and angular motions can and do happen in three dimensions, and therefore a full kinematic description requires the knowledge of all six degrees of freedom (6-DoF) as a function of time.

In gait analysis, the three-dimensional space is defined by a global reference frame, typically (but not obligatory) consisting of three orthogonal axes. The point of origin is arbitrary and will often be determined by the measurement system and form the origin of a 'lab based' reference frame. The definition of the three axes is also arbitrary, but in gait analysis the X-axis is typically chosen to be along the overall direction of travel (or long axis of the walkway in a lab). The Y-axis can be the lateral axis, or the vertical axis (as recommended by the International Society for Biomechanics, ISB; [10]) and the Z-axis will be the remaining axis. The signs are typically dictated by a right-handed triad: if positive X is the walking direction, then following ISB recommendations, positive Y will be up, and positive Z to the right. Different labs and different equipment manufacturers use different conventions and therefore it is important to specify them in any reporting.

Similar to the global reference frame, which is fixed to the earth (or the lab), we can define local reference frames which are fixed to body segments (Figure 1.3). This can also be done in many ways and it is therefore important to specify in any reporting (see ISB recommendations: [11,12]). It is very convenient that, once a global and a set of local reference frames are defined, full 6-DoF kinematics can be calculated using standard mathematical methods. Kinematics then becomes, in essence, a description of how the local reference frames move with respect to each other, and with respect to the global reference frame. For example, the forward speed of the head would be the translation of its local reference frame along the X-axis of the global reference frame. The hip abduction angle would be the rotation of the local reference frame of the thigh relative to that of the pelvis, around the X-axis of the pelvis local reference frame (if defined as in the ISB recommendations).

It is intuitive to think of rotations as being around three axes, and this often makes anatomical sense. For example, one axis would represent flexion–extension, one axis adduction–abduction and the third internal–external rotation. However, this

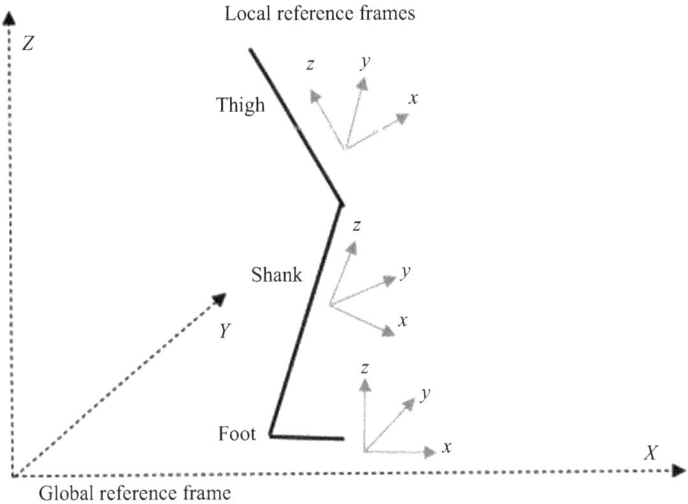

Figure 1.3 Reference frames

method, using the three so-called Euler angles, can generate mathematical problems due to the so-called gimbal lock effect. This happens when a rotation makes one axis effectively parallel to another, and therefore with only two different axes left, a full 3-D movement becomes impossible. This effect can be mimicked using a three-axis photo tripod head. One solution to the gimbal lock problem is to use quaternions, rather than Euler angles. Kinematics software will use quaternions 'under the bonnet'; most researchers will not actively use them for most applications. Once the calculations are done, the quaternion values (which are non-intuitive) can be unequivocally converted into easier-to-understand Euler values for reporting.

For a detailed account of kinematics, we refer to Chapter 3.

1.3 Overview of kinematic techniques used in gait analysis

Spatiotemporal variables are very easy to measure. Indeed, when no specialist equipment is available, they only require (inked) footprints on the ground for the spatial measures and a stopwatch for the temporal measures. In practice, it is of course desirable to use more advanced equipment to improve data quality and workflow, and to obtain full 6-DoF kinematics.

This book is entirely dedicated to such techniques and, in this introductory chapter, we therefore only provide a basic overview.

1.3.1 Video

Video cameras produce a series of stills, and therefore they fundamentally work like the systems pioneered in the nineteenth century by Muybridge and Marey (see Chapter 2).

The main advantages of video cameras are their availability and low cost, and even the ones in our phones are of sufficient standard to carry out basic gait analysis. In addition, video is non-invasive and easy to use. However, there are several disadvantages of using video for gait analysis. First, they produce 2D footage, so to yield 3D data at least two video cameras are needed; they need to be synchronised and calibrated. This is possible but not trivial. Second, most video cameras are limited in spatial resolution (the number of pixels) and temporal resolution (the number of frames per second); in many cameras, the user can choose a compromise, but only professional cameras will have both a high spatial and temporal resolution. Third, video footage needs to be processed to extract kinematic parameters. This typically requires the manual or semi-automatic determination of landmark points and tracking them over time, which is a time-consuming process. Despite its disadvantages, video can be a suitable technique for basic gait analysis, e.g., in athletic coaching, basic clinical assessment, animals or for student projects.

1.3.2 Camera-based motion capture systems

The most common technique to quantify gait kinematics is to use a dedicated motion capture system. These come in many configurations, but they usually consist of several cameras, in which each produce 2D footage, and reconstruct 3D data by combining the signals from the individual cameras. The user is usually not much involved in this process.

Most motion capture systems only track markers attached to the subject by the researcher. The choice of marker set is crucial and determines what variables can later be reliably calculated [13–16]. Individual markers are put on anatomical landmarks, so that their location can be tracked, and also clusters of three or more ‘tracking’ markers are placed on the segments, so that the segments’ 6D position (translations and rotations) can be unequivocally measured.

Two types of markers exist. Passive markers are highly reflective spheres that reflect light produced by the cameras (usually invisible infrared) back into the camera. The advantage is that these markers are relatively inexpensive, and many can be used on the body (50–100 is common, depending on the application). One disadvantage is that the motion capture system is naïve to the anatomical location of these markers, so this needs to be assigned by the researcher, usually in post-processing. The spatial resolution of these systems varies but sub-mm accuracy and frame rates of 100–500 Hz are typical. Active markers are powered units that produce uniquely coded signals which are then captured by the cameras. This has the advantage that the motion capture system ‘knows’ how to differentiate markers and, unlike with passive markers, they cannot be confused. Systems with active markers tend to have even higher spatial accuracy than those with passive markers, but if a lot of markers are used their temporal resolution drops, which does not happen with passive systems.

In most cases, several cameras are set up by the user to suit their needs. When collecting data in a relatively small and defined volume, a smaller number of cameras can suffice. Even though in principle two cameras suffice to get 3D data, in reality more are needed to deal with obscured markers and to improve reconstruction accuracy; therefore, six cameras are a practical minimum for gait analysis. When increased accuracy is required, a large volume needs to be measured, or dropouts (invisible markers) are to be minimised, many more cameras can be used. Once the cameras are set up, the system needs to be calibrated so that the relative position of the cameras to each other is known, and so that the global (lab) reference frame is known to the system. Practically, the latter is done by capturing a physical reference frame to define the three lab axes and their orientation, and the former by a dynamic process in which an object of precisely known length (usually two markers) is captured as it is moved within the entire measurement volume.

Systems exist in which multiple cameras come precisely mounted within one unit; in this case, the system is factory calibrated and can be positioned at will by the researchers, or even be moved while capturing data.

A recent development is markerless motion capture. In this technique, video signals are used to estimate pose, using complex algorithms, and calculate kinematic variables without the need to attach markers onto the body. This technique is well known from gaming consoles, which have also been used in gait research for non-critical purposes. However, recently, dedicated scientific systems with much improved reconstruction accuracy and good temporal resolution are becoming available. It is highly likely that such systems will gradually replace marker-based systems.

1.3.3 Non-camera-based motion capture systems

Systems based on electromagnetic tracking, rather than cameras, have very high spatial and temporal resolution. Most systems are wired, and are ideally suitable for small, confined movements (such as hand kinematics during fine manipulation tasks). However, systems are available that can track whole-body kinematics, but they are not very commonly used.

All of the above systems collect positional data: X, Y and Z positions and displacements of points.

From positional data, velocities, accelerations and jerk (the change in acceleration over time) can be calculated using numerical differentiation. This is mathematically very simple with the main challenge being that any noise in the positional data will be amplified during differentiation, and therefore low-pass filtering of the data is necessary.

Kinematic analysis can be carried out in a variety of software packages. Motion capture software usually has basic built-in analysis options, and further processing can be done by exporting data into a spreadsheet. However, dedicated biomechanics software can greatly increase the analysis and reporting possibilities as well as workflow efficiency, especially when working with large data sets. Many laboratories also use in-house developed software for very specific purposes. Chapter 6 outlines considerations for data analysis in more detail.

It is possible to obtain kinematics by using an approach which is the reverse of the method described above, i.e., by measuring acceleration and calculate velocity and displacement (position) by numerical integration. This is not trivial because the acceleration data measured will be the sum of gravity and actual acceleration, it requires the incorporation of an integration constant (which is usually not explicitly known) and any drift in the signal will have a great influence on the data. For this reason, this method typically involves the synchronous measurement of several types of raw data by inertial measurement units (IMUs). These data are as follows: accelerometry (for accelerations), gyroscopy (for relative rotation), magnetic field sensors (for absolute angular position) and global positioning system (GPS) (for absolute position). Even though the individual signals of these sensors are imperfect, data fusion algorithms can combine these signals into usable kinematic data. IMU-based systems are increasingly common and have the potential to be low-cost (even inexpensive smart phones contain a full set of IMU sensors).

1.4 Gait analysis – beyond kinematics

1.4.1 Measuring forces

Motions are caused by forces and in many cases, we need to measure these. This is the field of kinetics and is discussed in detail in Chapter 4. In gait analysis, where the feet are the only mechanical interaction with the environment, we are interested in the ground reaction forces (GRFs). These GRFs are equal in magnitude but opposite in direction to the forces exerted by the body upon the ground, but the use of GRFs is standard because these forces can directly be used as the 'external' forces in biomechanical modelling.

GRFs are measured by means of force plates. These should be mounted flush into the ground (or in a raised platform if need be) and allow measurement of 6-DOF kinetics, i.e., 3D forces and 3D moments. With this information, the point of application of the force can be calculated. Force plates can be based on strain gauge or piezo-electric technology (each with its own pros and cons) but both types will yield GRF data of very high spatial and temporal resolution for use in gait analysis.

1.4.2 Combining kinematics and kinetics – modelling

Having kinematic as well as kinetic data for a gait trial is highly useful; not only for the data as such, but the combination of both types of data opens up very useful modelling approaches. A detailed discussion on modelling is beyond the scope of this chapter, but we will briefly discuss the most common modelling technique in gait analysis: inverse dynamics analysis. Advanced biomechanical software packages will perform such analysis, but it is important to understand the underlying principles.

Inverse dynamics is used to calculate the forces and moments at the joints, rather than measure them directly and invasively, which would be impossible in research and clinical practice. Inverse dynamics allows, for example, to calculate the net moments produced by the hip, knee and ankle joints during walking. Combined with kinematic data (i.e., angular joint velocity), this allows to determine power and work produced by these joints.

Inverse dynamics is a form of multibody analysis, in which the human body is treated as a set of rigid segments (e.g., the thigh, the shank, etc.) which are each treated as 'free bodies', i.e., their movements are completely dictated by their inertial properties (mass and moment of inertia) and by the forces acting upon them. In live subjects, inertial properties can be determined using scanning techniques (e.g., DEXA: dual-energy X-ray absorptiometry), but in practice they are nearly always estimated using previously established anthropometric relationships. Whilst these relationships provide a rough approximation only, the error they induce is relatively small when the accelerations are small, as in walking.

The forces acting upon the body segments (called 'external forces') have four possible components, which together form the so-called Newton–Euler equations. The first force, which acts upon the CoM of the segments, is the force due to gravity. This force can be easily calculated by multiplying the mass of the segment with the gravitational acceleration constant. The second force includes the GRF acting on the feet (but can also include other external forces, e.g., forces acting upon crutches). This force is directly measured using a force plate. In addition, this force also includes the forces exerted by one segment onto the next (e.g., by the shank onto the thigh). The third force, due to inertia, acts on each body segment and is due to their mass being accelerated (and the rotational equivalent due to an angular acceleration of the moment of inertia). This force can be calculated based on the estimate of segment inertial properties, and their acceleration (from kinematics) using Newton's second law (force equals mass times acceleration). The fourth force is due to centrifugal and coriolis forces; it is small during gait and can in many cases be ignored.

1.4.3 Measuring pressures

Pressure plates are commonly used, either on their own or combined with force plates. Such plates allow the measurement of plantar (foot) pressures thereby, unlike force plates, showing which anatomical regions of the foot sole are subject to the highest loads. This information can be very valuable, e.g., in clinical research, particularly where the tissue of the foot is compromised and at risk of ulceration when subject to high or prolonged pressure during gait, or in footwear design.

By integrating the pressures over the total foot area, the force can be calculated, but this will be less accurate than when it is measured directly using a force plate. It also only has the vertical force component (because pressure is, by definition, perpendicular to the surface), and no moments. Therefore, if accurate and full 3D forces are needed, as for inverse dynamics, pressure plates will not suffice.

Pressure systems also exist as insoles, which allow the measurement of in-shoe pressures and thus the interface of the foot and the shoe, unlike plates which measure the interface between the shoe and the ground (albeit more accurately), when not walking barefoot. In-shoe pressure devices are particularly useful in a clinical setting, negating the need for the service user to visit the gait laboratory.

1.4.4 Measuring physiology

A number of additional techniques are commonly used in gait analysis, many of them measuring physiological aspects related to locomotion.

Electromyography (EMG) measures the electrical signal associated with muscle activation. This requires the insertion of electrodes into the muscles of interest or, much more commonly, the use of non-invasive electrodes attached to the skin which detects the activities of superficial muscles. Most systems are wireless and user-friendly. Electromyography enables the researcher to assess when muscles are active and can also help interpret the results from inverse dynamics (which are net joint moments but do not tell anything about the contribution of individual muscles). The main challenges associated with EMG are the presence of measurement noise and the fact that the absolute values measured have no direct relation to muscle output (e.g., force or power). Relative EMG magnitudes can be normalised, e.g., to maximal voluntary contractions (MVCs) or to the peak value observed during the experiment. Advanced signal processing (e.g., spectral analysis) is also used to gain an insight into the recruitment of muscle motor units. A very useful technique, it is also important to understand the limitations of EMG. It is worth mentioning a related technique, muscle mechanography (MMG), which measures the sound muscles produce when contracting, rather than the electrical signal when they are stimulated, but MMG is much less common than EMG.

Respirometry is a direct measure of energy consumption through oxidative pathways (used for sustained activities like gait). It does so by measuring oxygen consumption and usually, carbon dioxide expulsion. In practice, this technique requires the subject to wear a respiration mask and carry a unit containing measurement electrodes. In many cases, these units are wireless. The advantage of this technique is that it measures actual metabolic energy use. This is usually more relevant than mechanical energy use, measured using kinematics and kinetics, which is only a rough estimate of metabolic energy use. A disadvantage of respirometry is that it can be uncomfortable and interfere with gait.

A variety of additional techniques exist to measure or estimate physiological parameters. Of particular note are wrist-worn devices such as consumer-grade smart watches or research-grade units. These measure parameters such as heart rate or skin resistance (a measure for stress) and can provide useful experimental data in a cost-effective and user-friendly way and allow measurements to be captured in a clinical setting.

For details on physiological measurements for gait and on novel technologies, we refer to Chapters 5 and 7.

1.5 Introduction to healthy gait biomechanics

In this chapter, the gait cycle will be described using the Ranchos Los Amigos terminology [17], which consists of initial contact (approximately 0–2% of the gait cycle), loading response (approximately 2–10% of the gait cycle), mid-stance (approximately 10–30% of the gait cycle), terminal stance (approximately 30–50% of the gait cycle), pre-swing (approximately 50–60% of the gait cycle), initial swing (approximately 60–73% of the gait cycle), mid-swing (approximately 73–87% of the gait cycle) and terminal swing (approximately 87–100% of the gait cycle).

During normal gait, the ankle is positioned at approximately 90° at initial contact, with the knee flexed at approximately 5° and the hip at approximately 30°, initiating the first ankle rocker. The forward reach of the limb tilts the tibia upwards 15° and positions the heel as the lowest segment of the foot [3]. Ankle plantarflexion and eversion are initiated to decelerate the impact of the falling body. During loading response, the knee is flexed to approximately 20°, and 5° of ankle plantarflexion is required to reduce the heel rocker effect so that the tibia doesn't advance too rapidly [3]. If the ankle were to remain at 90°, the tibia would accompany the foot through its rapid arc of motion [3]. A second advantage of plantarflexion at this stage of the gait cycle is shock absorption.

During mid-stance, knee flexion decreases to approximately 15° with the hip moving towards extension, the full foot is in contact with the floor providing a stable position and the second rocker is initiated. Progression of the limb is continued by dorsi-flexion of the ankle moving from an initial 5° plantarflexed to 5° dorsi-flexed. This is followed by a rapid movement into dorsi-flexion of between 10° and 15° which continues through mid-stance. The ability to dorsi-flex the foot during mid-stance is essential to allow the forward rotation of the tibia and forward progression of the trunk over the stance limb.

As terminal stance begins, the femur continues to advance over the stable tibia, knee flexion is reduced and maximum knee extension occurs (approximately 5°) at approximately 40% of the gait cycle. The heel begins to rise, the sub-talar joint moves into an inverted position locking the mid-tarsal joints. The forefoot is now supporting the bodyweight. The metatarsal-phalangeal joints (MTPJs) and phalanges provide the third foot rocker allowing the body vector to advance for continued progression. Peak hip extension occurs at 50% of the gait cycle.

During pre-swing, passive knee flexion occurs and the hip moves into approximately 10° extension. In pre-swing, the MTPJs continue to extend combined with a change in the position of the ankle from a dorsi-flexed position at the end of terminal stance to a plantarflexed position at the end of pre-swing. The bodyweight is transferred from the trailing limb to the forward limb during this period of double foot to floor contact [3]. During the remainder of the swing phase, knee and hip flexion peak and the ankle is held at 90° to clear the floor.

We can identify mid-stance as a particularly difficult stage of the gait cycle; as the limb is in single support, stability is crucial. Of the five pre-requisites for normal gait [3], it is stability that also contributes to the other four factors [2]. It is the importance of stability that highlights mid-stance as being a crucial aspect of the gait cycle as it is at this stage of the gait cycle when the foot is in full contact with the floor, producing a stable alignment of the foot segment coupled with a virtually stationary and inclined shank at temporal mid-stance (TMST) which places the knee joint centre over the centre of the foot, providing a very stable base of support [2].

TMST occurs at 30% of the gait cycle; during TMST, the pelvis, the trunk and the head are directly over the foot, and the GRF appears to be vertical. The knee of the swing limb can just be seen anterior to the stance limb, and the heel of the swing limb can be seen posterior to the stance limb [18].

Another important aspect of normal gait is the ability to support one's own bodyweight and decelerating the downward velocity of the CoM in late stance phase; this is demonstrated in the GRF. This is often a significant issue for children with gait pathologies [19], and without corrective action, the limb would collapse into flexion. The GRF provides information concerning the magnitude, direction and point of application of the impact forces. The vertical GRF shows the least variability between and within subjects [20,21].

1.6 Uses and applications of gait analysis

1.6.1 Fundamental understanding of gait

Gait analysis has helped us and is continuing to do so, to unravel the fundamental biomechanics of human locomotion. Such knowledge, in addition to being of interest for its own sake, also helps us establish which features of gait are essential to effective locomotion, which is useful, for instance, if we are to assess pathological gait, design prostheses or bio-inspired robots.

Even though everyone has a unique gait (see Chapter 10 on the use of gait in forensics), its fundamental mechanics are shared. Normal human gait is characterised by its fluctuations in the body's CoM and the potential and kinetic energy of the CoM.

During walking, both forms of mechanical energy fluctuate mostly out of phase: when its position and therefore the potential energy is the highest (during mid-stance), its speed and therefore kinetic energy is the lowest, and vice versa. Therefore, these forms of energy can be exchanged which makes human walking very efficient compared to most other animals. This mechanism is called the inverted pendulum mechanism.

During running, the two forms of mechanical energy also fluctuate but they do so mostly in phase. They can therefore not be exchanged; however, running uses a different (somewhat less efficient) energy-saving mechanism, in which mechanical energy is stored in elastic structures during early stance, and partly released during late stance. This mechanism is called a spring-mass mechanism.

The functional distinction between walking and running, described above, does not always correspond to the kinematic distinction based on duty factor (see above), for example, in grounded running, which shows the dynamics of running at a duty factor greater than 0.5.

It is very interesting that, although inverted pendulum and the spring-mass mechanisms are functionally very different, they are performed by the same anatomy and can be thought of as different dynamic output behaviours resulting from different input conditions (forces and body configurations), as expressed in the spring-loaded inverted pendulum model.

In addition to CoM behaviour, gait is also impacted at a basic level by collisions. These are sudden changes in the body's CoM velocity vector, as observed during initial foot impact. Collisions are a major source for energetic losses in human gait.

It is also important to consider the effect of the surface on gait, for example, differences in kinematics are evident when walking overground compared to walking on a treadmill [22].

1.6.2 Athletic performance and injury prevention

Gait, and especially running, is crucial in many sports and its analysis can help increase performance and reduce the risk of injury, even though both are complex and multi-factorial. Basic gait analysis using inexpensive tools can assist coaches at all levels. The use of advanced techniques in the lab and the field can help optimise performance (e.g., by quantifying power produced at the whole-body or the joint level) and assess biomechanical risk factors for traumatic or overuse injury.

1.6.3 Footwear design

Shoes are used for many purposes. In our daily lives, they protect us from injury or cold; in sports, they become a piece of equipment; and they can also be a therapeutic intervention (like orthoses). In any case, footwear forms the mechanical interface between our body and the environment; it can therefore have a major effect on gait. Even though the passive mechanical properties of shoes can be easily measured, how they influence gait cannot and requires gait analysis; indeed, it is well established that different types of footwear trigger different responses in the body so the effect of, for example, cushioning cannot be predicted by studying the shoe alone. For example, rocker sole shoes, which have a rigid sole to limit movement of the foot joints during stance, are used for people who are at risk of developing a foot ulcer (e.g., people with diabetes or rheumatoid arthritis who are considered high risk). This footwear reduces movement of the plantar tissues and thereby reduces plantar pressures, with different designs having varied effects on plantar pressures [23].

1.6.4 Pathological gait

Gait analysis is very useful for the diagnosis, treatment and follow-up of gait-related conditions, whatever their basis, e.g., neurological (examples include cerebral palsy, Parkinson's disease gait and diabetic neuropathy) or mechanical (examples include overuse injuries or amputations). For an excellent overview on clinical gait, we refer to Chapters 8 and 9, and to Ref. [24].

1.6.5 Gait and ageing

Efficient and stable gait results from a complex interaction between neural control and mechanics. These aspects change with ageing; even in healthy ageing, we will experience a gradual reduction of muscle mass and function as well as perception (e.g., vision, vestibular, skin receptors), which affect gait. In addition, age-related diseases such as sarcopenia or osteo-arthritis further affect locomotor ability. Gait analysis can help quantify the changes in gait, and the effectiveness of treatment interventions.

1.6.6 Gait and prosthetics

Research and clinical work on prostheses requires a detailed understanding of gait. Gait analysis can help assess the quality of gait in prosthetic users, performance (e.g., in athletics) and assists the design of prostheses. Gait analysis in healthy people is instrumental in providing a baseline of understanding to which pathological gait can be compared. One example is that it has been demonstrated that human gait is primarily ankle-powered, and therefore powered ankle prostheses are a major focus in prosthetic design. Another example is that the roll-over pattern of a healthy human foot can be functionally represented as a 'roll-over shape' which can inspire prosthetics design.

1.6.7 Clinical applicability

There has been an increase in demand for clinical gait analysis services over the last decade, fuelled by growing evidence that gait analysis measurements are more repeatable than traditional physical examinations. Although the visual assessment of gait is crucial in clinical decision-making, accurate assessment of gait is difficult by eye alone, namely, due to the complexity and the speed at which the phases of the gait cycle change, especially when assessing patients with neurological disorders. Clinicians have utilised technology in a clinical setting (video, smart phones and smart watches) with varying success, but the quality and depth of the data are not comparable to that which is produced by a gait analysis laboratory. The relatively new emergence of IMUs has great potential for clinical use, negating the need for service users to visit gait laboratories, and giving clinicians on hand, readily accessible objective and repeatable gait data to enable them to plan treatment and review the efficacy of treatment interventions.

1.7 Conclusion

Gait analysis has many uses in fundamental research, clinical research and practice, and sports and exercise science (without even mentioning non-human gait). In all these cases, technology plays a key role. Huge technological progress has been made since the early pioneers of gait analysis, and new technologies are currently emerging at an ever-increasing pace. This allows not only for an improved quality of data but also makes gait analysis accessible to a much larger and non-specialist audience. This book will outline the past, present and future of these technologies and techniques in gait analysis.

References

[1] Inman VT and Ralston HJTF. (1981). *Human Walking*. Baltimore, MA: Williams & Wilkins.

[2] Owen E. (2016). Normal gait kinematics and kinetics. In: *Physical Therapy for Children with Cerebral Palsy; An Evidence-Based Approach* (Rahlin M, ed.). Chicago, IL: Publisher Slack Incorporated.

[3] Perry J, Burnfield JM and Cabico LM. (2010). Gait analysis: normal and pathological function. *J. Pediat. Orthoped.*, 12(6): 551.

[4] Rose J and Gamble JG. (2005). *Human Walking* (3rd ed.). Philadelphia, PA: Lippincott Williams and Wilkins.

[5] Winter DA. (1990). *Biomechanics and Motor Control of Human Movement* (2nd ed.). New York, NY: Wiley; 277 p.

[6] Enoka RM. (2008). *Neuromechanics of Human Movement*. Human Kinetics.

[7] Shumway-Cook A and Woollacott MH. (2017). *Motor Control: Translating Research into Clinical Practice* (5th ed.). Philadelphia, PA: Lippincott Williams & Wilkins.

[8] Biknevicius AR and Reilly SM. (2006). Correlation of symmetrical gaits and whole body mechanics: debunking myths in locomotor biodynamics. *J. Exp. Zool. Part A,* 305A: 923–934.

[9] Alexander RM. (1976). Mechanics of bipedal locomotion. In: *Perspectives in Animal Biology* (Spencer Davies P, ed.). Oxford: Pergamon; pp. 493–504.

[10] Wu G and Cavanagh PR. (1995). Isb recommendations for standardization in the reporting of kinematic data. *J. Biomech.* 28: 1257–1261.

[11] Wu G, Siegler S, Allard P, *et al.* (2002). Isb recommendation on definitions of joint coordinate system of various joints for the reporting of human joint motion – Part i: ankle, hip, and spine. *J. Biomech.,* 35: 543–548.

[12] Wu G, van der Helm FCT, Veeger HEJ, *et al.* (2005). Isb recommendation on definitions of joint coordinate systems of various joints for the reporting of human joint motion – Part ii: shoulder, elbow, wrist and hand. *J. Biomech.,* 38: 981–992.

[13] Leardini A, Benedetti MG, Catani F, Simoncini L and Giannini S. (1999). An anatomically based protocol for the description of foot segment kinematics during gait. *Clin. Biomech.,* 14: 528–536.

[14] Leardini A, Stebbins J, Hillstrom H, Caravaggi P, Deschamps K and Arndt A. (2021). ISB recommendations for skin-marker-based multi-segment foot kinematics. *J. Biomech.,* 125: 110581. doi:10.1016/j.jbiomech.2021.110581. Epub 2021 Jun 25. PMID: 34217032.

[15] Carson MC, Harrington ME, Thompson N, O'Connor JJ and Theologis TN. (2001). Kinematic analysis of a multi-segment foot model for research and clinical applications: a repeatability analysis. *J. Biomech.,* 34(10): 1299–1307.

[16] Davis RB III, Ounpuu S, Tyburski D and Gage JR. (1991). A gait analysis data collection and reduction technique. *Human Move. Sci.,* 10(5): 575–587.

[17] Gronley J and Perry JK. (1984). Gait analysis techniques. Rancho Los Amigos Hospital gait laboratory. *Physcial. Ther.*, 64(12): 1831–1838.

[18] Gibson T, Jeffery RS and Bakheit AM. (2006). Comparison of three definitions of the mid-stance and mid-swing events of the gait cycle in children. *Disabil Rehabil.*, 28(10): 625–628.

[19] Williams SE, Gibbs S, Meadows CB and Abboud RJ. (2011). Classification of the reduced vertical component of the ground reaction force in late stance in cerebral palsy gait. *Gait Posture [Internet],* 34(3): 370–373.

[20] Keller TS, Weisberger AM, Ray JL, Hasan SS, Shiavi RG and Spengler DM. (1996). Relationship between vertical ground reaction force and

speed during walking, slow jogging, and running. *Clin. Biomech.*, 11(5): 253–259.

[21] Cavanagh PR and Lafortune MA. (1980). Ground reaction forces in distance running. *J. Biomech.*, 13: 397–406.

[22] Chockalingam N, Chatterley F, Healy A, Greenhalgh A and Branthwaite H. (2012). Comparison of pelvic complex kinematics during treadmill and overground walking. *Arch. Phys. Med. Rehab.,* 92(12): 2302–2308.

[23] Healy A, Chatzistergos P, Needham R, Naemi R and Chockalingam N. (2013). Comparison of design features in diabetic footwear and their effect on plantar pressure. *Footwear Sci.,* 5(Supp. 1): S67–S69.

[24] Richards J. (2018). *The Comprehensive Textbook of Clinical Biomechanics* (2nd ed.). New York, NY: Elsevier.

Chapter 2
Gait analysis – a historical perspective

Tom Shannon[1]

2.1 Introduction

Homo Sapiens have evolved to walk with their body upright, neither leaning forwards or backwards with an even stride and arms swinging at the side, conventionally described as gait. Locomotion when in the upright position is facilitated by attempting to hold a constant centre of mass to achieve good balance and economy, expressed as unit energy consumed against distance travelled. The emergence of an evolutionary advantage through efficient bipedal locomotor capability continues to remain a research focus to correlate validated motion and applied forces with measurements of the morphology and anatomy of primates, early hominins, and humans.

Around four million years ago, our ancestors adapted through evolutionary development or mutation to a terrestrial life by walking using just their hind legs as evidenced by fossilised footprints discovered at Laetoli in Tanzania. The physiological and anatomical reconfiguration to an efficient upright bipedal posture saw the spine evolve as a slight sagittal curvature with a skull optimally placed on top, and longer legs, broader hips, feet with arches and a big toe aligned with the rest of the foot. When compared with our ape relatives, it is apparent that the human frame in terms of the skeleton, muscle attachments and posture is designed for walking and running as a normal activity, whereas contemporary sedentary behaviour and the impact of pathologies on individuals is not (Figures 2.1 and 2.2).

Kozma *et al.* [1] investigated the tri-dimensional (3-D) shape of hominoid pelves and hip extension performance. They postulated that humans with reduced extensor moments offer a greater degree of hip extension so sacrificed climbing capability for improved walking economy. Conversely, extant apes exhibit a crouched gait with a more limited hip extension and better extensor moments at the hip to match the optimum survival needs within their arboreal–terrestrial environments. The authors applied their findings to fossil pelves and concluded that early hominin gait capability differed from both humans and primates. 'Ardipithecus' achieved a compromise between human bipedal gait and climbing ability but the

[1]Centre for Biomechanics and Rehabilitation Technologies, Staffordshire University, Stoke on Trent, UK

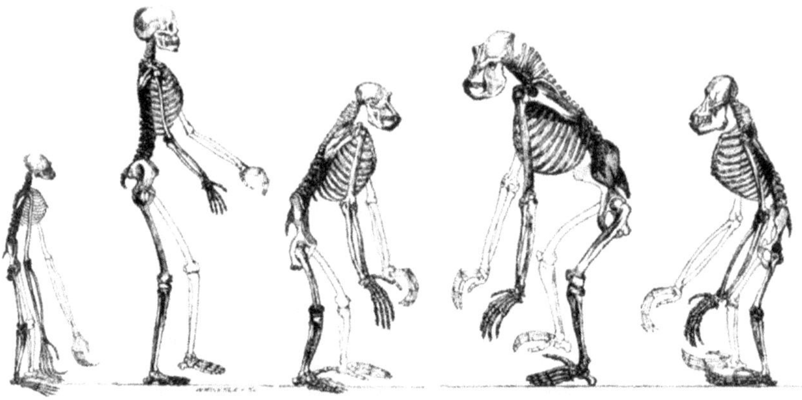

Figure 2.1 Ape and Homo sapiens skeletons

later ‘Australopithecus afarensis’ and ‘Australopithecus africanus’ exhibited reduced climbing advantage to achieve more efficient gait. Lovejoy [2] observed that ‘Australopithecus’ and modern humans evolved long and flexible lumbar spine adaptations that exhibited lordosis to place the head, trunk and arms vertically above the lower limb joint centres thus allowing the extension of both anatomical regions during upright walking and running.

The study and understanding of human and animal locomotion have been a topic of fascination to the arts and sciences for over a millennia. Cave drawings from the Palaeolithic era depicting animals and men in motion could have been an early attempt to record survival questions to escape predators and hunt for food. Greek philosophers (500–300 BC) were driven to observe, analyse and describe human motion in their attempts to seek harmony in the universe [3]. Their belief that truth could only be reached through logical thought at the expense of the experimental method did influence their conclusions.

Andriacchi and Alexander [4] suggested that the Greek philosophers still posed relevant questions when addressing the challenges of studying human locomotion even today. In particular, can our senses deceive us, so limiting our ability to observe and analyse complex 3-D motion? Do the experimental methods and measurement protocols contaminate the natural characteristics we are attempting to observe? The authors concluded that these remain valid concerns that need consideration when interpreting measurement outcomes from studied locomotion, especially when formulating any generalisations from specific experimental results.

An early example of the limitations of our physical senses to interpret what we are observing can be seen in the depiction of horses at the gallop in art. Generations of artists had portrayed the horse with both front legs extended forward and both hind legs extended rearward in a pose analogous to the ‘rocking horse’. Clearly, the temporal resolution limitations of the human eye were not capable of capturing the sequence of rapid limb movements of horses in motion until the invention of

Figure 2.2 Aelius Galenus (131–210 CE)

photography in the early nineteenth century. The later introduction of new methods of observation proved that at the gallop one hoof remains in contact with the ground except in the suspension when the legs are bent rather than extended.

The need to answer fundamental questions continues to drive advances in the study of human locomotion coupled with improvements in methodologies and technologies to enhance our understanding within a plethora of potentially symbiotic fields including medicine, ergonomics, animation and sporting performance to name but a few. This chapter provides an examination and analysis of the key historical factors and developments and the influence they played in the growth and wider acceptance of the field in order to predict and attempt to extrapolate the direction it may evolve in the coming decades.

2.2 The ancient world

The study of Biomechanics (from the Ancient Greek: *βίος* translated as 'life' and *μηχανική* as 'mechanics') has acted as an important adjunct to historical and contemporary anatomical and physiological investigations and basic research.

The field continues to be widely applied to describe the fundamentals of human motion in terms of well-understood mechanical principles.

Plato (428–348 BC) recognised that the head was the most divine part of the body with all other members to be its servants and to be its means of locomotion 'with which it might take hold and to be able to pass through all places, carrying on high the dwelling place of the most sacred and divine part of us' (Timaeus, 360 BC, translated by Benjamin Jowett). Hippocrates (460–370 BC) recognised that if a dislocation occurred in a leg joint, then the gait would be adversely affected and the body weight would be supported by the un-injured limb and on occasion by the use of a crutch. Aristotle [5] questioned the fundamental reasons why different species used a variety of ambulatory solutions to achieve motion and what were the fewest points necessary to achieve progression. In his work 'De Motu Animalium' (On the Movement of Animals), he described the bodies of animals in terms of mechanical systems so he could subject them to geometric analysis for the first time. Aelius Galenus (131–201 CE), a Roman citizen of Greek origin, published in his essay 'De Motu Musculorum' (On the Movements of Muscles) the differences between motor and sensory nerves, agonist and antagonist muscles and introduced the concept of tonus. Human dissection was discouraged during his lifetime but as a physician, he still taught that muscle contraction resulted from messages sent by the brain via the nerves that he described as 'animal spirits'.

2.3 The renaissance

No historical perspective would be complete without considering the contribution of the great Italian renaissance polymath, Leonardo da Vinci (1452–1519), particularly in his broad interests in the symbiosis of the sciences and the arts (Figure 2.3). Andrea Ancillao [6] described how da Vinci's principal objective was to study the proportions of the human body in his own words 'to represent man and the intentions of his soul, and the attitudes and movements of his limbs'. His approach was artistic to communicate a representation of the world that included man. He realised that modelling human motion was not a trivial task due to the interaction of the skeleton, muscles and tendons to achieve locomotion. His

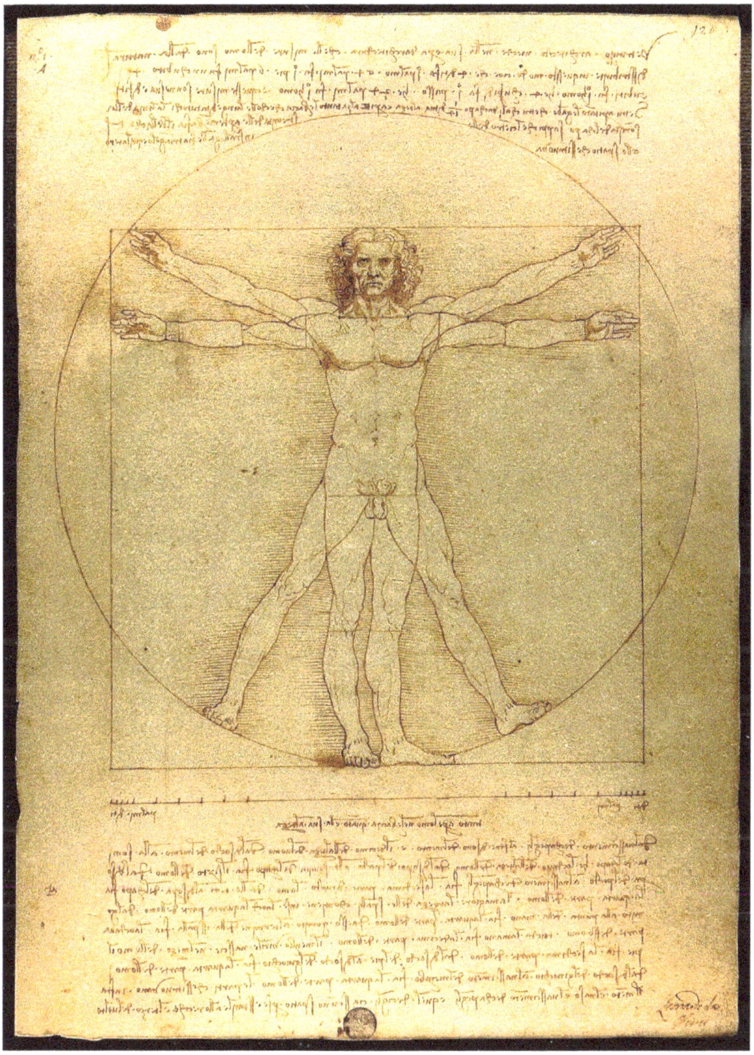

Figure 2.3 Leonardo da Vinci Vitruvian Man 1490

solution was to attempt to gain insight and to explain such processes through the dissection of cadavers with detailed drawings and notes. One of his most famous and readily recognised images is the 'Vitruvian Man' that contains some fundamental observations about the anatomy and physiology of the human body. Knowing that da Vinci had stated that 'The human foot is a masterpiece of engineering and a work of art', and as an example of his significant contribution, Jastifer and Toledo-Pereyra [7] reviewed his commentary and anatomical drawings of around the 30 cadaver dissections he performed. They concluded that his concepts of foot and ankle anatomy and mechanics have little difference from contemporary theories of proportion, statics and joint stability, sesamoid biomechanics and the structural support of the foot. da Vinci with his genius and artistic ability coupled with an inherent interest in human anatomy laid the foundation for the study of orthopaedics, biomechanics and the application of the clinical sciences that impacts research thinking and approaches even today.

Galileo Galilei (1564–1642) was a medical student before studying physics for which he is much better known in history. His clinical interests focussed towards the circulatory system also discovering the constancy of the period of a pendulum that he used to express quantitatively the pulse rate in terms of the length of the device arm to achieve synchrony. His contribution to biomechanics and the analysis of gait is to introduce the experimental method rejected by the ancient philosophers as a valid approach.

Contemporary scientists in seventeenth-century Italy embraced the new philosophical approach to science and medicine through appreciating the importance of applying the physical laws of nature for empirical experimentation, developing mathematical concepts and explanation of observations. The physicist, astronomer and personal medical consultant to the Queen of Sweden, Giovani Alfonso Borelli (1608–79), was no exception and was centred on and inspired by the rigorous analytical methods by Galileo particularly to biological questions. Piolanti *et al.* [8] identified that in his seminal work 'De Motu Animalium' (On the Motion of Animals) published in Rome in 1680, Borelli analysed the static and dynamic cases of structure, motion, balance and forces concerning almost all the principal joints of the human body (Figure 2.4). He studied the anatomy and biomechanics of the knee joint in detail, concluding that femoral condyles shift backwards during flexion to facilitate a wider range of motion. He published his observations that when the knee flexes, the lateral condyle moves rearward to a greater degree than the medial condyle, described today as medial pivoting. Borelli also established the centre of mass of a body by placing subjects on a rigid platform supported on a knife-edge. He measured the location centre of mass once the platform was in a position of balance.

2.4 The enlightenment and the nineteenth century

In 1791, Luigi Galvani (1737–98) published 'De Viribus Electridtatis' (Commentary on the Effects of Electricity on Muscular Motion), where he described how he had observed the effects of atmospheric electricity on dissected frog muscles (Figure 2.5). He observed that muscular contraction sometimes occurred when touched by a scalpel leading him to conclude that there existed 'in dwelling electricity which proceeded along the nerve'. Piccolino [9] suggests that Galvani's findings were the earliest explicit statement of the presence of electrical potential in muscles and nerves that laid the foundations of the new science of electro-physiology.

Brothers, Ernst Heinrich (1795–1878) and Eduard Friedrich Wilhelm (1806–71) Weber, collaborated on a number of physiological works before publishing 'Die Mechanik Der Menschlichen Gerverzeuge' (Mechanics of the Human Walking Apparatus) in 1836 [10] (Figure 2.6). In the work, they described and

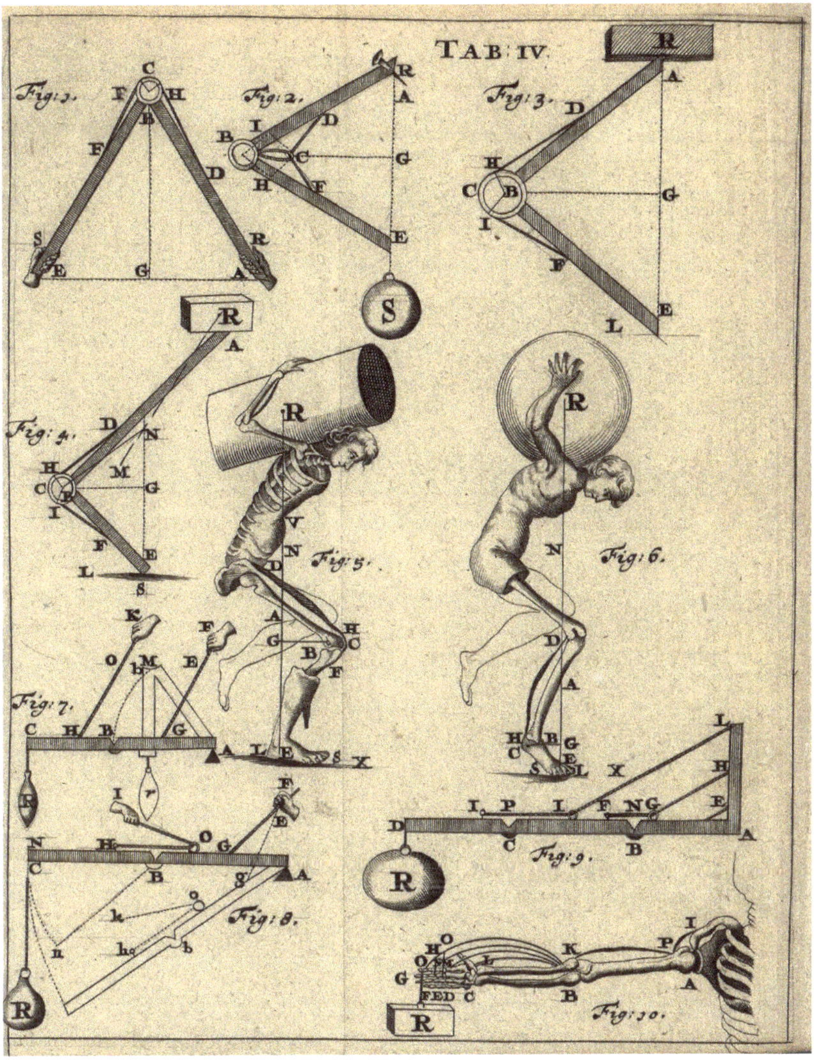

Figure 2.4 Illustration from De motu animalium, 1685, Giovanni Alfonso Borrelli (1608–1679)

illustrated their experiments showing the position of the limbs and the attitude of the limb segments at 14 different temporal events in the gait cycle. Using only a stop-watch, measuring tape and telescope, they attempted to demonstrate that there was a reliable correlation between walking speed, step length and cadence. Although their measures were quantitative, many of their conclusions were of qualitative conjecture resulting in much discussion, debate and improvements in understanding in the subsequent decades.

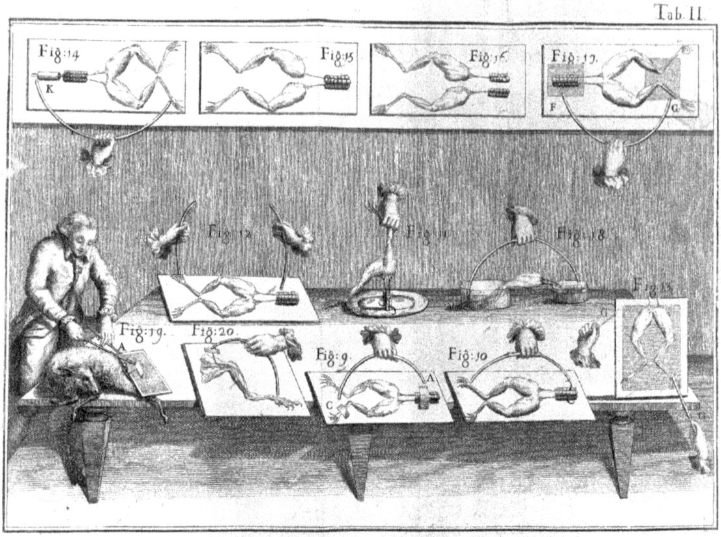

Figure 2.5 De Viribus Electridtatis, Luigi Galvani (1737–1798)

E. Harless initially used the same approach as the Weber brothers to derive absolute and relative lengths of the body and its segments but extended his research to include the absolute and relative location of the centres of mass in the sagittal plane of movable body segments. He published his results in an 1860 treatise 'The Static Moments of the Human Body Limbs' [11], based on the dissection of five male and two female cadavers [12], segmented along planes passing through the axes of rotation of each of the primary joints. The mass and the centres of mass were measured using scales and balance plates with the volume and density of each segment estimated using Archimedes' water displacement principle. Harless was the first to conclude that age and sex are significant influencers on the density of segments of the human body.

Étienne-Jules Marey (1830–1904) in his 'Introduction to La Methode Graphique dans les Sciences Experimentales' published in 1878 stated 'From the invisible atom to the celestial body lost in space, everything is movement.... It is the most apparent characteristic of life; it manifests itself in all functions, it is even the essence of several of them (Figure 2.7). All movement is the product of two factors: time and space; to know the movement of a body is to know the series of positions that it has occupied in space in a series of successive moments'. Since Isaac Newton (1642–1727) had first published his 'Philosophiæ Naturalis Principia Mathematica' (Mathematical Principles of Natural Philosophy) in 1687, there had been significant academic discourse if the motion of the human body was subject to the same laws as nature. Richard Baker [13] concluded that Marey was convinced

Figure 2.6 Ernst Heinrich (1795–1878) and Eduard Friedrich Wilhelm (1806–1871)

that it was and devoted his career in physiology that by making careful and appropriate measurements he could deduce how Newtonian mechanics could be applied to an approach that could be described as the first gait analysis (Figure 2.8a and b).

Marey collaborated with his student, Gaston Carlet (1849–92), on his early studies of human gait (Figure 2.9). Carlet first recorded the 'butterfly pattern' of the ground reaction forces exerted by the foot during walking from his development of a shoe with three pressure transducers built into the sole. Carlet's thesis, published in 1872, described various experimental techniques concluding with an essentially accurate and still valid description of the gait cycle of typically developed adults [14].

Interest in motion in the mid-nineteenth century was directed principally towards quantifying how a horse moved and solving a persistent debate about if there was an instant if all four hooves were off the ground when trotting. Marey

Figure 2.7 Étienne-Jules Marey (1830–1904). Courtesy of Collection BIU Health Medicine, Open License

approached the problem by adapting Carlet's ground reaction force transducers and fitting them to the distal portion of a horse's leg. Marey's results demonstrated conclusively that there was a short period when none of the hooves were in contact with the ground with several artists modifying their pictures of moving horses to be a more accurate representation of reality. Although accepted as truth in Europe, the argument still raged in the United States of America between a former Governor of California and racehorse owner, Leland Stanford, and the

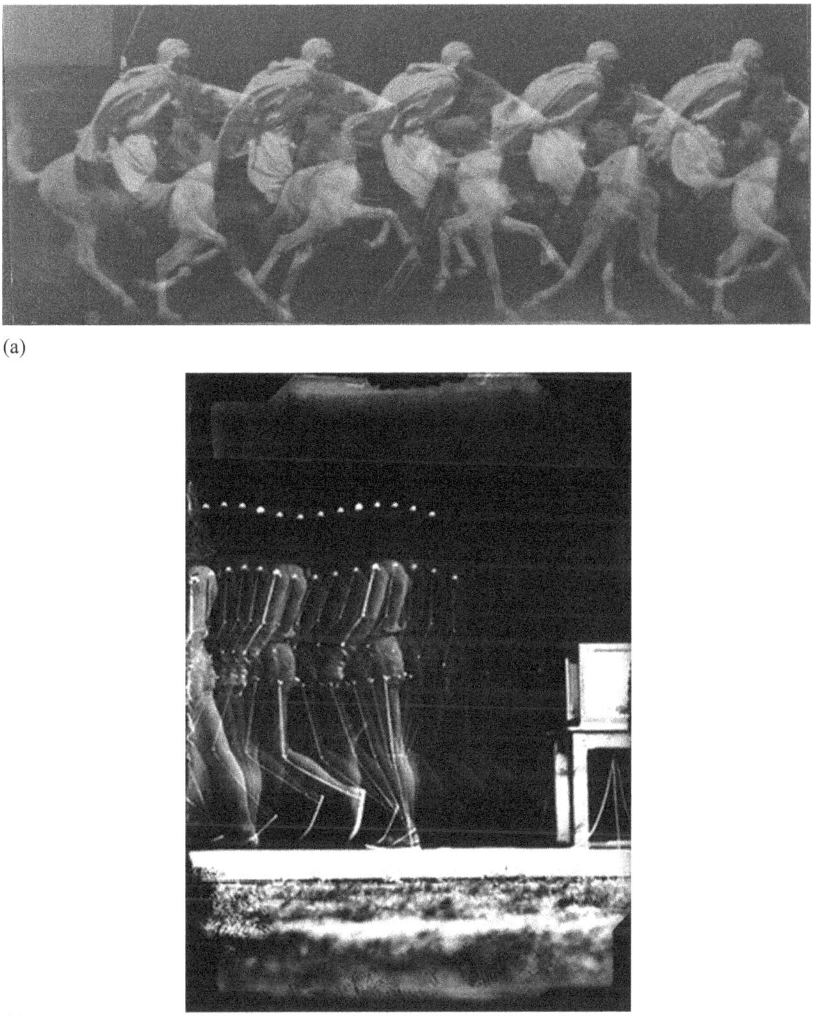

(a)

(b)

Figure 2.8 (a) Étienne-Jules Marey Arab Horses at the Gallop. (b) Étienne-Jules Marey Chronophotography of Homme-qui-marche (1890–91)

wealthy sporting community on the East Coast, all fuelled by a buoyant and partisan press corps.

The first photograph 'View from the Window at Le Gras' taken in 1826 or 1827 by Joseph NicÉphore NiÉpce gave scientists access to a new method to instantly record the motion of humans and animals (Figure 2.10). A proponent of the new and later technology was an Englishman who had migrated to North America to become a well-known photographer, Eadweard Muybridge (1830–1904) (Figures 2.11–2.13). As an inveterate showman and self-publicist, he quickly

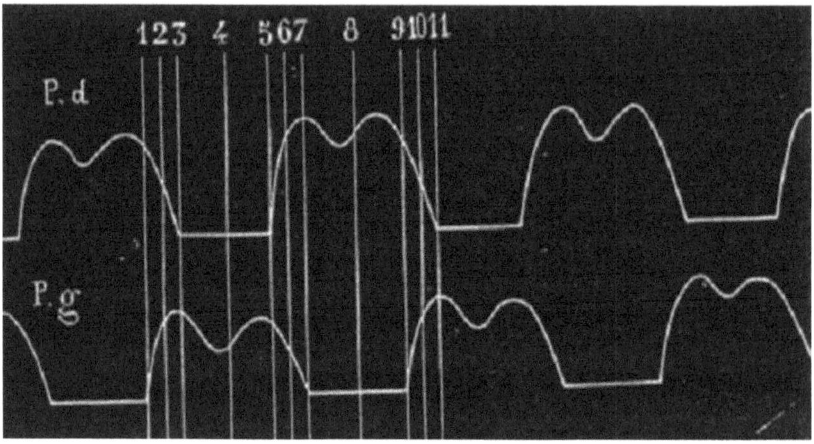

Figure 2.9 Gaston Carlet (1849–92). Ground reaction forces visualised [14]

Figure 2.10 Joseph NicÉphore NiÉpce, View from the Window at Le Gras (1826–27)

established himself as a landscape photographer through producing dramatic views of both Yosemite and San Francisco. His reputation as being an adventurous and progressive photographer led him to work as both a war and official government photographer.

Figure 2.11 Eadweard Muybridge (1830–1904)

Leland Stanford achieved great wealth as the builder of the Central Pacific railroad investing heavily in developing the greatest racing stable on the West Coast where he acquired and utilised the latest scientific research. He knew of Marey's work and needed a way to confirm independently what he also believed was accepted truth so as to defeat his nemeses on the East Coast. In 1872, Stanford hired Muybridge who developed a system of multiple cameras with trip wire shutter releases to capture horse motion at speed. The project took 5 years in what proved to be a turbulent period in his life as in 1874 he shot his young wife's lover but was subsequently acquitted of the crime as a 'justifiable homicide'. After the trial, Muybridge set out for an expedition to Central America but returned in 1878 to use his cameras to prove conclusively that a galloping horse lifts all four hooves off the ground. The effect of his discovery was seismic and widely published with Muybridge touring extensively in America and Europe to present his photographic series to astonished audiences. However when Stanford published 'The Horse in

Figure 2.12 Eadweard Muybridge, The Horse in Motion (1878)

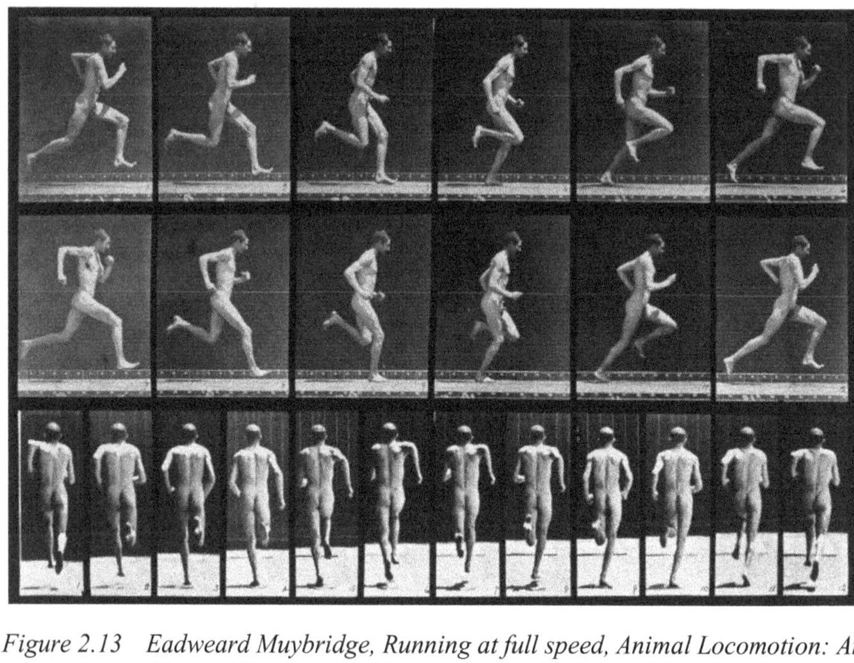

Figure 2.13 Eadweard Muybridge, Running at full speed, Animal Locomotion: An Electro-Photographic Investigation of Consecutive Phases of Animal Movement 1872–1885, Royal Academy of Arts

Motion', Muybridge's contribution was barely mentioned with the relationship between the two quickly disintegrating.

Muybridge's most enduring and influential body of work 'Animal Locomotion' (1884–87) was when he was sponsored and funded by the University of Pennsylvania. The project became a systematic and wide-ranging study of the science of motion and the emerging art form of photography through the creation of a series of sequential plates, each imaging the phases of one animal movement. Muybridge also extensively documented human subjects engaged in walking, running, descending staircases as well as engaging in activities such as boxing, fencing, weight lifting and wrestling. In the 3 years he spent working on the project, Muybridge took over 20,000 photographs, ruthlessly editing the final selection to 781 plates. During his time in Pennsylvania, he also worked on the use of clock-work motors to fire camera shutters in sequence so as to remove the dependence on the subject's motion to trigger an exposure. Muybridge also extended the sensitivity of his capture of human and animal motion by using rear and angled views so pre-empting the concept to acquire 3-D imagery. He also invented the first machine to project moving photographic images, and his 'zoopraxiscope' could be used to animate entire sequences of photographs, so anticipating the rise of cinema.

Marey and Muybridge met in Paris in 1881, and although the former was inspired by the achievements of the latter, he quickly realised that the Muybridge's camera solution prevented consistent measurement of gait as each plate was captured from a slightly different angle. To solve this weakness, Marey developed the 'chronophotograph' that used a shutter enabling several images to be captured on a single photographic plate. A limitation of the new technique was that taking reliable measurements from overlapped images was challenging. It was at this time he started to work with a new student, Georges Demeney (1850–1918), and together they started identifying joint anatomical landmarks using different types of markers. The work resulted in exquisite photographs of both typical and pathological gait patterns where it was possible to extract meaningful measurements.

2.5 The twentieth century

Marey limited his investigations to two dimensions (2-D) only and it was left to the German mathematician Otto Fischer (1861–1917) to conduct and achieve the first 3-D gait analysis (Figures 2.14 and 2.15). His name will always be linked with Wilhelm Braune (1831–92), a Professor of Anatomy, who died during the seminal work but still recognised as a collaborative author in 'Der gang des menschen (1900)' (The Human Gait) [15]. The researchers used two pairs of cameras in darkness set for continuous exposures with the subject fitted with Geissler tubes strapped to his lower limb segments. A tuning fork was used to make and break the electrical circuits to provide accurate timing of the flashes of light. The process required the accurate determination of each of the camera positions, achieved by projecting a mesh of known dimensions during a separate exposure. Points were measured from the images acquired from each of the cameras on the respective side

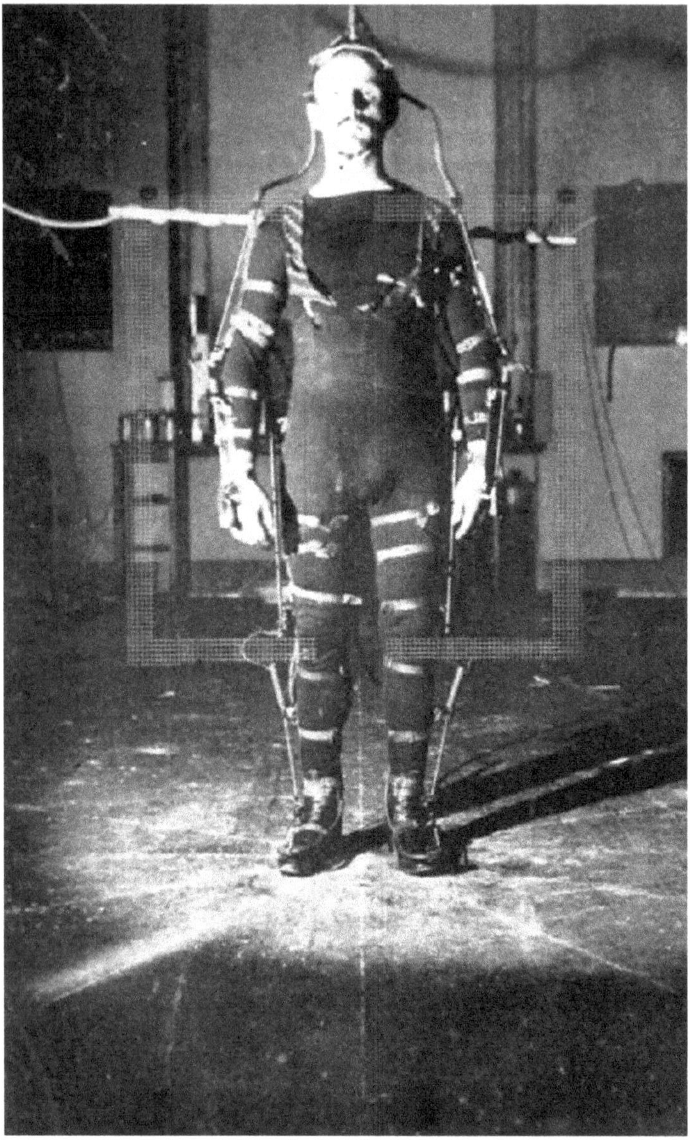

Figure 2.14 Otto Fischer (1861–1917), Wilhelm Braune (1831–1892). Instrumented subject of the first three-dimensional movement analysis.

of the subject, resulting in a full reconstruction of each point in 3-D space with a precision of the order of 1 mm. Once the point coordinates were identified, the joint centres could then be calculated. As it was not possible to derive the internal or external rotations of the limbs, Fischer assumed that the elbow, wrist and knee joint centres were medial to the joint markers by half the joint width. For the hips and

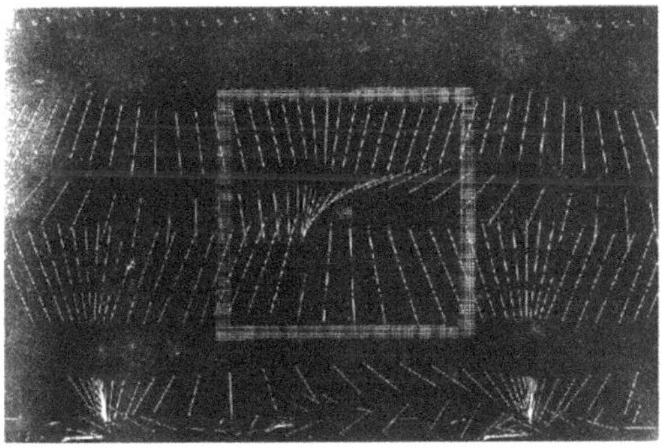

Figure 2.15 Otto Fischer (1861–1917), Wilhelm Braune (1831–1892). Results of the first three-dimensional movement analysis [15].

shoulders, he took the measurements from both sides to determine the rotation of the trunk and pelvis in the transverse plane. Although the hand-plotted results looked superficially similar to Marey's photographs, they were a more accurate representation as all parallax artefacts had been addressed in the 3-D analysis. Fischer used the results to calculate the trajectory of the centre of mass of each body segment and used the inverse dynamics mathematical approach to calculate the lower limb joint moments during the swing phase of the subject's gait.

Nikolai Aleksandrovich Bernstein, a neurophysiologist, modified and built on the work of Braune and Fischer [16,17] in the Soviet Union in the 20s and 30s with a focus on identifying the small movements when undertaking manual tasks such as cutting metal with a chisel and hammering in order to introduce ergonomic changes to improve worker productivity. His research included a series of experiments that studied the development of walking as humans matured and the gait among patients diagnosed with a neurological disorder.

The work of Braune, Fischer and Bernstein can be best described in mechanical terms as the study of 'kinematics' which is concerned only with the characteristics or properties of the motion of an object without considering the forces that cause the motion. This is of course only half the story so biomechanical analysis of humans must also incorporate the study of 'kinetics' that is concerned with analysing the effects of forces and torques on the motion of objects having mass.

Demeny and Marey modified Carlet's earlier work on in-shoe pressure measurement to develop a vertical component ground reaction force plate [18] based on a pneumatic principle that they synchronised with their 'chronophotograph' to investigate both the kinematics and kinetics of gait. The limitation of the approach was that in order to understand fully the kinetics of walking, it was important that all three axial components of ground reaction needed to be simultaneously acquired.

As a rehabilitation specialist, Jules Amar (1879–1935) faced with mass casualties in France during and after the Great War was the first to develop a three-component force plate [19]. His work was driven to meet an immediate clinical need to quantify improvement among his injured patients rather than to satisfy any scientific curiosity

[20]. Notable work on the technology continued in the United States with Herbert Elftman [21] developing a three-component force plate in 1938 at Columbia University but also looking at the pressure distribution under the foot and postulating the forces, moments and changes in energy during walking. Don Cunningham and Wayne Brown achieved a full six-component force plate using strain gauge technology at the University of California at Berkley in the early 50s [22] with similar designs soon being developed in Europe, notably at Strathclyde University in Glasgow, Scotland.

The first commercial force plate with specifications suitable for biomechanical analysis and based on a piezo-electric principle was manufactured by Kistler Instrumente AG in Switzerland in 1969 first for the surgeon, Dr S.M. Perren, then at the Laboratorium für experimentelle Chirurgie in nearby Davos. In 1976, Advanced Mechanical Technology, Inc. constructed the first commercially available strain gauge force plate for gait analysis at the biomechanics laboratory of the Boston Children's Hospital.

2.6 Post Second War

Just as in Jules Amar's time, the end of the Second World War in 1945 and the major improvements in battlefield casualty evacuation led to the return of tens of thousands of amputees with differing physical problems all trying to re-build their lives in a civilian environment. In response and at the urging of the National Research Council, the Surgeon General's office (later the Veteran's Administration) created the Committee on Prosthetic Devices. A team of around 40 scientists and clinicians were assembled within a dedicated biomechanics laboratory at Berkeley, in collaboration with faculty from the Department of Engineering, and on the San Francisco campus within the Department of Anatomy.

Orthopaedic surgeon, Verne T. Inman (1905–80) headed the team and engineer, Howard D. Eberhart (1906–93) was interestingly Inman's patient after he was badly injured resulting in his foot being amputated by his future colleague (Figure 2.16).

The team quickly discovered that basic research in the biomechanics of locomotion including the phasic electrical activity of appropriate muscles was needed if they were to make considered design changes to the unsatisfactory prosthetic devices of the period. Over time, the basic research, as well as their clinical application in a collaborative environment populated by a diverse group of surgical, scientific and engineering specialities established high standards and major publications in the fields of the study of human locomotion, the mechanisms of pain and innovative prosthetic design. As a group, they worked on the fundamental principle, first elucidated by Jules Amar, that an understanding of typically developed walking was an essential precursor before studying amputee gait. Ebehart, Inman and Bresler [23] stated ' . . . it is obvious that any improvement – either in surgical and physiotherapeutic procedures or in braces and prostheses – must rest upon an accurate knowledge of the functional characteristics of the normal locomotor system'. A serious limitation in performing routine gait analysis in a clinical setting was the reality that the exercise took the Berkeley group approximately 500 man-hours, eventually reduced to 250 man-hours with experience, to undertake

Figure 2.16 Verne T. Inman (1905–80) and Howard D. Eberhart (1906–93)

the 14,000 numerical calculations by hand, with 72 curves plotted and 24 curves with applied graphical differentiation, all form a single stride. Clinical gait analysis would never become a useful tool until processing times were drastically reduced.

Mary Pat Murray (1925–84) was a renowned physical therapist who began her professional career at the Curative Workshop of Milwaukee in 1945 and later at the Veterans Administration Medical Centre in Wood, Wisconsin in 1948 (Figure 2.17). She was a renowned teacher of kinesiology at Marquette University between 1954 and until her death but is also remembered as an internationally respected researcher with a focus on the measurement of normal and pathological gait patterns.

In 1964, Dr Murray created and was appointed Chief of the Kinesiology Research Laboratory at the Veterans Administration Medical Centre in Milwaukee, where over the next 20 years she adapted a century-old interrupted photography technique by placing a large angled mirror over a walkway to capture sagittal and transverse planes onto the same photographic plate. Although the resulting imagery looked superficially similar to those created by Braune and Fischer, Murray did not attempt a full coordinate point reconstruction but some key measures were corrected for parallax. She used the approach to measure whole movement patterns of normal human gait from infancy to old age. Murray *et al.* [24] studied the gait patterns of 60 walking men, including in their work to measure the kinematics of

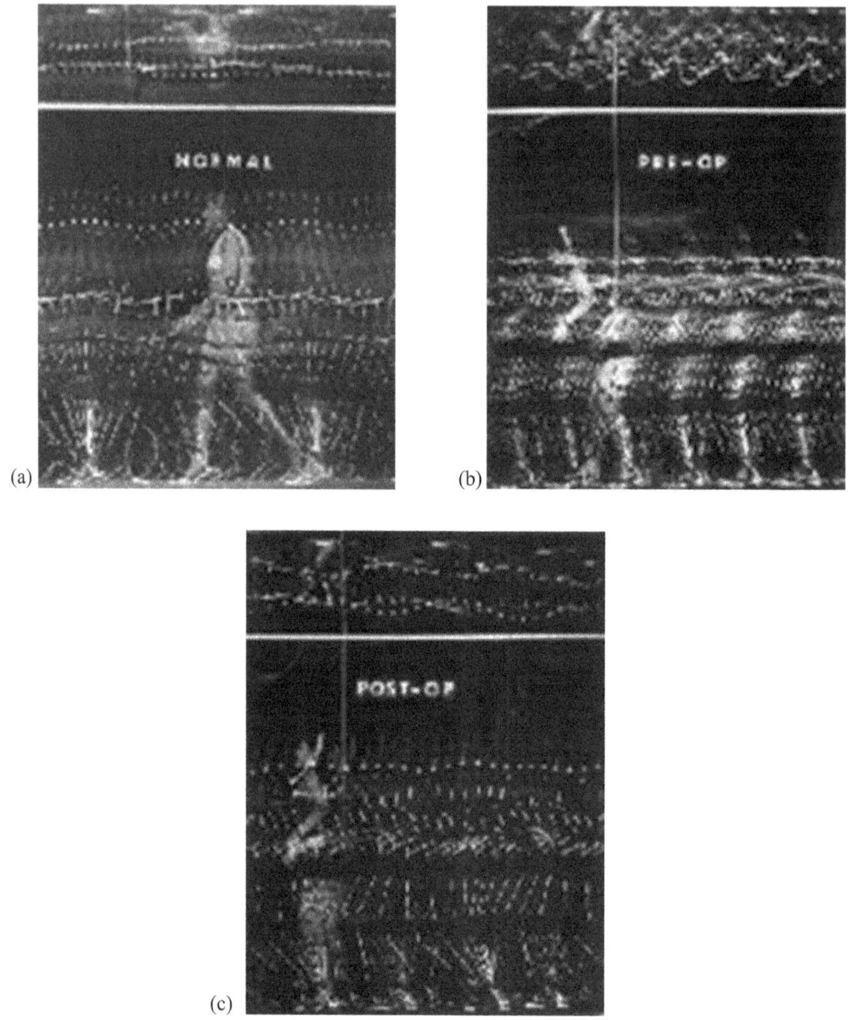

Figure 2.17 Mary Pat Murray (1925–84), study of 60 walking men (a, b, c) [24]

body segments in multiple planes of normal and pathological gait. The results although appearing crude are still regarded as a foundation for many subsequent studies across the world with patterns that are very similar to those obtained using contemporary technologies.

The study of kinematics required a method to quantify anatomical joint rotations. Fortunately, Leonhard Euler (1707–83) used spherical geometry to develop his 'rotation theorem' published in 1775. One outcome of this work was the definition of 'Euler angles' that are still used routinely to describe the orientation of a rigid body with respect to a fixed coordinate system or a mobile frame of reference.

A very well-understood adjunct technology was the goniometer that is a simple instrument that can measure an angle or be used to facilitate a limb joint to be moved to a pre-determined angular position. The name of the device derives from two Greek words 'gōnia', translated as angle, and 'metron', as a measure. In the post-war period, researchers started connecting electrical angle measuring and recording equipment to these mechanical devices to create the 'electrogoniometer'. In 1959, the Karpovich brothers [25] published their conclusions that the technology could easily capture multiple gait cycles and quickly display analogue graphs of motion without any need for any further manual data reduction calculations. In 1976, Bajd and Stanic *et al.* [26] published their design of a gait analysis system based on six precision potentiometers that provided reliable, inexpensive temporal-dependent hip, knee and ankle measures of both legs in the sagittal plane.

In 1971, Larry Lamoureux, an engineer with the University of California at Berkeley, used goniometers applied to a metal exo-skeleton to measure Eular angles at the hip and one-dimensional angles at the knee and ankle [27]. By 1980, Edmund Chao of the Mayo Clinic in Rochester, Minnesotta [28] had developed a tri-axial goniometer based on a gyroscopic principle and capable of generating Eular angles. The approach was dogged with cross-talk problems and like all other implementations introduced physical measurement artefact, inherent mechanical morphological inflexibility and was challenged when attempting to correlate the surface measures with actual skeletal joint rotations. Even today, goniometers have a place in general physical therapy clinics but do not have a wide acceptance in gait analysis environments. Chao's use of 3-D joint angle descriptors remains widely employed today.

2.7 Electromyography

Six decades after Luigi Galvani published his discovery that electrical stimulation could initiate a muscle contraction, Emil du Bois-Reymond (1818–96) established in 1849 that it was also possible to record the changes in electrical potential during a voluntary muscle contraction. Marey made the first actual recording in 1890, also introducing the term 'electromyography'. The technique is widely used today for evaluating the electrical potentials in skeletal muscle cells when electrically or neurologically activated to detect abnormalities, activation levels or recruitment order as an important element in the biomechanical analysis of human movement [29]. Early developments of clinical gait analysis were forged by two of Verne Inman's students, physician, Jacquelin Perry MD (1918–2013), and surgeon, David Sutherland MD (1923–2006) (Figures 2.18 and 2.19). Perry was employed by Rancho Los Amigos National Rehabilitation Centre in Downey, California and Sutherland first by the San Francisco Shriner's Hospital for Crippled Children and later was employed by the Children's Hospital of San Diego. In the 1960s and 1970s, the application of electromyography was the dominant clinical analysis technology, as being an analogue signal, it was much easier to acquire than any kinematic measures. Both Perry and Sutherland recognised the limitations in using

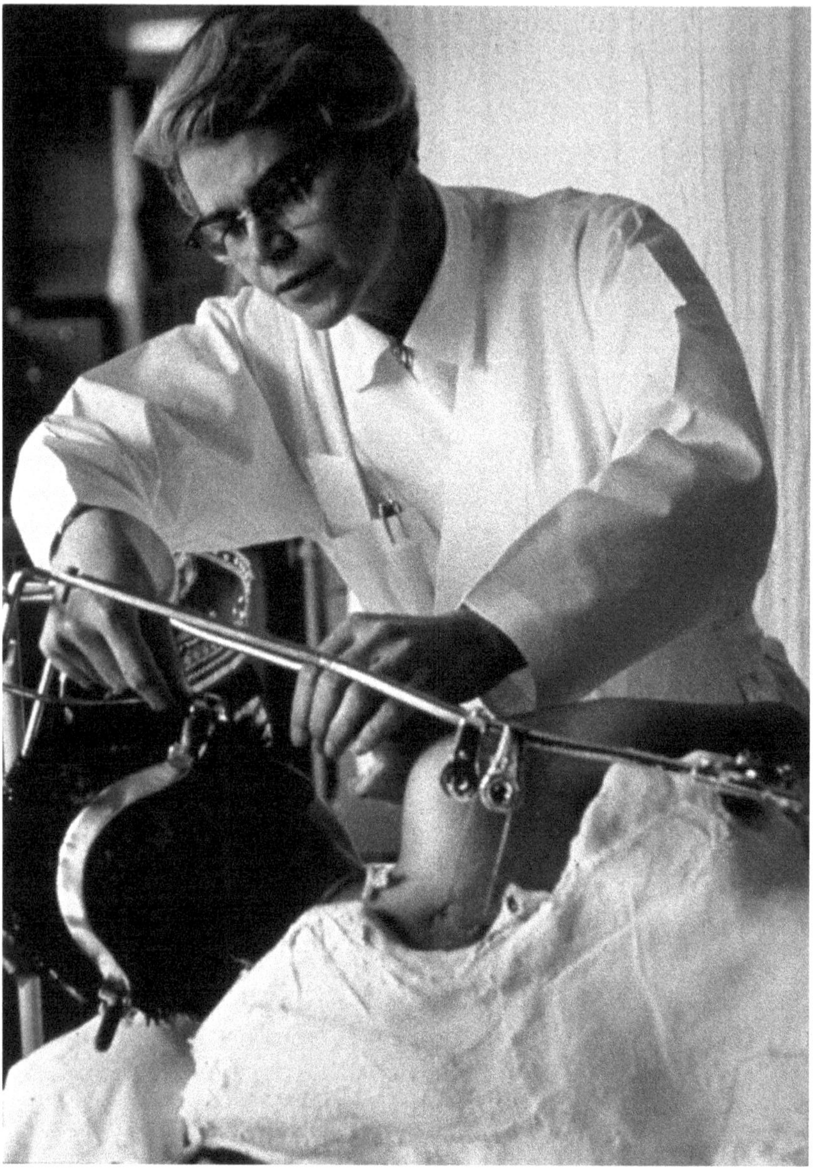

Figure 2.18 Jacquelin Perry, MD (1918–2013)

electromyography alone to diagnose and treat presenting aetiologies and embarked on a research to incorporate kinematic and kinetic measurement methodologies into their clinical practices. Perry concentrated on developing observational protocols [30] as well as applying some instrumented methods [31] to measure simple temporal–spatial gait characteristics to complement the routine clinical electromyographic acquisition.

Figure 2.19 David H. Sutherland, MD (1923–2006)

A casual review of a technical journal took Sutherland on a very different path [32,33] after he saw an advertisement for a 'Vanguard Motion Analyzer' (Figure 2.20). The equipment was designed to back-project cine film for simple frame-by-frame measurement of selected points to derive 2-D coordinates. He was also aware of the work of the Lockheed Aircraft Corporation, Missiles and Space Company in Sunnyvale, California who were using the equipment to derive pitch, yaw and roll by applying mathematical formulae to the coordinates from the film outputs of two or more cameras. Sutherland visited the facility and convinced John Hagy, Richard Oyama and later Cecil Keller to act as volunteers at the Shriner's Hospital, bringing with them two high-speed cameras, generously loaned by the Company. Sutherland and his Lockheed volunteers were able to create a system to add kinematics to the electromyography publishing their initial results in 1967 [34]. The early technique remained very time consuming requiring the manual recording of the 2-D coordinates and using a slide rule to complete the trigonometric

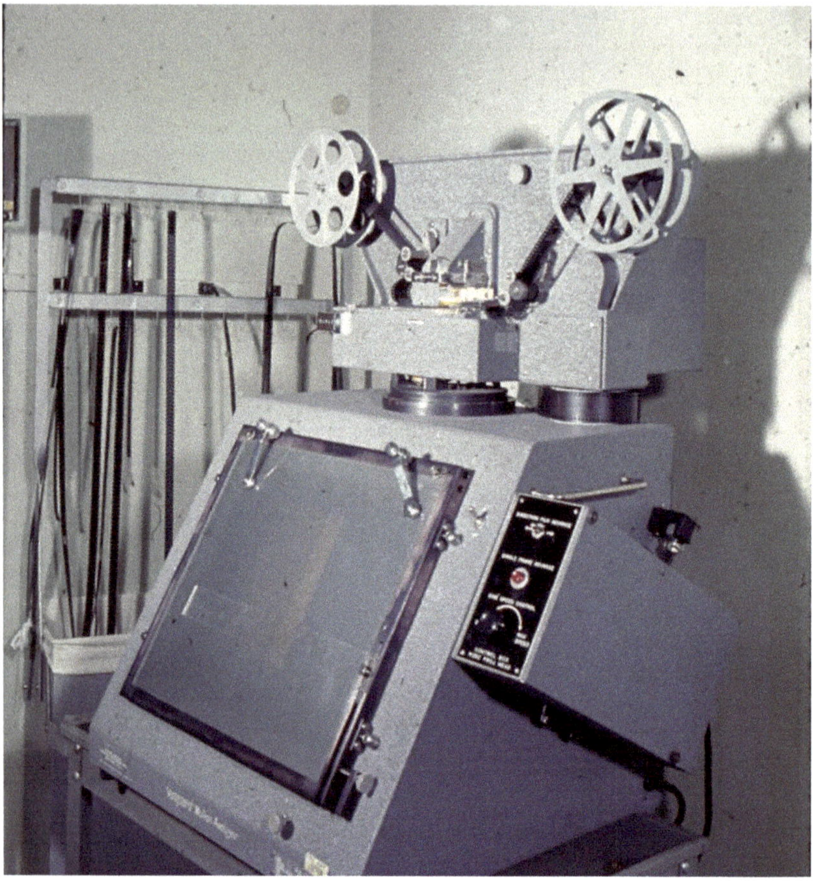

Figure 2.20 Vanguard motion analyser

calculations. Later improvements included the incorporation of electronic hand digitisers and an early computer to store and perform the reconstruction calculations but the approach remained very labour-intensive and not feasible as a general clinical tool. By 1972, the researchers reported that acquisition would take an average of 20 minutes to complete and a single investigator could derive five joint angles in approximately 2 hours. Despite the practical technical limitations, Sutherland and his team remained committed to the potential for clinical gait analysis, publishing papers on crouch gait [35], gait analysis in cerebral palsy [36], the role of the ankle plantar flexors [37] and the pathomechanics of gait in Duchenne muscular dystrophy [38]. Others could see the possibilities of this emerging science but very few were willing to invest in the levels of time and effort required to extract 3-D clinical motion measurements until there were significant improvements in the technique most likely through developments in computational automation.

In Europe between the wars, Richard Scherb (1880–1955) concentrated on studying phasic muscle activity during gait cycles. He developed a manual treadmill using an endless canvas fitted around drums and moving on a wooden board. Scherb used a palpation technique [39] to evaluate muscle tone together with electrodes under the feet to acquire the phases of contact during gait cycles. He used the outcomes to draw conclusions about gait abnormalities and to plan post-procedure recovery protocols [40]. In 1960, Willy Teillard founded a gait analysis laboratory within the University Children's Hospital in Basel, taken over by another surgeon, Jürg Baumann in 1964, following a 2-year fellowship with David Sutherland and Sheldon R. Simon MD. He continued his work on correlating electromyography with cine film [41,42] but remains best known for his images of movements of the foot taken from underneath a transparent plate during the stance phase.

2.8 The development of motion analysis systems

In 1967, E. Hans Furnée, PhD, of the Faculty of Applied Physics, Technical University, Delft, The Netherlands began the development of a motion analysis system based on television cameras that were becoming affordable for out-of-studio applications (Figure 2.21). His experiment used passive, circular paper markers, impregnated with a chemical that would glow brightly when illuminated by ultraviolet light, then applied to surface anatomical landmarks on a cat running on a treadmill. Using an early digital computer, he and his researchers were able to

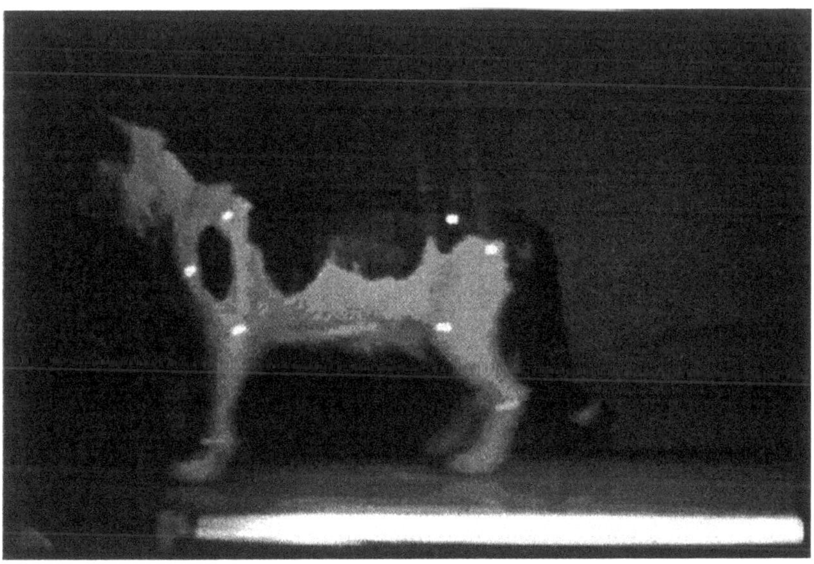

Figure 2.21 E. Hans Furnée. Measurement of animal joint angles (1967) [43]

derive multiple joint angles from the animal [43]. The light source was strobed to minimise motion smear in each video frame with the technological achievements remaining a window into the widespread future adoption of the photo-electronic method to the measurement of the kinematics of human and animal gait. Furnée continued the development in the form of the PRIMAS system [44–46] from the Motion Laboratory of his University with further innovations including the application of circle fitting algorithms to identify marker centres in real-time.

Professor John Paul, PhD (1927–2013), of Strathclyde University was also interested in the measurement of kinematics using a pair of orthogonal and synchronised cine-film cameras with a locally constructed ground reaction force plate (Figures 2.22 and 2.23). He became aware of Hans Furnée's work when he published his 1967 hybrid instrumentation paper in Stockholm [47]. Professor Paul allocated two research students to develop a two-camera system to achieve the 3-D reconstruction of motion. In 1972, Mick Jarrett and Brian Andrews commenced their PhD research to develop a functioning system based around a Digital Equipment Corporation (DEC) Programmed Data Processor-12 (PDP-12) minicomputer (Figure 2.24). Brian Andrews did not complete his studies at that time but did return as a staff member in 1980. He completed further development work by interfacing the bespoke motion analysis equipment to a PDP-11 minicomputer [48–50]. Professor Paul stated in a letter to David Sutherland in 2002 [32] that in order to improve marker definition and to minimise image smear, Andrews had investigated the early use of camera shutters and believed he was the first to employ strobed infrared light sources. Both methods would gain wide acceptance in future research and commercial motion capture systems. Professor Paul, due to his research output and of that from a number of his former students across the

Figure 2.22 Professor John Paul (1927–2013)

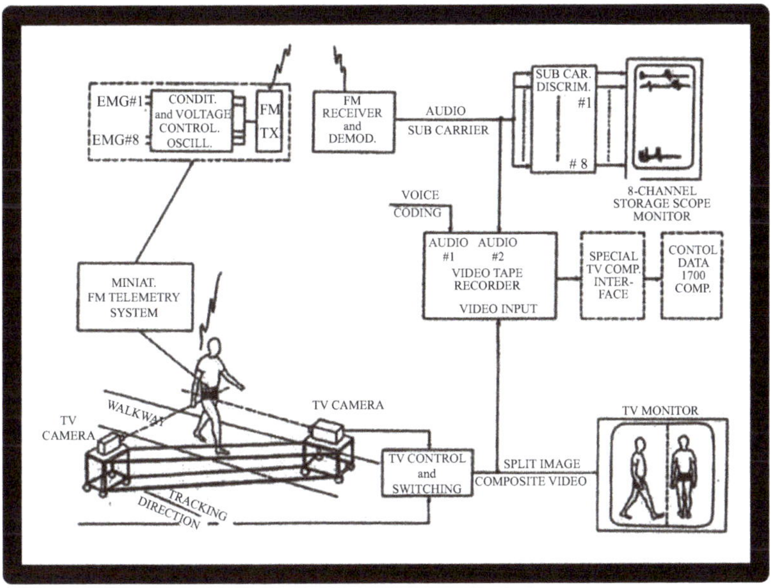

Figure 2.23 David A. Winter, CINTEL System (1969) [56]

Figure 2.24 Digital Equipment Corporation PDP-11/40

world, played a very important part in the later rise of clinical motion analysis. As an aside, Professor Brian Andrews was one of the doctoral supervisors of the author.

Jarret completed his studentship and was later employed jointly by the University and the limb-fitting surgeon, Professor George Murdoch (1920–2004) at Dundee to develop a second system locally. Julian Morris DPhil at the Nuffield Orthopaedic Centre, University of Oxford became aware of the Strathclyde work in 1973 and during 1974–75 started working with Jarret to introduce 3-D automated gait analysis technology at their respective hospitals. Morris arranged for Malcolm Herring, a very experienced electronics designer in Oxford, to redesign Jarret's prototype with three units being built for Strathclyde, Dundee and Oxford.

2.9 The introduction of the computer

The practical implementation of the emerging motion analysis technology was wholly dependent on the commensurate development of affordable minicomputers and flexible retro-reflective material for those designs that were reliant on the application of passive markers to palpated surface anatomical landmarks.

The most common early minicomputer available to engineers and physics graduates of the late 1970s to late 1980s was the DEC PDP series. DEC was founded by Ken Olsen and Harlan Anderson who as engineers at the Massachusetts Institute of Technology on early designs launched their first product in 1961. During the following decade, they produced machines aimed towards the laboratory marketplace, culminating in 1964 with the introduction of the PDP-8, still regarded as the foundation design across the world. The PDP-11 was a later generic 16-bit computer that incorporated a bespoke backplane structure marketed as UNIBUS which supported all peripherals through memory mapping. New devices could be easily added by simply installing a hardware interface board into the backplane and then installing software that could read and write to the mapped memory. The ease of interfacing spawned an enormous market in third-party add-ons which of course included bespoke motion capture interface boards. The design was later expanded to allow paged physical memory, memory protection features, multi-tasking and time-sharing. By the end of 1996, over 600,000 units had been sold. The ubiquitous design also included display terminals and incorporated an operating system called RSX-11 that supported a wide variety of programming languages.

2.10 Retro-reflectivity

Retro-reflectivity is the physical phenomenon exhibited by a device or material that reflects radiation, usually light, back to its source with minimal scattering (Figure 2.25a and b). Unlike a planar mirror placed exactly perpendicular to a radiation wave front with a zero angle of incidence, a retro-reflector will exhibit a wide angle of incidence or acceptance angle. The acceptance angle of a single

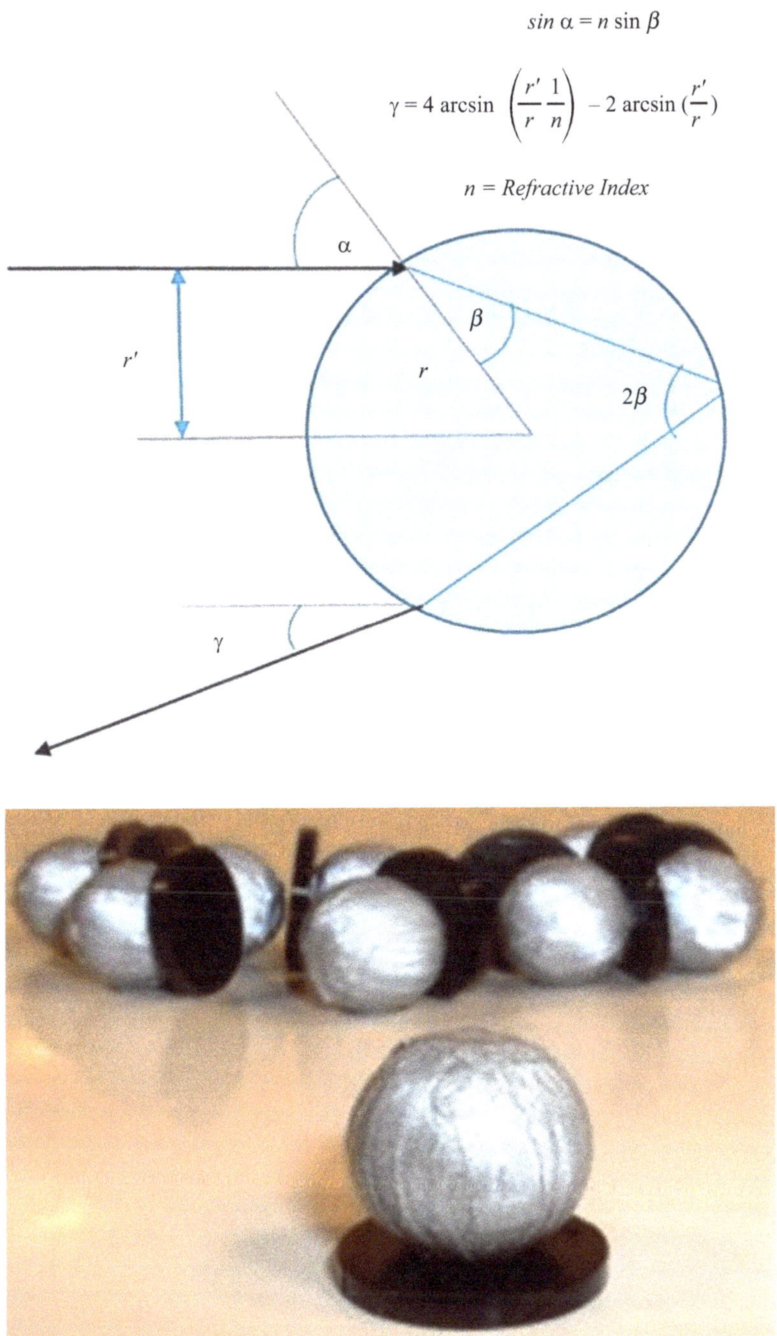

(a)

(b)

Figure 2.25 *(a) Retro-reflectivity equation. (b) Retro-reflective motion capture markers. (Source: Vicon Motion Systems Ltd)* retro-reflector or material supporting an array of retro-reflectors is the incident angle at which a reflected ray is present. A common type of retro-reflector consists of multiple miniature spherical, high refractive index optical elements arranged uniformly upon a flexible and reflective surface. One commercial example is 3M™ Scotchlite™ Sheeting # 7610 [51] that has been successful when used to coat passive optical motion capture markers. The material has retro-reflectivity characteristics 900 times greater than a white paper and still at 700 times at $\pm$ 45° so exhibiting a wide acceptance angle.

2.11 The commercialisation of optical motion capture technologies

Morris resigned from the Nuffield Orthopaedic Centre in 1977 to take up a new appointment in the industry as the Technical Director of Oxford Medical Systems, a subsidiary of the Oxford Instruments Group. Although his company was focussed towards products for cardiology applications, Morris believed there was a potential market for an automated 3-D gait analysis system. He hired Malcolm Herring, Graham Klyne and Annabel MacLeod to create commercial versions of Jarett's hardware under licence to the researcher and the Universities of Strathclyde and Dundee. Klyne and Morris designed and wrote the 3-D photogrammetric software drawing on the published ideas of many others including the seminal biomechanical mathematical achievements of Dr Herman J. Woltring (1943–92). In 1983–84, the Oxford Instruments Group decided to dispose of the biomechanics business which was sold to Morris and a number of co-founders including the development staff and the author to later become Oxford Metrics Ltd with the product VICON (from Video-Convertor) (Figure 2.26).

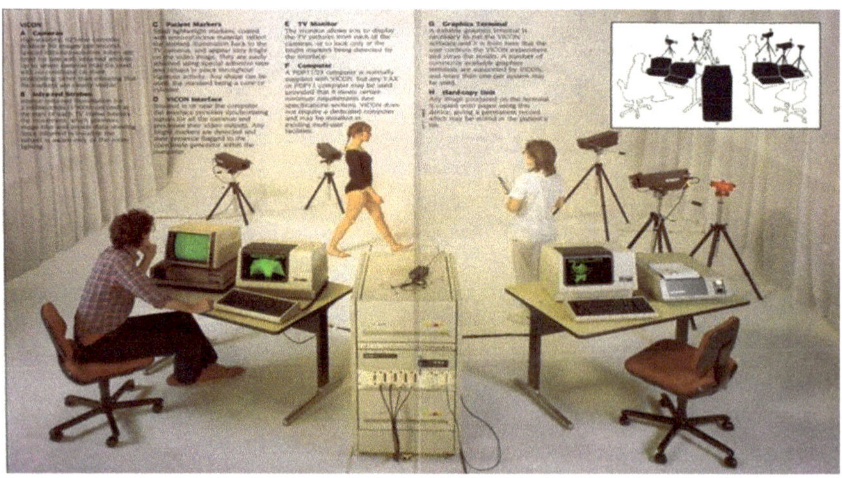

Figure 2.26 First-generation gait analysis system (1978) (Source: Vicon Motion Systems Ltd)

The first VICON system was shipped in 1980 to Eric Radin, MD in West Virginia to study the walking and running gait of sheep on a treadmill with the second being supplied to Sheldon Simon MD at the Boston Children's Hospital a few months later. Following a visit to Radin's laboratory in 1983, Sutherland with two colleagues, Ed Biden DPhil and Marilynn Wyatt PT purchased a VICON system in 1984. Biden wrote custom clinical applications software for the San Diego Gait Laboratory before undertaking a cross-validation exercise by comparing the cine film results and those of the new system using data captured from typically developed individuals before transitioning to the video camera method. The early VICON systems were still dependent on tracking and labelling in 2-D, which remained challenging with Dr Simon stating that in his opinion there was no improvement in processing time between cine film and video measurement methods.

Michael Whittle, MD, PhD completed a master's degree in biomechanics after completing a post-internship surgical research. As a Royal Air Force medical officer, he was seconded to National Aeronautics and Space Administration (NASA) in Houston for 3 years to supervise six medical experiments on the Skylab Space Station programme. Whittle returned to the United Kingdom before taking over from Morris at the Nuffield Orthopaedic Centre. He used his unique experience as both a scientist and clinician to develop 3-D motion capture software to help further embed the gait analysis laboratory as an essential function within the Hospital as it remains to this day. Until his retirement, Whittle held the Cline Chair of Rehabilitation Technology at the University of Tennessee at Chattanooga. He is a co-author of a book entitled, 'Whittle's Gait Analysis', which is now in its fifth edition [52].

Another early European commercial system developer, formed in 1986, is Bioengineering Technology Systems (BTS) located in Milan, Italy. Like VICON, the technology originates from academic research with significant contributions of the Bioengineering Centre of the Pro Juventute Foundation and the Politecnico di Milano. The engineering contributions of Giancarlo Ferrigno PhD, Professor Antonio Pedotti and Professor Aurelio Capozzo were fundamental to the development of the ELITE system [53–55]. The company's focus from its roots continues to be developing turnkey clinical gait analysis solutions through combining kinematics, kinetics and electromyographic acquisition.

Significant work was also occurring in North America in parallel to the European experiences. The chief surgeon of the Shriners Hospital in Winnipeg, Canada, Robert K. Greenslaw, MD was aware of Inman and Sutherland's work co-opting engineer David Winter, PhD, PEng (1930–2012) of the University of Manitoba to develop a gait laboratory for his hospital. The results of the early work submitted in 1969 and published in 1970 by Dinn *et al.* [56] were followed by a further description of their technology in 1972 [57]. Later publications concentrated on biomechanics and clinical gait analysis studies that became cornerstones of the emerging science [58–60]. Winter's book 'Biomechanics and Motor Control of Human Movement' [61] requires reading together with Whittle's *et al.* work [52] for any student of the subject.

In 1982, Dr John Greaves founded the Motion Analysis Corporation, based on his earlier work in tracking microscopic marine plankton. The company provides 3-D passive optical motion capture systems for a wide range of applications, including use in clinical gait laboratories and sports analysis research facilities. In collaboration with the Cleveland Clinic Foundation and in conjunction with Chester Tylkowski, MD, then at the Human Motion Laboratory within the University of Florida at Gainesville, they developed their ORTHOTRAK software. Tylkowski first became interested in gait analysis when working as a fellow with Dr Simon at the Boston Children's Hospital in 1978–79 and has since opened a number of clinical gait laboratories across the United States.

2.12 The rise of clinical gait analysis

In 1981, orthopaedic surgeon, James R. Gage, MD established a gait analysis laboratory within the Newington Children's Hospital, Connecticut using bespoke technology developed by United Technologies Research Corporation, co-located in the city (Figure 2.27). The laboratory routinely performed post-operative gait analysis from among patient cohorts who had undergone pre-operative analyses. The research provided the avenue for the publication of dedicated clinical papers [62–66] and a seminal work entitled 'Gait analysis in cerebral palsy' [67] in 1991, now in its second edition [68] in 2009. Following Gage's move to Gillette Children's Hospital, Minnesota in 1990, the laboratory continued under Peter DeLuca, MD as Medical Director with senior scientific and clinical support by Roy B. Davis, PhD and Sylvia Õunpuu, MSc [69], a former student of David Winter. The laboratory later moved to the Children's Hospital of Connecticut with many clinical papers subsequently published from both the Minnesota and Connecticut centres to this day. Gage's greatest contribution remains as an unswerving advocate of the employment of routine pre- and post-operative gait analysis in patients with cerebral palsy.

DEC designed a replacement for their PDP series of minicomputers with an improved 32-bit computer series named VAX sometimes referred to as 'super-minis'. The equipment supported the new concept named 'X Window' that permitted access to multiple activities on a single terminal. VAX computers were supplied throughout the 1980s with over 400,000 sold. In parallel, Oxford Metrics developed their second-generation equipment sold as VICON-VX which separated the acquisition unit from the computer with data transfer achieved via an early network protocol. The unit contained a single board PDP computer with camera and network interface electronics installed onto a DEC UNIBUS backplane. A prototype created by Malcolm Herring, Annabel MacLeod and the author proved to be commercially viable with over 70 systems sold between 1988 and 1993. Much of the acceptance of the new system can be credited to Dr Andrew Dainis and Doug McGuire of the National Institute of Health (NIH), Bethesda, Maryland, who in 1987 collaborated to develop software to more efficiently process raw digitised video data into 3-D coordinates. The researchers formed a company called

Figure 2.27 James R. Gage MD

ADTECH to commercially exploit their software through awarding distribution rights to Oxford Metrics to market their AMASS software as part of the VX system. Dainis and McGuire also described the C3D binary format which continues to be the default file structure that is independent of a commercial offering. A typical C3D file stores both the 3-D reconstruction of the position of palpated kinematic points in space and synchronised analogue kinetic and electromyography (EMG) information from a gait analysis trial. Other features included applying circle fitting algorithms to pixels outlining markers in video images to derive 2-D digitised

Figure 2.28 CCD camera, Willard Boyle and George Smith (1969)

marker centres; minimising tangential and radial distortions due to lens non-linearity; automated 3-D reconstruction; and tracking of un-identified 2-D marker locations in digitised video data (Figure 2.28).

In 1994, Dainis designed a now widely adopted, quick volume calibration technique in terms of camera location/orientation and image scaling that did not rely on knowing the location of multiple markers by an alternate and time consuming, direct linear transformation method. This major innovation eliminated a chief source of inaccuracy in motion capture systems and reduced calibration times from hours to minutes.

2.13 The introduction of solid-state camera technologies

Another significant source of inaccuracy in early motion capture systems was due to a reliance on tube video cameras (VIDICON and silicon diode arrays developed in the 50s) whose 2-D image was subject to distortion with time and changes in ambient temperature. A successful gait analysis was dependent on all cameras reaching a stable temperature before undertaking a volumetric calibration, so limiting the practical usefulness of the method in a busy clinic. VIDICON tubes were popular in the 1970s and 1980s but were rendered obsolete by the availability of inherently stable solid-state image sensors in the form of 'Charge Coupled Devices' (CCD), invented in 1969 by Willard Boyle and George Smith of Bell Labs, and later 'Complementary Metal Oxide Semiconductor' (CMOS) sensors.

In 1993, Oxford Metrics announced their third-generation VICON 370 System which was based on the emerging Microsoft ® Windows Operating System,

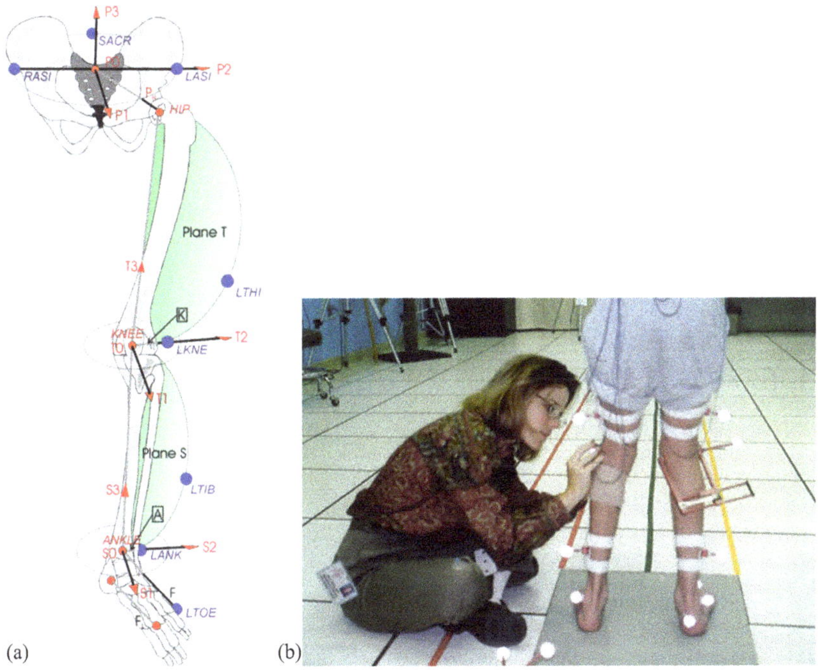

(a) (b)

Figure 2.29 (a) Helen Hayes marker set. (b) Helen Hayes marker set applied to a patient. (Source: Vicon Motion Systems Ltd)

resident on a low-cost personal computer (PC). Realising that its future lay in clinical gait analysis, the company responded by incorporating Clinical Manager software based on the state-of-the art gait models of the period.

The story of motion capture is far from over with the entry of many new companies and new generations of systems constantly being launched with clinical gait laboratories now available to patients across the world. The engineering problem to reliably capture and track points in space has long been solved with research effort continuing to be focussed on achieving a better understanding of the association between the precise measurements of the location of bony landmarks on the skin surface and the actual location of the underlying skeletal structures during motion, commonly described as the 'soft tissue artefact' so elegantly described by Sutherland [32] as trying '...to measure the movements of a broomstick within a gunny sack' (Figure 2.29a and b).

2.14 The development of unified biomechanical models

Murali Kadaba PhD became interested in gait analysis after he was appointed as a research scientist within the Helen Hayes Rehabilitation Hospital in West Haverstraw, New York. His interest was further piqued after visits to Dr Perry at Rancho Los Amigos and Dr Sutherland in San Diego. Dr Kadaba received an NIH grant in 1984 to investigate the reproducibility and reliability of gait data [70], later focussing on the numerical representations of kinematic and kinetic data for pattern recognition in spastic diplegia [71]. A secondary outcome was that under his leadership, a team within the hospital created bespoke clinical software to fill a void where laboratories were then wholly dependent on generalist motion capture software. The software included a detailed description and protocol when placing lower limb markers in a consistent way later renamed as the widely accepted 'Helen Hayes marker set'. Other marker sets including the Cleveland Clinic Foundation version and the later Helen Hayes–Newington gait model were introduced and incorporated into commercial offerings. This important initiative meant that the ready availability of a family of closely related and widely used biomechanical models for gait analysis could be used by multiple laboratories to exchange clinical experience based on a unified measurement protocol. Gait and movement analysis societies now exist in the United States and Europe with conferences arranged throughout each year. Stephen J. Stanhope, PhD formerly of the NIH Biomechanics Laboratory, has been a strong advocate of inter-laboratory reliability and a need to quantify system kinematic performance. His achievement has been to lift the standards of clinical gait analysis laboratories in the United States with similar initiatives in the European Union and other countries across the world.

Roy B. Davis *et al.* [69] described the clinical application of movement analysis as to collect information to aid in the understanding of the aetiology of presenting gait abnormalities and to assist in treatment decision-making. To this day, the interpretation of acquired kinematic and kinetic data remains in the hands of inter-disciplinary teams with substantial knowledge of normal and pathologic gait where characteristics are measured, abnormalities are identified, causes are postulated and treatments are proposed.

Many researchers have proposed techniques to identify joint centres and axes functionally through fitting the data to an idealised joint model that also incorporates some form of soft tissue artefact compensation. This work continues and like the earlier innovations is rapidly gaining acceptance through close collaborative efforts between clinicians, scientists and engineers in laboratories, academia and industry across the world.

2.15 Wider applications of motion capture

The need to quantify sporting performance has resulted in the development of cameras with significant improvements in quantum efficiency, speed and resolution that are now capable of working outside the laboratory to track fast and subtle motions with consistency. Access to this new technology has meant that the motion of the Mars Rover over desert terrain could be measured together with many other new and exciting applications such as tracking swarms of quadcopters and even measuring the locomotion of kangaroos (Figure 2.30).

Since the release of the film *Titanic* in 1997, there has been an explosion in the use of motion capture to enhance films, games, virtual reality environments, music

Figure 2.30 Mars Rover. Image Courtesy of NASA and California Institute of Technology (Caltech)

Figure 2.31 Paddington Bear. Image courtesy of the Press Association

videos and television that have brought many stories to life, thought almost impossible before the ready availability of the technology to the entertainment sector. With it has brought many challenges including the need to capture the motion of multiple characters simultaneously, in large volumes and with high camera counts. The sector has also demanded improved subtlety in the measurement of the motion of hands, feet and facial expressions with a need to provide a real-time reconstruction of the animated characters who may or may not be humanoid (Figures 2.31–2.34).

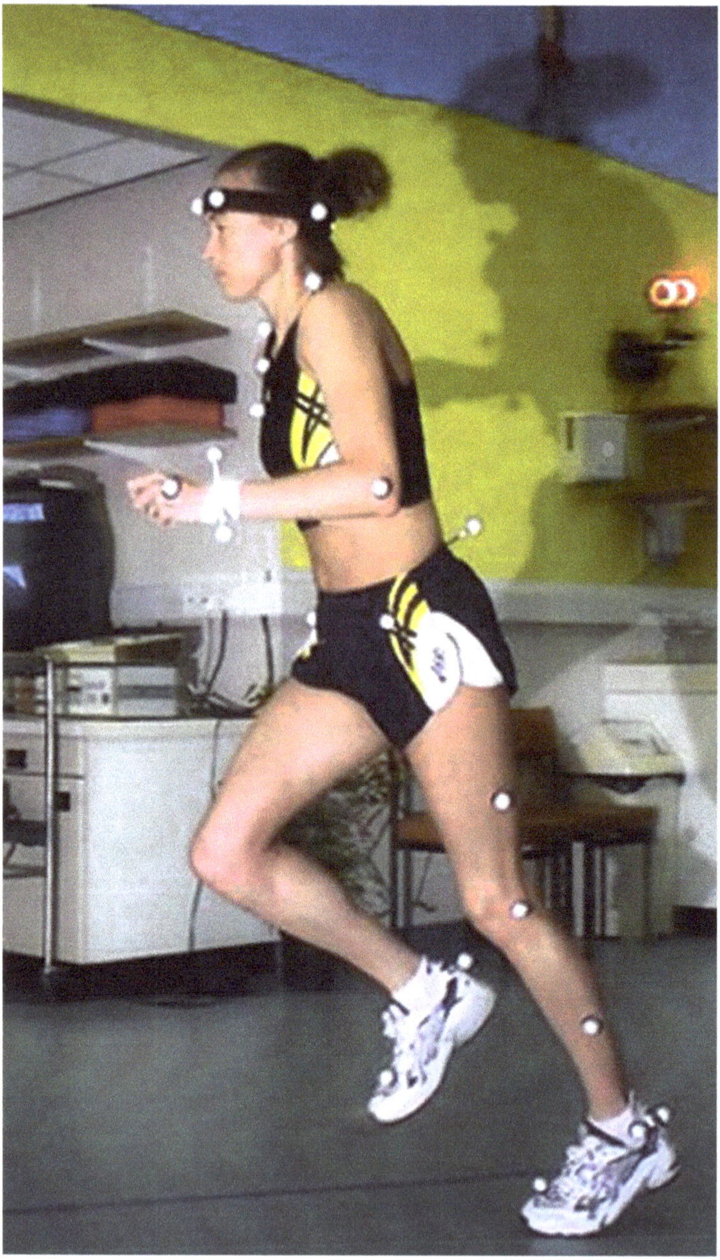

Figure 2.32 Measuring sporting performance (Source: Vicon Motion Systems Ltd)

Figure 2.33 Vicon 370 system (Source: Vicon Motion Systems Ltd)

Figure 2.34 Vicon 370 camera (Source: Vicon Motion Systems Ltd)

The demands of life sciences including clinical gait analysis and sport; engineering including ergonomics; and the entertainment sector have resulted in a synergy that has driven and improved the core technologies for the benefit of all. In just one example, the life sciences' need to continuously improve skeletal modelling has benefited animators to create ever-more realistic characters, whilst their requirements have directly improved how the technologies are applied and data are presented within the clinical gait and biomechanics communities.

References

[1] Kozma E.E, Webb N.M., Harcourt-Smith W.E.H., *et al.* (2018). Hip extensor mechanics and the evolution of walking and climbing capabilities in humans, apes, and fossil hominins. *Proc Natl Acad Sci*, 115(16): 4134–4139; DOI:10.1073/pnas.1715120115

[2] Lovejoy C.O. (2004). The natural history of human gait and posture I: Back and pelvis. *Gait Posture*, 21: 95–112.

[3] Lorini, G., Bossi, D. and Specchia, N. (1992). *The concept of movement prior to Giovanni Alfonso Borelli*, 23.

[4] Andriacchi T.P. and Alexander E.J. (2000). Studies of human locomotion: past, present and future. *J Biomech*, 33: 1217–1224.
[5] Aristotle. On the Gait of Animals, Written 350 BCE, Translated by A. S. L. Farquharson.
[6] Ancillao A. (2017). Modern functional evaluation methods for muscle strength and gait analysis. In *Springer Briefs in Applied Sciences and Technology*. New York, NY: Springer. ISBN 978-3-319-67437-7.
[7] Jastifer J. R. and Toledo-Pereyra L.H. (2012). Leonardo da Vinci's foot: historical evidence of concept. *J Invest Surg*, 25(5): 281–285. DOI: 10.3109/08941939.2012.725011.
[8] Piolanti N., Polloni S., Bonicoli E., Giuntoli M., Scaglione M. and Indelli P. F. (2018). Giovanni Alfonso Borelli: the precursor of medial pivot concept in knee biomechanics. *Joints*, 6(3): 167–172. doi:10.1055/s-0038-1675164.
[9] Piccolino M. and Galvani L. (1997). Animal electricity: two centuries after the foundation of electrophysiology. *Trends Neurosci*, 20(10): 443–448.
[10] Weber W. and Weber E. (1991). *Mechanics of the Human Walking Apparatus*. Translated by Maquet P. and Furlong R. Berlin: Springer-Verlag.
[11] Harless E. (1860). The static moments of human limbs (in German). Treatises of the Math.-Phys. *Class of the Royal Acad Sci Bavaria*, 8: 69–96 and 257–294.
[12] Drillis R., Contini R. and Bluestein M. (1964). Body segment parameters; a survey of measurement techniques. *Artif Limbs*, 8: 44–66.
[13] Baker R. (2007). The history of gait analysis before the advent of modern computers. *Gait Posture,* 26: 331–342.
[14] Carlet G. (1872). Essai experimental sur la locomotion humaine, etude de la marche. *Ann Sci Nat: Zool*; 15.
[15] Braune W. and Fischer O. (1987). The human gait. Translated by Maquet P and Furlong R. Berlin: Springer-Verlag.
[16] Jansons H. (1992). Bernstein: the microscopy of movement. In: Cappozzo A, Marchetti M and Tosi V, editors. *Biolocomotion: A Century of Research Using Moving Pictures.* Rome: International Society of Biomechanics.
[17] Bongaardt R. and Meijer O.G. (2000). Bernstein's theory of movement behaviour: historical development and contemporary relevance. *J Motor Behav,* 32(1): 57–71.
[18] Marey E.-J. (1883). De la mesure dans les differents acts de la locomotion. *C R Acad Sci Paris*, 97: 820–825.
[19] Amar J. (1916). Trottoir dynamographique. *C R hebdomadaires seances Acad Sci,* 163: 130–133.
[20] Amar J. (1920). The Human Motor. London: George Routledge and Sons.
[21] Elftman H. (1938). The measurement of the external force in walking. *Science*, 88: 152–153.
[22] Cunningham D.M. and Brown W.G. (1951). Two devices for measuring the forces acting on the human body during walking. *Proc Soc Exp Stress Anal,* 9: 75–90.

[23] Klopsteg P.E. and Wilson P.D. (1954). *Human Limbs and Their Substitutes.* New York, NY: McGraw-Hill.
[24] Murray M.P., Drought A.B. and Kory R.C. (1964). Walking patterns of normal men. *J Bone Joint Surg*, 46-A: 335–360.
[25] Karpovich P.V. and Karpovich G.P. (1959). Electrogoniometer: a new device for study of joints in action. *Fed Proc*, 8:79.
[26] Bajd T., Stanic U., Kljajic M. and Trnkoczy A. (1976). On-line electro-goniometric gait analysis. *Comp Biomed Res*, 9(5): 439–444.
[27] Lamoureux L. (1971). Kinematic measurements in the study of human walking. *Bull Prosthet Res*, 10(15): 3–8.
[28] Chao E.Y.S. (1980). Justification of triaxial goniometer for the measurement of joint rotation. *J Biomech*, 13: 989–1006.
[29] Reaz, M.B.I., Hussain, M.S. and Mohd-Yasin, F. (2006). Techniques of EMG signal analysis: detection, processing, classification and applications (Correction). *Biol Proc Online;* 8: 163.
[30] Perry J. (1992). Gait analysis. Thorofare: SLACK.
[31] Perry J. (1974). Clinical gait analyzer. *Bull Prosthet Res*, Fall: 188–192.
[32] Sutherland D.H. (2002). The evolution of clinical gait analysis. Part II kinematics. *Gait Posture,* 16(2): 159–179.
[33] Sutherland D.H. and Hagy J. (1972). Measurement of gait movements from motion picture film. *J Bone Joint Surg*, 54A: 787–797.
[34] Sutherland D.H., Schottstaedt E.R., Vreeland R.W., Oyama F.R. and Hagy J. L. (1967). Measurement of movements and timing of muscle contraction from movie film. *J Bone Joint Surg*, 49A: 1248–1249.
[35] Sutherland D.H. and Cooper L. (1978). The pathomechanics of progressive crouch gait in spastic diplegia. *Orthop Clin North Am*, 9: 143–154.
[36] Sutherland D.H. (1978). Gait analysis in cerebral palsy. *Dev Med Child Neurol*, 20(6): 807–813.
[37] Sutherland D.H., Cooper L. and Daniel D. (1980). The role of the ankle plantar flexors in normal walking. *J Bone Joint Surg*, 62(3): 354–363.
[38] Sutherland D.H., Olshen R.A., Cooper L., *et al.* (1981). The pathomechanics of gait in Duchenne muscular dystrophy. *Dev Med Child Neurol*, 23: 3–22.
[39] Basile M., Coco M., Buscemi A. and Petralia M. (2018). Brief review on the historical evolution of kinematics and its application in clinical practice. *Acta Med Mediterran,* 34: 911.
[40] Scherb R. (1927). Ein Vorschlag zur kinetischen Diagnostik in der Orthopädie. *Verhd Dtsch Orthop Gesell. Beilageheft der Z Orthop Chir*, 48: 462–472.
[41] Baumann J. and Hanggi A. (1977). A method of gait analysis for daily orthopaedic practice. *J Med Eng Technol*, 2: 86–91.
[42] Baumann J. and Baumgartner R. (1974). Synchronised photo-optical and EMG gait analysis with radiotelemetry. In: Proceedings of the Biotelemetry II Second International Symposium.

[43] Furnée E.H., Halbertsma J., Klunder G., *et al.* (1974). Proceedings: automatic analysis of stepping movements in cats by means of television system and a digital computer. *J Physiol*, 240(2): 3P–4P.

[44] Furnée E.H. (1989). TV/computer motion analysis systems: the first two decades. Delft: Delft University of Technology.

[45] Furnée E. (1990). Opto-electronic movement measurement systems: aspects of data acquisition, signal processing and performance. In: *Otto Bock Foundation International Symposium*. Berlin, Germany: Mecke Druck and Verlag, Duderstadt.

[46] Furnée E.H. and Jobbágy Á. (1993). Precision 3-D motion analysis system for real-time applications. *Microprocessors Microsyst*, 17(4): 223–231.

[47] Furnée E.H. (1967). Hybrid instrumentation in prosthetics research. In: *Proceedings of the 7th International Conference on Biomedical and Bioinformatics Engineering*, Stockholm.

[48] Andrews B.J., Jarrett M.O. and Paul J.P. (1975). On-line kinematic data acquisition. In: *B.E.S. Conference*, Edinburgh, Scotland.

[49] Jarrett M.O., Andrews B.J. and Paul J.P. (1976). A television–computer system for the analysis of human locomotion. In: *I.E.R.E. Golden Jubilee Conference. An exhibition on Application of Electronics in Medicine*. Southampton University, Southampton, England.

[50] Paul J.P., Jarrett M.O. and Andrews B.J. (1974). Quantitative analysis of locomotion using television. In: *World Congress I.S.O,* Montreux.

[51] http://multimedia.3m.com/mws/media/169286O/3mtm-scotchlitetm-high-gain-reflective-sheeting-7610.pdf?fn=7610.pdf. Accessed January 2020.

[52] Levine D., Richards J. and Whittle M. (2012). Whittle's Gait Analysis (5th ed.). London: Churchill Livingstone.

[53] Ferrigno G. and Pedotti A. (1985). ELITE: a digital dedicated hardware system for movement analysis via real-time TV signal processing. *IEEE Trans Biomed Eng*, 32(11): 943–950.

[54] Cappozzo A., Leo T. and Pedotti A. (1975). A general computing method for the analysis of human locomotion. *J Biomech*, 8(5): 307–320.

[55] Cappozzo A. (1984). Gait analysis methodology. *Hum Mov Sci*, 3: 27–50.

[56] Dinn D.F., Winter D.A. and Trenholm B.G. (1970). CINTEL–computer interface for television. *IEEE Trans Comput*, C-19: 1091–1095.

[57] Winter D.A., Greenlaw R.K. and Hobson D.A. (1972). Television–computer analysis of kinematics of human gait. *Comp Biomed Res*, 5: 498–504.

[58] Letts R.M., Winter D.A. and Quanbury M. (1975). Locomotion studies as an aid in clinical assessment of childhood gait. *CMA J*, 112:1091–1094.

[59] Winter D.A., Quanbury A.O., Hobson D.A., *et al.* (1974). Kinematics of normal locomotion—a statistical study based on TV data. *J Biomech*, 7(6): 479–486.

[60] Winter D.A. and Yack H.J. (1987). EMG profiles during normal human walking: stride-to-stride and inter-subject variability. *Electroencephalogr Clin Neurophysiol,* 67(5): 402–411.

[61] Winter D.A. (2009). *Biomechanics and Motor Control of Human Movement* (4th ed.). New York, NY: Wiley.

[62] Gage J.R., Fabian D., Hicks R. and Tashman S. (1984). Pre- and post-operative gait analysis in patients with spastic diplegia: a preliminary report. *J Pediatr Orthop*, 4: 715–725.

[63] Gage J.R., Perry J., Hicks R.R., Koop S. and Werntz J.R. (1987). Rectus femoris transfer to improve knee function in children with cerebral palsy. *Dev Med Child Neurol*, 29: 159–166.

[64] Gage J.R. (1990). Surgical treatment of knee dysfunction in cerebral palsy. *Clin Orthop*, 253: 45–54.

[65] Gage J.R. (1993). Gait analysis. An essential tool in the treatment of cerebral palsy. *Clin Orthop*, 288: 126–134.

[66] Gage J.R. (1994). The role of gait analysis in the treatment of cerebral palsy. *J Pediatr Orthop*, 14(6): 701–702.

[67] Gage J.R. (1991). *Gait analysis in cerebral palsy.* New York, NY: MacKeith Pres.

[68] Gage J.R., Schwartz M.H. , Koop S.E. and Novacheck T.F. (2009). *Identification and Treatment of Gait Problems in Cerebral Palsy* (2nd ed.). New York, NY: MacKeith Press.

[69] Davis R.B., Õunpuu S., DeLuca P.A. and Romness M.J. (1998). Clinical gait analysis and its role in treatment decision-making. *Med Gen Med,* 1(1) [formerly published in Medscape Orthopaedics & Sports Medicine eJournal 2(5), 1998]. Available at: http://www.medscape.com/viewarticle/440148

[70] Kadaba M.P., Ramakrishnan H.K., Wootten M.E., Gainey J., Gorton G. and Cochran G.V. (1989). Repeatability of kinematic, kinetic, and electromyographic data in normal adult gait. *J Orthop Res*, 7(6): 849–860.

[71] Kadaba M.P., Ramakrishnan H.K., Jacobs D., Chambers C., Scarborogh N. and Goode B. (1995). Pattern recognition in spastic diplegia. *Biomecanica*, 3(4): 49–58.

Chapter 3

Gait analysis – kinematics

Robert A. Needham[1]*, Jonathan K. Sinclair*[2]*, Michael Lawson*[1]*, Roozbeh Naemi*[1] *and Nachiappan Chockalingam*[1]

3.1 Introduction

Modern biomechanical movement analysis is undertaken using motion capture (camera) based systems. Simple 2-D dimensional analyses which are now almost completely limited to field-based analyses require as little as one camera, whereas 3-D examinations of human movement can require many cameras depending on the movement and space requirements of the examination. From a 3-D analysis standpoint, it is important to consider several factors to optimise the quality of the kinematic data (on which this chapter focusses) that you collect and maximise its usefulness in both clinical and sports settings.

Inertial measurement units (IMUs) are often employed in field-based analysis of walking and running as they are lightweight, easily transportable and have low power requirements [1]. IMUs contain either or a combination of the following sensors: accelerometers, gyroscopes and magnetometers. Accelerometers measure the linear acceleration of an object and gyroscopes measure the rotational acceleration, and together they can quantify the frequency and intensity of movement in all three planes of motion [2,3]. Magnetometers, however, measure the ambulatory position and orientation of an object and are primarily employed within IMUs for azimuth determination and correction of gyroscope drift in the horizontal plane [4,5]. This chapter will also explore how commonly used IMU parameters within walking and running analysis provide information on either the quantity, rate, or quality of locomotion.

3.2 Optoelectronic stereophotogrammetric marker-based systems

Data capture of human movement no longer uses video-based recordings to perform kinematic analyses in 3-D, rather optoelectronic stereophotogrammetric

[1]Centre for Biomechanics and Rehabilitation Technologies, Staffordshire University, Stoke on Trent, UK
[2]Reader in Sport Exercise & Nutritional Sciences, School of Sport & Health Sciences, Faculty of Allied Health & Wellbeing, University of Central Lancashire, Preston, UK

systems are now widely used to capture human movement by tracking the trajectories of retroreflective markers attached to the body itself. For 3-D dimensional analyses, it is necessary to position these retroreflective markers onto specific anatomical landmarks in order to represent specific body parts or to give their correct term 'segments'. The method and selection of anatomical landmarks/retroreflective markers used to represent segmental movement in 3-D are termed a marker set.

3.3 Soft tissue artefact

The displacement of retroreflective markers away from anatomical landmarks during dynamic tasks such as walking and running is a general concern when using marker-based systems due to the influence on segmental angle calculations. As a consequence of soft tissue artefact (STA), there are two types of error associated with changes in the position of markers. The relative movement between markers that define a rigid segment describes relative error, whereas absolute error refers to the movement of a marker against the anatomical landmark it is overlying. An applicable estimation of bone pose position and joint/segmental kinematics via advancements in kinematic and mathematical modelling have provided techniques that could compensate for the degree of STA experienced during movement [6–9].

3.4 Kinematic modelling of the pelvis and lower limbs

The simplest 'marker-set' for the quantification of segmental kinematics simply requires the fixation of retroreflective markers to specific anatomical landmarks at the proximal and distal ends of different segments. This allows the segment itself to be delineated using a linear line joining the two markers and joint angles quantified using either the internal or external angle created at the adjoining point to two segments – this process, however, is rarely used for 3-D kinematics due to its limitations in quantifying non-sagittal rotations and does not have its own technical term due to its simplicity. Another commonly adopted marker set in 3-D kinematic analyses is the Helen Hayes technique, in which anatomical landmarks are positioned bilaterally onto the acromion process, lateral humorous, anterior superior iliac spine, sacrum, greater trochanter, lateral femoral epicondyle, lateral malleoli, fifth metatarsal and calcaneus alongside wands that are positioned onto the thigh and shank segments to allow non-sagittal rotations to be calculated. The Helen Hayes marker set was widely used, although it had deficiencies in relation to the identification of joint centres which were identified by projecting medially from the lateral anatomical landmarks.

The most commonly utilised marker set in contemporary 3-D examination of human movement is the calibrated anatomical systems technique (CAST) [10]. The CAST method requires retroreflective markers to be positioned onto specific anatomical landmarks on the body segments in order to define the segment anatomy (i.e. anatomical frame) and tracking markers to allow the cameras to track the movement of each segment (i.e. technical frame). Tracking markers can be positioned arbitrarily onto the body segments and can be mounted directly to the skin, onto wands or onto rigid cluster plates. The CAST methodology requires each segment to be defined at its proximal and distal ends as well as medial and lateral aspects, therefore anatomical markers are typically (although not exclusively) positioned bilaterally onto the acromion process, iliac crest, anterior superior iliac spine, posterior superior iliac spine, greater trochanter, medial/lateral femoral epicondyles, medial/lateral malleoli, fifth metatarsal head and first metatarsal head.

The advantages of the CAST method are that it allows the segments to be modelled in six degrees of freedom, provided that it has a technical frame and an anatomical frame. This means that the interaction and relation between each segment in relation to the segment about which it proximally articulates with (i.e. a joint) can be quantified. Six degrees of freedom means that every segment in the body is able to move in all three directions, i.e. forwards and backwards, up and down and side to side as well as being free to rotate about all three anatomical planes, i.e. flexion and extension, abduction and adduction and also internal and external rotation; when the three movements are added to the three rotations, this gives six ways in which a segment can move which are termed degrees of freedom. It is necessary to have a sufficiently complex marker set, i.e. CAST, to allow six degrees of freedom in relation to the modelled segments in order to identify rotational motions independently in each of the three planes and obtain an accurate representation of joint movement that is essential to clinical practice. Whilst it is evident that more complex six degrees of freedom models are probably not necessary in order to measure joint kinematics in the sagittal plane (i.e. flexion/extension), coronal and transverse plane joint kinematics are extremely important in terms of being related to the aetiology of injuries, so it is very important in clinical practice that these rotations be measured accurately and managed through various orthotic and bracing techniques.

In the CAST marker system, with markers positioned onto the medial and lateral aspects of segments at its proximal and distal ends as well as the position of a cluster plate onto the segment, this allows the segment anatomy to be identified in relation to the position of the cluster itself. The positions of both anatomical and cluster (i.e. tracking markers) in relation to one another are identified by obtaining what is commonly known as a static calibration trial. This is a relatively straightforward process whereby the participant being examined using 3-D motion capture stands without movement (i.e. static) in an anatomical position within the motion capture space to allow a very short duration capture of all of the experimental markers. When a 3-D analysis of human movement is being undertaken, the software that is being used will transform the coordinates of the markers from what is typically referred to as the laboratory coordinate system to a segment coordinate system.

Although the CAST marker technique remains the gold standard in terms of the available surface mounted marker techniques available within the biomechanics literature, like all approaches whether good or bad it is not without its own limitations. First and foremost, this technique is reliant on palpable landmarks being available at the medial and lateral aspects of each segment, which presents a significant challenge when attempting to delineate the proximal end of the thigh segment. Whilst there is a clearly identifiable landmark on the lateral aspect of the proximal thigh at the greater trochanter, there is no palpable medial landmark at which to position a marker. This made the identification of the hip joint centre problematic for researchers, practitioners and clinicians involved in human movement. The hip joint centre landmark is critical in human movement analyses as it allows the determination of the anatomical reference frame of the femur. Accurately locating the hip joint centre is required to accurately quantify hip and knee joint kinematics in gait analyses.

A number of techniques currently exist for the estimation of the hip joint centre. The most commonly adopted method is the anatomical method of Bell *et al.* [11] whereby the position of the hip joint centre is estimated relative to anatomic pelvic landmarks. Specifically, the hip joint centre is identified using this technique at a point 14% of the anterior superior iliac spine breadth medially, 19% posteriorly and 30% distally from the ipsilateral anterior superior iliac spine landmark. More recently, a functional technique has also been developed that estimates the centre of rotation of the proximal aspect of the thigh segment, i.e. the hip joint centre relative to the pelvis using 3-D kinematic data. Finally, a projection technique method has also been developed that places the hip joint centre at one-quarter of the distance from the ipsilateral to the contralateral greater trochanter. Previous in-vivo and fluoroscopic investigations examining the accuracy of the various methods of identifying the hip joint centre have generally shown that the functional approach is superior in comparison to predictive methods. However, as the functional method relies on individuals being able to exhibit an almost unrestricted range of motion at the hip joint in order to create the movement required to generate the functional centre of joint rotation, this approach is not frequently adopted in analyses concerning pathological gait. Thus, the question regarding hip joint centre accuracy remains unclear within the biomechanical modelling literature.

However, in addition to accuracy, reliability of joint kinematics is also important in practice as the majority of analyses are typically undertaken with the aim of comparing information pre- and post-intervention. Whilst the literature is unclear regarding the most accurate method of identifying the hip joint centre location, there is more clarity regarding the most reliable approach. In a reliability investigation undertaken by Sinclair *et al.* [12] examining the test–retest efficacy of the above methods, the results showed significant test–retest differences in the coronal and transverse planes using the functional and projection techniques. Although strong test–retest intraclass correlations were observed for all techniques, the anatomical method was associated with the highest levels of reliability.

Similarly, a further concern associated with the CAST technique is that markers are positioned onto the skin, with the goal of quantifying movement of the underlying bones. Whilst there is little alternative to this process without using invasive methods for the positioning and identification of anatomical landmarks, a key concern that remains is the positioning of markers onto the shoe in order to quantify foot kinematics as there is likely to be movement of the foot itself within the shoe relative to the positions of the markers on the shoe outsole. Previous analyses have attempted to quantify the errors that are associated with placing markers onto the foot by making small holes in the shoes to allow the simultaneous measurement of foot kinematics using both skin and shoe-mounted markers. Sinclair *et al.* when modelling the foot as a single rigid segment showed that intraclass correlations between the two methods were highest in the sagittal plane and lowest in the transverse plane and that the shoe-mounted markers significantly underestimated several important coronal and transverse plane ankle kinematic parameters. In addition, when undertaking the same analysis using a multi-segment foot model concerning the calcaneus, midfoot and metatarsus, Sinclair *et al.* found firstly that strong correlations were observed between the two methods at the calcaneus-tibia. However, at the more distal foot articulations, much lower correlations were found, i.e. the midfoot-calcaneus, metatarsus-midfoot and metatarsus-calcaneus. Significant differences in discrete kinematic parameters were also observed between the skin and shoe-mounted markers, at the midfoot-calcaneus, forefoot-midfoot and forefoot-calcaneus articulations. The results led to the conclusion that that shoe-mounted markers do not fully represent true foot movement and should therefore be interpreted with caution during the examination of multiple-segment foot kinematics.

The proper identification of anatomical landmarks is essential for the correct interpretation of joint kinematics. One of the principal sources of measurement error in three-dimensional kinematic analyses is the definition of the anatomical joint centre about which segmental rotations occur. As outlined previously, the definition of the joint centre is considered at the midpoint between the medial and lateral anatomical landmarks at that specific joint location – for example, the ankle joint centre is positioned at the midpoint between the medial and lateral malleolus markers. This is a relatively straightforward process provided that of course there are medial and lateral landmarks to position the anatomical landmarks (see above issues in relation to the hip joint centre). However, although it has not received attention on the scale of the hip joint centre issues, there has been debate regarding the anatomical landmarks required to identify the position of the knee joint centre as there are several landmarks available around the knee joint that has been adopted. Previous work by Thewlis *et al.* has shown that delineating the knee joint centre via the commonly utilised femoral epicondyles, tibial ridges and femoral condyle landmarks had a significant influence on the resultant knee joint moments in the coronal and sagittal planes. Similarly, Sinclair *et al.* also on examining the femoral epicondyles, tibial ridges and femoral condyle methods as well as the functional identification of the knee joint centre, investigated the test–retest reliability of the different methods of delineating the knee joint centre. Their results showed that there were significant differences in coronal and transverse planes angulation using the tibial ridge and functional methods, and that the strongest test–retest intraclass correlations were observed for the femoral epicondyle and femoral condyle configurations. This led to their conclusion that these techniques for the estimation of the knee joint centre are currently the most reliable modalities.

3.5 Pelvis alternative models

According to the International Society of Biomechanics (ISB) guidelines [13], the coordinate system of a pelvis segment is defined by skin-mounted markers placed on the anterior and posterior landmarks of the ilium. This retroreflective marker configuration is associated with the conventional models, which is an approach commonly employed in clinical gait analysis [14]. Skin-mounted markers attached to the anterior superior iliac spines (ASISs) displace more during gait due to STA than markers attached to the posterior superior iliac spines (PSISs) (70 mm vs. 35 mm, respectively) [15,16]. In young adults with a regular body mass index (BMI – 22.8 $\pm$2.7 kg/m^2), the effect of STA on pelvis kinematics during gait is reported to be less than 1° [15]. However, the effect of STA on pelvis kinematics would be more of an issue for those individuals with a higher BMI. Due to excessive soft tissue in the abdominal region, alternative pelvis models are available considering the concerns of using the ASIS. With advancements in biomechanical software, this has presented an opportunity to use virtual markers to support the creation of a segment and aid in the tracking of segmental movement [15]. For instance, markers positioned on the iliac crests (ICs) in combination with anatomical markers can create a technical coordinate system (TCS). The ASIS markers are no longer required following the recording of a static trial, as the TCS creates the position of the ASISs during the dynamic trials [17]. Another approach is to disregard the ASIS and IC landmarks and to assess the movement of the pelvis using the sacral region alone or in combination with the PSIS landmarks [18,19].

3.6 Kinematic modelling of the spine and trunk

3.6.1 Two-dimensional modelling of the trunk and spine

The planar projection of a virtual line between two adjacent markers on spinous processes can be projected onto a global reference frame that provides reference to a relative angle to the vertical or horizontal. Using this technique, planar projection angles between multiple adjacent segments can provide an assessment of intersegmental movement of the spine. Alternatively, the entire trunk region can be modelled as a rigid segment using a virtual line between C7 and the sacrum. However, analysing intersegmental movement of the spine would be more appropriate for gait analysis, since it has been found that spine segments laterally flex towards the opposite side at initial contact in comparison to segments above T7 [20]. This provokes the question on the traditional technique of modelling the thoracic segment of the spine below the level of T6 [21]. The attachment of retroreflective markers on the spinous processes of the vertebrae provides a measure of sagittal and frontal movement of spine segments [20,22,23]. It is not possible to measure transverse plane movement if spinous processes are used due to the collinearity of the markers. While markers are not directly attached to the back surface, a transverse plane angle of the entire trunk region is measured by projecting a line between the left and right acromion process (outlines shoulder girdle) and a line between the left and right PSISs.

Although upper limb movement associated with this modelling technique has been shown to introduce movement artefact [24]. Using the planar projection technique, the relative angle between defined line segments provides a measurement on postural classifications such as kyphosis (the projection angle of two segments that are not adjacent to each other) and lordosis (relative angle between two-line segments defined by markers attached at L1, L3 and L5).

3.6.2 Three-dimensional modelling of the thoracic region

A minimum of three non-collinear markers are required to define the position and orientation of a body segment in 3-D space. For the thorax model defined by the ISB [13], markers attached to spinous processes define the proximal (C7) and distal (T8) ends of the segment, while the sternum markers provide the non-linear landmarks. A similar marker configuration is noted for the Plug-in-Gait thorax module although T10 defines the distal end of the thorax. While thorax angles during gait do not seem to be influenced if T8 or T10 is employed, it has been recommended that T2 is used to define the proximal end since C7 will represent movement in the cervical region [25,26].

3.6.3 3-D modelling of the lumbar region

Unlike the thoracic region, there are no accessible landmarks in addition to the spinous processes in the abdominal region to define a 3-D lumbar segment. Although invasive and not suitable for routine clinical use, indwelling bone pins inserted into L1 to L5, and the spinous process of S1, have quantified intersegmental movement of the lumbar vertebrae and pelvis during gait [27]. A technical frame referred to as a marker cluster is attached to each bone pin and consists of at least three markers that are positioned in a non-linear configuration. A parallel technique is the use of a skin-mounted 3-D cluster that can be attached over a spinous process. A comparison of angle data between the bone pin and skin-mounted approach has noted a similar range of motion values and kinematic waveforms in the sagittal and frontal plane [21]. Structural factors of a skin-mounted 3-D cluster and the invasive nature of using indwelling bone pins may explain differences in transverse plane kinematics. For example, pin movement may be influenced by constraints of soft tissues along with the mild discomfort associated with indwelling bone pins that impact the perceived action of walking. For the skin-mounted 3-D cluster technique, inertial perturbations, and individual participant characteristics such as skin elasticity and body composition can introduce a 'wobble effect' during dynamic movements. Portus *et al.* [28] observed less perturbation during a high-impact task when the 3-D cluster was attached to a semi-rigid structure in comparison to a rigid base structure. Whilst limitations are known, a systematic review on 3-D kinematics of the lumbar spine during gait using marker-based systems reported that skin-mounted 3-D clusters were shown to be a reliable technique when consistently employed in the same movement analysis laboratory [29]. Indeed, a $<1°$ difference across gait cycles has been reported when using a 3-D cluster to assess lumbar spine angles [29].

Another technique to define a 3-D lumbar segment is to place one marker on either side of the markers that are attached to spinous processes. Typically, the lumbar region can be defined as one rigid segment (L1–L5) or divided into an upper (L1–L3) and lower (L3–L5) segment. However, it has been speculated that the movement of these lateral markers that are attached to the back surface will be affected by greater morphological factors such as skin elasticity and body composition, which may increase relative error. In a recent study, Needham *et al.* [30] compared two kinematic modelling approaches that quantify 3-D lumbar movement. Model one used an individual marker configuration, with markers attached over the spinous processes of L1 and L5, in addition to one marker attached onto the back surface on either side of L1 [31]. Model two was a skin-mounted 3-D cluster consisting of three markers positioned in a non-linear configuration on a semi-rigid that was attached over the spinous process of L3 [21]. The kinematic waveform profiles of the two models were similar in all three planes of motion, which was interesting since model one measures the entire lumbar region, whereas model two measures a specific region (L3). However, a greater range of motion was noted in the transverse plane for the skin-mounted 3-D cluster technique, which is opposite to the view on relative error. It was assumed that a greater soft tissue artefact would be associated with model one due to the independence and location of the markers, which would potentially lead to a greater segment range of motion. Although a semi-rigid base plate would have ensured the 3-D cluster structure was potentially less susceptible to excessive perturbation, the authors stated that the movement of the lateral sides of the base plate could have been influenced by the para-spinal musculature, which may have increased axial rotation angles. Yet, the range of motion values reported in previous studies [21,30] is comparable with Konz *et al.* [32], where a similar 3-D cluster was attached over L3. The only difference being the structural dimensions of the 3-D cluster, whereby Konz *et al.* [32] used a smaller rigid base that can attach over the spinous process of L3. Further comparisons on the impact of 3-D cluster structural factors were not possible since Konz *et al.* [32] did not provide relative or segment angle time-series data. Indeed, the lack of detail provided on materials and structural dimensions is a major limitation of using hand-built 3-D clusters from a study replication standpoint [29]. Therefore, it has been recently advised that 3-D printed clusters offer external validation between gait laboratories [33,34].

3.7 Inertial measurement units

Similar to optical motion analysis systems, IMU sensors quantify locomotion by applying specific algorithms to the IMU data that identify patterns which correspond to gait events, e.g. initial foot contact and terminal foot contact. This allows for the stance and swing phases of gait to be distinguished and subsequent parameters of interest to be calculated (stance time, swing time, stride length). These algorithms will vary depending on the type of sensor setup used, but there have been several methods validated which have employed either a single IMU sensor [35] or a system of multiple sensors mounted in different anatomical locations [36].

The review focusses on the practical aspects and logistics of data collection for each parameter by discussing the variations in the setup procedures and how they have been employed in applied research studies.

3.8 Quantifying kinematic parameters related to walking and running

3.8.1 Number of steps and cadence

Counting the number of steps a person makes is one of the simplest forms of analysis, and it is either presented as the total number of steps in any particular time period or as the average number of steps per minute (cadence). Its simplicity lies in that a sensor only must detect the shock or the impact force transmitted to the body during a foot strike with no requirement to measure the intensity or the temporal characteristics of the foot strike. For this reason, step counting has been conducted since 1965 [37] and was initially measured with a spring-levered pedometer sensor mounted to the hip. However, due to the mechanical design of these pedometers, they have often been found to underestimate the number of steps at lower walking speeds [38,39]. More recently, newer piezoelectric pedometers have been developed which are similar to accelerometers as it can measure the intensity of force and can accurately detect foot strikes at lower speeds [40]. One problem that remains though is that the sensors are uni-axial, so if the sensor is tilted, then the accuracy can become compromised. This is seen with people who are severely overweight or pregnant due to the increase in waist size and wobbling mass of the upper body which causes misalignment of the sensor and more sensor oscillations [41,42].

As a result, methods using tri-axial accelerometers have been validated when the sensor is mounted at the ankle, thigh and waist [43]. The ability to count steps in all three axes of motion theoretically means that the orientation of the sensor should not affect its accuracy. This has led to the development of step counting applications for smartphones and other devices using the inbuilt tri-axial accelerometers [44,45]. These developments, therefore, allow for the physical activity profiles to be conveniently measured on the general population; however, the validity of these applications can be affected by the location of the smartphone on the user during locomotion and the step counting algorithms can differ between different applications [46].

The methods to identify foot contact within walking and running are the same. The accuracy of the methods employed in walking though differs when used for running analysis. When using uni-axial accelerometers, the error increases when the running speed is >10 km/h [47]. This is due to an increase in multi-planar motions, and therefore methods using tri-axial accelerometers are needed to accurately detect ground contact during running [47].

3.8.2 Stride length, speed and distance

The calculation of average walking speed in the field can be done easily with a stopwatch when the distance covered is known. When the distance is not known, stride length and number of strides can be used to estimate walking speed [48]. A stride is considered as two consecutive steps on the same limb and stride length is the distance between these two steps. Multiplying the stride length by the number of strides will give the total distance covered and then dividing the distance covered over time will give the average walking speed. Identification of a step, as previously mentioned, with an IMU is simple. However, the calculation of stride length is more complex.

There have been a variety of methods developed to measure stride length, yet they all are based initially on identifying foot contact through the protocols previously discussed. Amongst these methods are human gait models, which estimate stride length by measuring the vertical displacement of the centre of mass with an inverted pendulum model [49,50]. The benefit of this model is that it only requires one tri-axial accelerometer placed at the centre of mass to analyse both limbs. They do require additional anthropometric information though, and the calculation procedures are complex when a highly accurate measurement is required [51]. Another approach is the direct integration method, which does not require anthropometric data. This is because it uses an added gyroscope to measure sensor and limb orientation with respect to the global coordinate system and then integrates the sensor acceleration, from the start and end of the stride, to estimate the stride length [52]. However, for bilateral analysis, an accelerometer and gyroscope are needed to be attached to each foot, and the accuracy of this method is heavily dependent on the correct identification of sensor orientation. Gyroscopes are susceptible to accumulative drift error and thus enhanced signal processing methods such as zero-velocity update (ZUPT) are required to estimate this error within each gait cycle and improve accuracy [53–55].

There is a clear biomechanical distinction between walking and running in that running has periods of no ground contact referred to as the double float phase [56]. The presence of this phase in running means that the inverted pendulum models used to calculate stride length, distance and speed in walking cannot be applied directly to running. Instead, running is modelled as a spring-mass system where, during foot contact, the lower limbs (spring) absorb the energy from the upper body (mass) and then release the energy to propel the body upward and forward [57]. This model is utilised in the direct integration method, proposed by Yang *et al.* [58] to identify the start and end points of the stride and subsequently calculate stride length, distance and speed at running speeds of 2.5–3.5 m/s. Additionally, Potter *et al.* [54] found that the inclusion of the ZUPT processing method allowed for accurate measurement up to running speeds of 6.4 m/s but IMU ranges needed to be at least 50*g* for accelerometers and 1,000 deg/s for gyroscopes.

Running speed has also been accurately estimated with a single IMU through the artificial neural network method. This method uses the acceleration profile of the gait cycle at a known speed during a controlled calibration run and then uses the same profile during an uncontrolled run to estimate the speed [59]. The benefit of this method is that it can provide bilateral analysis through a single trunk-mounted accelerometer [60], however, it requires a pre-calibration procedure on a treadmill before field-based use.

One area of consideration that is unique to running speed estimation is that accelerated running is biomechanically different from steady-state running as the body leans forward [56,61]. As a result, models that have been validated for steady-state running speed have difficulties when applied to accelerated running such as sprinting. Parrington *et al.* [62] tested whether a direct integration method as described above could accurately measure peak velocity and average 10 m split time during a 100 m sprint. However, unlike previous methods that used a direct integration model, the authors used a centre of mass-mounted IMU containing an accelerometer and gyroscope. Despite the strong correlations of average 10 m split velocities from 10 m to 100 m and peak velocity, the model had large errors and poor correlation in the first acceleration phase from 0 m to 10 m. Caution should be taken when using IMUs for speed estimation during accelerated running.

3.8.3 Stance and swing

Identification of initial foot contact (IFC) and terminal foot contact (TFC) allows for the distinction between the two phases of gait (stance and swing), and subsequently, parameters such as stance time and swing time can be calculated. There have been several validated multiple sensor systems that use accelerometers and/or gyroscopes to detect IFC and TFC when mounted at either the foot [63], shank [64] or thigh [65]. For unilateral limb comparison though, sensors need to be placed on each leg which increases set up time, cost and requires data integration procedures. Instead, single sensor systems have also been developed mounted close to the centre of mass that can provide unilateral comparisons [66,67]. These methods use linear acceleration peaks within the anterior/posterior axis to detect IFC and TFC and peaks within the medial/lateral axis to identify which foot is in contact with the ground. McCamley *et al.* [68] also showed that IFC and TFC are identifiable in the vertical axis with enhanced signal processing techniques. Single sensor methods, therefore, may be a better option for field-based analysis.

The double float phase is characterised within running-based research studies as flight time [69]. Its inclusion within the gait cycle, along with an increased speed of locomotion, decreases the percentage of time spent in stance from 62% for walking to 31% in moderate running and 22% for sprinting [70]. Similar to walking, the measurement of this phase along with stance and swing time requires the identification of IFC and TFC. Due to the increased movement of soft tissues during running, there is more noise within the IMU signal and thus identifying IFC and TFC can be difficult at higher speeds [71]. The mounting site is important in ensuring signal noise is reduced and there have been methods validated either mounting IMUs on the foot [72], tibia [35,73,74] or at the centre of mass [69,75,76]. Accelerometers and gyroscopes have both been validated in these mounting sites for low to moderate running speeds, however, only gyroscopes have been found to accurately identify IFC and TFC at speeds greater than 5.7 m/s [71].

IMUs have also been integrated with commercially available GPS devices to provide analysis into running kinematics for novice and professional running athletes. The validity of upper trunk-mounted IMUs on the anterior aspect in a heart rate belt [77,78], and on the posterior aspect in a specially designed vest has been assessed for stance times [79]. The positioning of these devices caused an underestimation of stance time at lower running speeds, due to greater vertical oscillations of the trunk, which decreased as speed increased until 4.4 m/s. Despite this, the authors suggested that the accuracy of these devices was acceptable for field-based use and the minimal set up time makes them appealing for athletes wanting to conduct this type of analysis.

3.9 Application of kinematic parameters related to walking and running

3.9.1 Number of steps and cadence

The total number of steps is a parameter that quantifies the quantity of locomotion as more steps are associated with more time a person has spent walking [80]. It is most commonly employed for research purposes within the clinical industry to observe the differences in the amount of physical activity that is undertaken by people with debilitating physical conditions such as osteoarthritis [81], osteoporosis [82] and multiple sclerosis [83]. Through combining this data with other clinical tests to assess a patient's quality of life, clinicians have been able to set physical output targets, in the form of the number of steps per day, in intervention programmes [83]. In addition, providing regular feedback to patients on the number of steps undertaken has been shown to be a useful motivational tool in intervention programmes of overweight hypertensive patients to prevent cardiovascular disease [84]. Aoyagi and Shephard [85] also proposed that equipping just 5% of the general public with the means to count steps and meet targets could reduce national health care costs by 3.7%, which displays the important value that this parameter has in the clinical industry.

Cadence, on the other hand, is a parameter that quantifies the rate of locomotion as more steps per minute reflect a faster pace of walking and vice versa. It is typically employed in a similar manner as the number of steps within the clinical industry to analyse the rate of the physical output, which individuals with debilitating health conditions undertake [84]. However, cadence is also important in the prevention of skeletal diseases caused by bone mineral density loss. Cadences of over 122 $\pm$ 10 steps/min corresponded to peak vertical accelerations and peak loading rate values recommended to maintain bone health (>4.9g and 43 BW/s, respectively) [86]. Therefore, the regular use of this measure, as a proxy measure of peak loading rates, is essential for older adults at risk of osteoporosis.

Cadence is of interest within applied running research, but it is often expressed as stride frequency or stride rate (cadence divided by 30) in the literature, and there have been several studies that have analysed the effect of manipulating this parameter. Stride frequency has been shown to influence oxygen consumption, and novice runners have naturally less optimal stride frequencies than experienced runners [87]. Consciously increasing stride frequency has also been linked to reducing the risk of injury as it was shown to decrease the tibial acceleration experienced at the point of ground contact [88]. Measuring cadence in running is therefore potentially useful for wanting to improve running efficiency and remain injury-free.

Both the total number of steps and cadence can provide insight into the quantity and rate of locomotion; however, on their own, they do not assess the quality of locomotion. This is because these parameters view the human body as a non-animated mass and cannot differentiate between the motions of the individual limbs. The inclusion of cadence though in a controlled setting where quantity and rate are measured through treadmill-derived distance and speed, as seen so frequently in the literature, shows that cadence can also provide a basic level of insight into the quality of locomotion.

3.9.2 Stride length, speed and distance

Stride length is a parameter that quantifies the quality of locomotion as it provides insight into how a person walks. When combined with the number of steps, both the quantity and rate of locomotion can be quantified through distance and speed, respectively. The use of these three parameters could be appealing for clinical practitioners as they quantify the three aspects of locomotion. However, there is a lack of applied research with patients of pathological conditions due to large errors observed when earlier methods, that have been validated with healthy subjects, are applied to patients with an altered walking gait [89]. Consequently, there have been recent developments in data processing procedures for patients with pathological conditions [90,91], but research in the application of these methods is needed to understand how longitudinal measurement of stride length can be used by clinicians.

The analysis of stride length during running is potentially useful in the prevention and treatment of running-related injuries. Heiderscheit *et al.* [92] found that increasing stride frequency and decreasing stride length resulted in less mechanical energy absorbed at the hip and knee joints which may reduce the risk of injury. Yet, there is a limited amount of applied running-related research that has demonstrated the value of employing IMU-derived stride length, speed and distance. A recent study included IMU-derived stride length with other biomechanical parameters of running to determine whether kinematic changes could be detected during a fatiguing run with IMUs, however, it was unclear whether stride length was able to detect this alone [93].

3.9.3 Stance and swing

Stance time can be separated into periods of single limb support and double limb support by identifying the time between IFC and TFC of each limb and calculating the difference. It represents the ability of the lower limbs to support the human body during locomotion [94]. Alternatively, swing time, which is the time between TFC and IFC, represents the motor control ability of a person to maintain an evenly balanced walking gait [95]. Stance and swing phase parameters are of interest for hemiplegic patients where both neuromuscular and motor control ability is affected. It has been reported that patients who have suffered a stroke spend more time in double support stance phase and total swing time is also increased [65,95]. Also, unilateral comparisons have shown the stance time of the affected limb to be less than the unaffected [94]. The analysis of the two phases of gait allows practitioners assess to what degree a person's neuromuscular capability has become impaired and has also been used in the analysis of other pathological conditions that cause large impairments in walking ability such as Parkinson's disease [96]. However, this is only to a certain extent as they fail to show the effect of only discrete changes in a person walking gait as seen after an anterior cruciate ligament reconstruction [97].

The stance phase in running can differ from walking as the anatomical location of IFC is determined by individual running technique and running speed and therefore it can vary between the heel, midfoot or forefoot. The type of IFC can be identified within the pattern of the accelerometer data by the timing and intensity of the accelerometer peaks throughout the foot contact [69]. Also, it has been suggested that, in 'heel strikers', the intensity of the first vertical acceleration peak during the foot contact corresponds to the maximum loading endured and the intensity of the second vertical acceleration peak corresponds to the maximum thrust produced [75]. This added information from the accelerometer data during the stance phase allows extra insight into the running performance, and it is widely used by track and field coaches to analyse their athletes. Furthermore, correlations have been observed between shorter stance times and a faster sprinting performance [98] which is due to a greater application of force during foot contact [99]. There is a popular focus within training programmes to increase force production during ground contact and therefore the measurement of foot stance time can offer insight into the progress of this.

Swing time has also been suggested to be an indicator of running performance. Naturally, as speed increases swing time has shown to decrease, but unlike stance time there is no correlation between swing time and sprinting performance [99]. Instead, swing time has been linked to running economy in long-distance running as athletes that employed foot stride kinematics that increased swing times and minimised stance time have a greater running economy [100]. This, therefore, indicates that analysing swing time could also be used by track and field coaches with long-distance athletes. However, the level of athlete needs to be considered as this correlation does not appear to be consistent with elite level runners where non-biomechanical factors have a greater effect on the running economy [101].

Similar to stride length, stance time and swing time both quantify the quality of locomotion. This is shown by their ability to detect walking impairments in the clinical setting and the associations they have with performance-related parameters during running. Stance and swing times can be calculated for the individual limbs, so therefore they provide a more detailed insight into the quality of locomotion than cadence does. When compared to stride length, stance and swing time assessment normally does not require long pre-calibration procedures, and hence they are a more practical option for practitioners to conduct analysis of quality of locomotion in the field.

IMUs can accurately quantify the quantity, rate and quality of locomotion without the assistance of other technology. With regards to quantity, the number of steps is the simplest and easiest form of measuring quantity, and methods that use a tri-axial accelerometer are the most accurate to count steps in both walking and running. Likewise, with regard to the rate of locomotion, analysing the number of steps per minute (cadence) is the most practical method. Quality of locomotion can also be analysed with cadence when the speed of locomotion is known, however, stance and swing times offer a more detailed view into the characteristics of the individual limbs. The three aspects of locomotion can also be measured with stride length, and stride length derived speed and distance, but research into the practical application of these parameters derived from IMUs is lacking.

There has been a substantial amount of research showing that there is valuable information to be gained by quantifying all these three aspects of locomotion that can be used within the clinical industry to help prevent and rehabilitate people with pathological conditions. A similar impact is seen within the sporting industry using parameters of quality of locomotion to enhance physical performance. Albeit, more insight into the practical use of employing this type of analysis could be gained, in both industries, if sensor setup procedures and error reduction techniques were improved. In the clinical industry, the development of algorithms that use the smartphone-based IMUs is a potential avenue to gain more frequent analysis of locomotion. The same can be said for the development of algorithms that use the data from IMUs based on the GPS tracking devices as they are already well established and heavily used within the sports industry. The more time patients and athletes spend under observation is going to give the practitioner a clearer picture of their locomotion ability and therefore improve the practitioner's capacity to detect abnormalities correctly and monitor intervention programmes.

References

[1] Fong, D.T. and Chan, Y. (2010). The use of wearable inertial motion sensors in human lower limb biomechanics studies: a systematic review. *Analysis*, *10*(12), 11556–11565. https://doi.org/10.3390/s101211556

[2] Lobo, J. and Dias, J. (2007). Relative pose calibration between visual and inertial sensors. *The International Journal of Robotics Research*, *26*(6), 561–575.

[3] Mathie, M.J., Coster, A.C., Lovell, N.H. and Celler, B. (2004). Accelerometry: providing an integrated, practical method for long-term, ambulatory monitoring of human movement. *Physiological Measurement*, *25*(2), 1–20. https://doi.org/10.1088/0967-3334/25/2/R01

[4] Picerno, P., Cereatti, A. and Cappozzo, A. (2008). Joint kinematics estimate using wearable inertial and magnetic sensing modules. *Gait & Posture*, *28*(4), 588–595. https://doi.org/10.1016/j.gaitpost.2008.04.003

[5] Raffin, E., Bonnet, S. and Giraux, P. (2012). Concurrent validation of a magnetometer-based step counter in various walking surfaces. *Gait & Posture*, *35*(1), 18–22.

[6] Cappello, A., Stagni, R., Fantozzi, S. and Leardini, A. (2005). Soft tissue artifact compensation in knee kinematics by double anatomical landmark calibration: performance of a novel method during selected motor tasks. *IEEE Transactions on Bio-Medical Engineering*, *52*(6), 992–998. https://doi.org/10.1109/TBME.2005.846728

[7] Cappozzo, A., Cappello, A., Della Croce, U. and Pensalfini, F. (1997). Surface-marker cluster design criteria for 3-D bone movement reconstruction. *IEEE Transactions on Bio-Medical Engineering*, *44*(12), 1165–1174. https://doi.org/10.1109/10.649988

[8] Lahkar, B.K., Rohan, P.-Y., Assi, A., *et al.* (2021). Development and evaluation of a new methodology for Soft Tissue Artifact compensation in the lower limb. *Journal of Biomechanics*, *122*, 110464. https://doi.org/10.1016/j.jbiomech.2021.110464

[9] Lu, T.W. and O'Connor, J.J. (1999). Bone position estimation from skin marker co-ordinates using global optimisation with joint constraints. *Journal of Biomechanics*, *32*(2), 129–134. http://www.ncbi.nlm.nih.gov/pubmed/10052917

[10] Cappozzo, A., Catani, F., Croce, U.D. and Leardini, A. (1995). Position and orientation in space of bones during movement: anatomical frame definition and determination. *Clinical Biomechanics (Bristol, Avon)*, *10*(4), 171–178. http://www.ncbi.nlm.nih.gov/pubmed/11415549

[11] Bell, A.L., Brand, R.A. and Pedersen, D.R. (1989). Prediction of hip joint centre location from external landmarks. *Human Movement Science*, *8*(1), 3–16. https://doi.org/10.1016/0167-9457(89)90020-1

[12] Sinclair, J., Taylor, P. and Hobbs, S.J. (2014). The test–retest reliability of three different hip joint centre location techniques. *Movement & Sport Sciences*, *83*(1), 31. https://doi.org/10.3917/SM.083.0031

[13] Wu, G., van der Helm, F.C.T., (DirkJan) Veeger, H.E.J., *et al.* (2005). ISB recommendation on definitions of joint coordinate systems of various joints for the reporting of human joint motion—Part II: shoulder, elbow, wrist and hand. *Journal of Biomechanics*, *38*(5), 981–992. https://doi.org/10.1016/j.jbiomech.2004.05.042

[14] Baker, R. (2006). Gait analysis methods in rehabilitation. *Journal of Neuroengineering and Rehabilitation*, *3*(1), 4. https://doi.org/10.1186/1743-0003-3-4

[15] Hara, R., Sangeux, M., Baker, R. and McGinley, J. (2014). Quantification of pelvic soft tissue artifact in multiple static positions. *Gait & Posture*, *39*(2), 712–717. https://doi.org/10.1016/j.gaitpost.2013.10.001

[16] Rozumalski, A., Schwartz, M.H., Novacheck, T.F., Wervey, R., Swanson, A. and Dykes, D.C. (2007). Quantification of pelvic soft tissue artifact. *Gait and Clinical Movement Analysis Society*, 10–11.

[17] Kisho Fukuchi, R., Arakaki, C., Veras Orselli, M.I. and Duarte, M. (2010). Evaluation of alternative technical markers for the pelvic coordinate system. *Journal of Biomechanics*, *43*(3), 592–594. https://doi.org/10.1016/j.jbiomech.2009.09.050

[18] Frigo, C., Rabuffetti, M., Kerrigan, D.C., Deming, L.C. and Pedotti, A. (1998). Functionally oriented and clinically feasible quantitative gait analysis method. *Medical & Biological Engineering & Computing*, *36*(2), 179–185. http://www.ncbi.nlm.nih.gov/pubmed/9684457

[19] Vogt, L., Pfeifer, K., Portscher, M. and Banzer, W. (2001). Influences of nonspecific low back pain on three-dimensional lumbar spine kinematics in locomotion. *Spine*, *26*(17), 1910–1919. http://www.ncbi.nlm.nih.gov/pubmed/11568705

[20] Syczewska, M., Oberg, T. and Karlsson, D. (1999). Segmental movements of the spine during treadmill walking with normal speed. *Clinical Biomechanics (Bristol, Avon)*, *14*(6), 384–388. http://www.ncbi.nlm.nih.gov/pubmed/10521619

[21] Needham, R., Naemi, R., Healy, A. and Chockalingam, N. (2016). Multi-segment kinematic model to assess three-dimensional movement of the spine and back during gait. *Prosthetics and Orthotics International*, *40*(5), 624–635. https://doi.org/10.1177/0309364615579319

[22] Chockalingam, N., Dangerfield, P., Ahmed, E.-N., Rahmatalla, A. and Cochrane, T. (2003). Movement Analysis of Scoliotic subjects: A Screening and Evaluation Tool? Leeds, UK: *British Scoliosis Society*.

[23] Frigo, C., Carabalona, R., Dalla Mura, M. and Negrini, S. (2003). The upper body segmental movements during walking by young females. *Clinical Biomechanics (Bristol, Avon)*, *18*(5), 419–425. http://www.ncbi.nlm.nih.gov/pubmed/12763438

[24] Nguyen, T.C. and Baker, R. (2004). Two methods of calculating thorax kinematics in children with myelomeningocele. *Clinical Biomechanics*, *19* (10), 1060–1065. https://doi.org/10.1016/j.clinbiomech.2004.07.004

[25] Armand, S., Sangeux, M. and Baker, R. (2014). Optimal markers' placement on the thorax for clinical gait analysis. *Gait & Posture*, *39*(1), 147–153. https://doi.org/10.1016/j.gaitpost.2013.06.016

[26] Leardini, A., Biagi, F., Merlo, A., Belvedere, C. and Benedetti, M.G. (2011). Multi-segment trunk kinematics during locomotion and elementary exercises. *Clinical Biomechanics*, *26*(6), 562–571. https://doi.org/10.1016/j.clinbiomech.2011.01.015

[27] MacWilliams, B.A., Rozumalski, A., Swanson, A.N., *et al.* (2013). Assessment of three-dimensional lumbar spine vertebral motion during gait with use of indwelling bone pins. *The Journal of Bone and Joint Surgery. American Volume*, *95*(23), e1841–8. https://doi.org/10.2106/JBJS.L.01469

[28] Portus, M.R., Lloyd, D.G., Elliott, B.C. and Trama, N.L. (2011). Kinematic perturbation in the flexion-extension axis for two lumbar rigs during a high impact jump task. *Journal of Applied Biomechanics*, *27*(2), 137–142. http://www.ncbi.nlm.nih.gov/pubmed/21576722

[29] Needham, R., Stebbins, J. and Chockalingam, N. (2016). Three-dimensional kinematics of the lumbar spine during gait using marker-based systems: a systematic review. *Journal of Medical Engineering & Technology*, *40*(4), 172–185. https://doi.org/10.3109/03091902.2016.1154616

[30] Needham, R., Healy, A. and Chockalingam, C. (2021). Quantifying three-dimensional lumbar kinematics during gait using optoelectronic motion capture: a comparison of two kinematic models. In *Studies in Health Technology and Informatics*. Bristol: IOP Press; pp. 276–277.

[31] Crosbie, J., Vachalathiti, R. and Smith, R. (1997). Patterns of spinal motion during walking. *Gait and Posture*, *5*(1), 6–12. https://doi.org/10.1016/S0966-6362(96)01066-1

[32] Konz, R.J., Fatone, S., Stine, R.L., Ganju, A., Gard, S.A. and Ondra, S.L. (2006). A kinematic model to assess spinal motion during walking. *Spine*, *31*(24), E898–E906. https://doi.org/10.1097/01.brs.0000245939.97637.ae

[33] De Mits, S., Willems, T., Needham, R., Palmans, T., Danneels, L. and Chockalingam, N. (2021). Assessment of three-dimensional movement of the spine and pelvis during routine clinical assessment. *Studies in Health Technology and Informatics*, *280*, 274–275.

[34] Willems, T., De Mits, S., Needham, R., Danneels, L. and Chockalingam, N. (2021). Coordination pattern and variability in a flexion movement control test used in clinical assessment. *Studies in Health Technology and Informatics*, *280*, 272–273.

[35] Purcell, B., Channells, J., James, D. and Barrett, R. (2006). Use of accelerometers for detecting foot-ground contact time during running. *BioMEMS and Nanotechnology II*, *6036*, 603615. https://doi.org/10.1117/12.638389

[36] Strohrmann, C., Harms, H., Kappeler-setz, C. and Tröster, G. (2012). Monitoring kinematic changes with fatigue in running using body-worn sensors. *IEEE Transactions on Information Technology in Biomedicine*, *16*(5), 983–990. https://doi.org/10.1109/TITB.2012.2201950

[37] Hatano, Y. and Tudor-Locke, C. (2001). Pedometer-assessed physical activity: measurement and motivations. In *American College of Sports Medicine Annual Meeting*, Baltimore, MD.

[38] Crouter, S.E., Schneider, P.L., Karabulut, M. and Bassett, D.R. (2003). Validity of 10 electronic pedometers for measuring steps, distance, and energy cost. *Medicine and Science in Sports and Exercise*, *35*(8), 1455–1460. https://doi.org/10.1249/01.MSS.0000078932.61440.A2

[39] Schneider, P.L., Crouter, S.E., Lukajic, O. and Bassett, J.D.R. (2003). Accuracy and reliability of 10 pedometers for measuring steps over a 400-m walk. *Medicine and Science in Sports and Exercise*, *35*(10), 1779–1784.

[40] Lee, J.A., Williams, S.M., Brown, D.D. and Laurson, K.R. (2015). Concurrent validation of the Actigraph gt3x+, Polar active accelerometer, Omron HJ-720 and Yamax Digiwalker SW-701 pedometer step counts in lab-based and free-living settings. *Journal of Sports Sciences*, *33*(10), 991–1000. https://doi.org/10.1080/02640414.2014.981848

[41] DiNallo, J.M., Downs, D.S. and Le Masurier, G. (2012). Objectively assessing treadmill walking during the second and third pregnancy trimesters. *Journal of Physical Activity & Health*, *9*(1), 21–28.

[42] Feito, Y., Bassett, D.J., Thompson, D. and Tyo, B. (2009). Body Mass Index affects accelerometer counts during walking: 1869. *Medicine & Science in Sports & Exercise*, *41*(5), 156–157.

[43] Fortune, E., Lugade, V., Morrow, M. and Kaufman, K. (2014). Validity of using tri-axial accelerometers to measure human movement – Part II: step counts at a wide range of gait velocities. *Medical Engineering & Physics*, *36*(6), 659–669. https://doi.org/10.1016/j.medengphy.2014.02.006

[44] Costa, P.H.V., de Jesus, T.P.D., Winstein, C., Torriani-Pasin, C. and Polese, J.C. (2020). An investigation into the validity and reliability of mHealth devices for counting steps in chronic stroke survivors. *Clinical Rehabilitation*, *34*(3), 394–403. https://doi.org/10.1177/0269215519895796

[45] Douma, J.A.J., Verheul, H.M.W. and Buffart, L.M. (2018). Feasibility, validity and reliability of objective smartphone measurements of physical activity and fitness in patients with cancer. *BMC Cancer*, *18*(1), 1052. https://doi.org/10.1186/s12885-018-4983-4

[46] Funk, M.D., Salazar, C.L., Martinez, M., *et al.* (2019). Validity of smartphone applications at measuring steps: does wear location matter? *Journal for the Measurement of Physical Behaviour*, *2*(1), 22–27. https://doi.org/10.1123/jmpb.2018-0025

[47] Rowlands, A.V, Stone, M.R. and Eston, R.G. (2007). Influence of speed and step frequency during walking and running on motion sensor output. *Medicine & Science in Sports & Exercise*, *39*(4), 716–727.

[48] Yang, S. and Li, Q. (2012). Inertial sensor-based methods in walking speed estimation: a systematic review. *Sensors*, *12*(5), 6102–6116.

[49] Alvarez, D., González, R.C., López, A. and Alvarez, J.C. (2006). Comparison of step length estimators from wearable accelerometer devices. In *Annual International Conference of the IEEE Engineering in Medicine and Biology – Proceedings*, pp. 5964–5967. https://doi.org/10.1109/IEMBS.2006.259593

[50] Zijlstra, W. and Hof, A.L. (2003). Assessment of spatio-temporal gait parameters from trunk accelerations during human walking. *Gait & Posture*, *18*(2), 1–10.

[51] Zijlstra, A. and Zijlstra, W. (2013). Trunk-acceleration based assessment of gait parameters in older persons: a comparison of reliability and validity of four inverted pendulum based estimations. *Gait & Posture*, *38*(4), 940–944.

[52] Sabatini, A., Martelloni, C. and Scapellato, S. (2005). Assessment of walking features from foot inertial sensing. *IEEE Transactions on Biomedical Engineering*, *52*(3), 486–494.

[53] Ojeda, L. and Borenstein, J. (2007). Non-GPS navigation for security personnel and first responders. *Journal of Navigation*, *60*(3), 391–407. https://doi.org/10.1017/S0373463307004286

[54] Potter, M.V, Ojeda, L.V, Perkins, N.C. and Cain, S.M. (2019). Effect of IMU design on IMU-derived stride metrics for running. *Sensors*, *19*(11), 2601. https://doi.org/10.3390/s19112601

[55] Rebula, J., Ojeda, L., Adamczyk, P. and Kuo, A. (2013). Measurement of foot placement and its variability with inertial sensors. *Gait & Posture*, *38*(4), 974–980.

[56] Novacheck, T. F. (1998). The biomechanics of running. *Gait & Posture*, *7*(1), 77–95.

[57] Blickhan, R. (1989). The spring-mass model for running and hopping. *Journal of Biomechanics*, *22*(11–12), 1217–1227. https://doi.org/10.1016/0021-9290(89)90224-8

[58] Yang, S., Mohr, C. and Li, Q. (2011). Ambulatory running speed estimation using an inertial sensor. *Gait & Posture*, *34*(4), 462–466. https://doi.org/10.1016/j.gaitpost.2011.06.019

[59] Herren, R., Sparti, A., Aminian, K. and Schutz, Y. (1999). The prediction of speed and incline in outdoor running in humans using accelerometry. *Medicine and Science in Sports and Exercise*, *31*(7), 1053–1059.

[60] Song, Y., Shin, S., Kim, S., Lee, D. and Lee, K.H. (2007). Speed estimation from a tri-axial accelerometer using neural networks. In *2007 29th Annual International Conference of the IEEE Engineering in Medicine and Biology Society*, pp. 3224–3227. https://doi.org/10.1109/IEMBS.2007.4353016

[61] Kugler, F. and Janshen, L. (2010). Body position determines propulsive forces in accelerated running. *Journal of Biomechanics*, *43*(2), 343–348.

[62] Parrington, L., Phillips, E., Wong, A., Finch, M., Wain, E. and MacMahon, C. (2016). Validation of inertial measurement units for tracking 100 m sprint data. *ISBS – Conference Proceedings Archive*, *34*(1), 324.

[63] Boutaayamou, M., Schwartz, C., Stamatakis, J., *et al.* (2015). Development and validation of an accelerometer-based method for quantifying gait events. *Medical Engineering and Physics*, *37*(2), 226–232. https://doi.org/10.1016/j.medengphy.2015.01.001

[64] Kotiadis, D., Hermens, H.J. and Veltink, P.H. (2010). Inertial gait phase detection for control of a drop foot stimulator. *Medical Engineering and Physics*, *32*(4), 287–297.

[65] Lau, H. and Tong, K. (2008). The reliability of using accelerometer and gyroscope for gait event identification on persons with dropped foot. *Gait & Posture*, *27*(2), 248–257. https://doi.org/10.1016/j.gaitpost.2007.03.018

[66] Bugané, F., Benedetti, M.G., Casadio, G., *et al.* (2012). Estimation of spatial-temporal gait parameters in level walking based on a single accelerometer: validation on normal subjects by standard gait analysis. *Computer Methods and Programs in Biomedicine*, *108*(1), 129–137. https://doi.org/10.1016/j.cmpb.2012.02.003

[67] Little, C., Lee, J.B., James, D.A. and Davison, K. (2013). An evaluation of inertial sensor technology in the discrimination of human gait. *Journal of Sports Sciences*, *31*(12), 1312–1318. https://doi.org/10.1080/02640414.2013.779739

[68] McCamley, J., Donati, M., Grimpampi, E. and Mazzà, C. (2012). An enhanced estimate of initial contact and final contact instants of time using

lower trunk inertial sensor data. *Gait & Posture*, *36*(2), 316–318. https://doi.org/10.1016/j.gaitpost.2012.02.019

[69] Wixted, A.J., Billing, D.C. and James, D.A. (2010). Validation of trunk mounted inertial sensors for analysing running biomechanics under field conditions, using synchronously collected foot contact data. *Sports Engineering*, *12*(4), 207–212. https://doi.org/10.1007/s12283-010-0043-2

[70] Mann, R.A. and Hagy, J. (1980). Biomechanics of walking, running, and sprinting. *The American Journal of Sports Medicine*, *8*(5), 345–350.

[71] Bergamini, E., Picerno, P., Pillet, H., Natta, F., Thoreux, P. and Camomilla, V. (2012). Estimation of temporal parameters during sprint running using a trunk-mounted inertial measurement unit. *Journal of Biomechanics*, *45*(6), 1123–1126. https://doi.org/10.1016/j.jbiomech.2011.12.020

[72] Benson, L., Clermont, C., Watari, R., Exley, T. and Ferber, R. (2019). Automated accelerometer-based gait event detection during multiple running conditions. *Sensors*, *19*(7), 1483. https://doi.org/10.3390/s19071483

[73] Heiden, T. and Burnett, A. (2008). Determination of heel strike and toe-off in the running stride using an accelerometer: application to field-based gait studies. *ISBS – Conference Proceedings Archive*, *1*(1).

[74] Sinclair, J., Hobbs, S.J., Protheroe, L., Edmundson, C.J. and Greenhalgh, A. (2013). Determination of gait events using an externally mounted shank accelerometer. *Journal of Applied Biomechanics*, *29*(1), 118–122. https://doi.org/2011-0260 [pii]

[75] Auvinet, B., Gloria, E., Renault, G. and Barrey, E. (2002). Runner's stride analysis: comparison of kinematic and kinetic analyses under field conditions. *Science and Sports*, *17*(2), 92–94. https://doi.org/10.1016/S0765-1597(02)00122-3

[76] Lee, J.B., Mellifont, R.B. and Burkett, B.J. (2010). The use of a single inertial sensor to identify stride, step, and stance durations of running gait. *Journal of Science and Medicine in Sport*, *13*(2), 270–273. https://doi.org/10.1016/j.jsams.2009.01.005

[77] Adams, D., Pozzi, F., Carroll, A., Rombach, A. and Zeni, J. (2016). Validity and reliability of a commercial fitness watch for measuring running dynamics. *Journal of Orthopaedic & Sports Physical Therapy*, *46*(6), 471–476. https://doi.org/10.2519/jospt.2016.6391

[78] Backes, A., Skejø, S.D., Gette, P., *et al.* (2020). Predicting cumulative load during running using field-based measures. *Scandinavian Journal of Medicine & Science in Sports*, *30*(12), 2399–2407. https://doi.org/10.1111/sms.13796

[79] Lawson, M., Needham, R.A., Chockalingam, N. and Naemi, R. (2021). The validity of the GPS-based accelerometer to measure foot stance characteristics during running. In *XXVIII Congress of the International Society of Biomechanics*.

[80] Tudor-Locke, C. and Bassett, D.R. (2004). How many steps/day are enough? *Sports Medicine*, *34*(1), 1–8.

[81] Winter, C.C., Brandes, M., Mueller, C., *et al.* (2010). Walking ability during daily life in patients with osteoarthritis of the knee or the hip and lumbar spinal stenosis: a cross sectional study. *BMC Musculoskeletal Disorders*, *11*(233), 1–7. https://doi.org/10.1186/1471-2474-11-233

[82] Dohrn, I.-M., Hagströmer, M., Hellénius, M.-L. and Ståhle, A. (2016). Gait speed, quality of life, and sedentary time are associated with steps per day in community-dwelling older adults with osteoporosis. *Journal of Aging & Physical Activity*, *24*(1), 22–31.

[83] Motl, R.W., Pilutti, L.A., Learmonth, Y.C., Goldman, M.D. and Brown, T. (2013). Clinical importance of steps taken per day among persons with multiple sclerosis. *PLoS One*, *8*(9), e73247. https://doi.org/10.1371/journal.pone.0073247

[84] Stefani, L., Mascherini, G., Scacciati, I., De Luca, A., Maffulli, N. and Galanti, G. (2013). Positive effect of the use of accelerometry on lifestyle awareness of overweight hypertensive patients. *Asian Journal of Sports Medicine*, *4*(4), 241–248.

[85] Aoyagi, Y. and Shephard, R.J. (2011). A model to estimate the potential for a physical activity-induced reduction in healthcare costs for the elderly, based on pedometer/accelerometer data from the Nakanojo Study. *Sports Medicine (Auckland, N.Z.)*, *41*(9), 695–708. https://doi.org/10.2165/11590530-000000000-00000

[86] Rowlands, A.V, Schuna, J.M., Stiles, V.H. and Tudor-Locke, C. (2014). Cadence, peak vertical acceleration, and peak loading rate during ambulatory activities: implications for activity prescription for bone health. *Journal of Physical Activity & Health*, *11*(7), 1291–1294. https://doi.org/10.1123/jpah.2012-0402

[87] de Ruiter, C.J., Verdijk, P.W.L., Werker, W., Zuidema, M.J. and de Haan, A. (2014). Stride frequency in relation to oxygen consumption in experienced and novice runners. *European Journal of Sport Science*, *14*(3), 251–258. https://doi.org/10.1080/17461391.2013.783627

[88] Clarke, T.E., Cooper, L.B., Hamill, C.L. and Clark, D.E. (1985). The effect of varied stride rate upon shank deceleration in running. *Journal of Sports Sciences*, *3*(1), 41–49.

[89] Hundza, S.R., Hook, W.R., Harris, C.R., *et al.* (2014). Accurate and reliable gait cycle detection in Parkinson's disease. *IEEE Transactions on Neural Systems and Rehabilitation Engineering*, *22*(1), 127–137.

[90] Nozaki, Y. and Watanabe, T. (2019). A basic study on detection of movement state in stride by artificial neural network for estimating stride length of hemiplegic gait using IMU. *2019 41st Annual International Conference of the IEEE Engineering in Medicine and Biology Society (EMBC)*, pp. 3151–3154. https://doi.org/10.1109/EMBC.2019.8856352

[91] Watanabe, T., Miyazawa, T. and Shibasaki, J. (2018). A study on IMU-based stride length estimation for motor disabled subjects: a comparison under different calculation methods of rotation matrix. In *2018 IEEE EMBS*

International Conference on Biomedical & Health Informatics (BHI), pp. 70–73. https://doi.org/10.1109/BHI.2018.8333372

[92] Heiderscheit, B.C., Chumanov, E.S., Michalski, M.P., Wille, C.M. and Ryan, M.B. (2011). Effects of step rate manipulation on joint mechanics during running. *Medicine & Science in Sports & Exercise*, *43*(2), 296–302. https://doi.org/10.1249/MSS.0b013e3181ebedf4.Effects

[93] Marotta, L., Buurke, J.H., van Beijnum, B.-J.F. and Reenalda, J. (2021). Towards machine learning-based detection of running-induced fatigue in real-world scenarios: evaluation of IMU sensor configurations to reduce intrusiveness. *Sensors*, *21*(10), 3451. https://doi.org/10.3390/s21103451

[94] Gaviria, M., D'Angeli, M., Chavet, P., Pelissier, J., Peruchon, E. and Rabischong, P. (1996). Plantar dynamics of hemiplegic gait: a methodological approach. *Gait & Posture*, *4*(4), 297–305. https://doi.org/10.1016/0966-6362(95)01055-6

[95] von Schroeder, H.P., Coutts, R.D., Lyden, P.D. and Nickel, V.L. (1995). Gait parameters following stroke: a practical assessment. *Journal of Rehabilitation Research and Development*, *32*(1), 25.

[96] Salarian, A., Russmann, H., Vingerhoets, F.J.G., *et al.* (2004). Gait assessment in Parkinson's disease: toward an ambulatory system for long-term monitoring. *IEEE Transactions on Biomedical Engineering*, *51*(8), 1434–1443.

[97] Patterson, M.R., Delahunt, E., Sweeney, K.T. and Caulfield, B. (2014). An ambulatory method of identifying anterior cruciate ligament reconstructed gait patterns. *Sensors*, *14*(1), 887–899. https://doi.org/10.3390/s140100887

[98] Farley, C.T. and Ferris, D.P. (1998). Biomechanics of walking and running: center of mass movements to muscle action. *Exercise and Sport Sciences Reviews*, *26*(1), 253–285.

[99] Weyand, P.G., Sternlight, D.B., Bellizzi, M.J. and Wright, S. (2000). Faster top running speeds are achieved with greater ground forces not more rapid leg movements. *Journal of Applied Physiology*, *89*(5), 1991–1999.

[100] Santos-Concejero, J., Tam, N., Granados, C., *et al.* (2014). Stride angle as a novel indicator of running economy in well-trained runners. *The Journal of Strength & Conditioning Research*, *28*(7), 1889–1895.

[101] Santos-Concejero, J., Oliván, J., Maté-Muñoz, J.L., *et al.* (2015). Gait-cycle characteristics and running economy in elite Eritrean and European runners. *International Journal of Sports Physiology and Performance*, *10*(3), 381–387.

Chapter 4
Gait analysis – kinetics

Claudia Giacomozzi[1], *Aoife Healy*[2] *and Nachiappan Chockalingam*[2]

4.1 Introduction

Kinetics examines the relationship between the forces acting on the body and how these forces affect motion. Ground reaction force, joint reaction force and muscle force are the major forces involved in human locomotion. This chapter will focus on ground reaction force; force provided by a supporting horizontal surface.

The center of mass of a complex system made of various material bodies moves only under the action of external forces, while each single body may be subjected to both internal forces – exchanged with the other bodies of the system – and external forces. Bodyweight force is one of the external forces that the body exchanges with the environment. In the human body, the internal forces acting on a body segment are largely determined by the action of the muscles and depend on the joint constraints, which are generally complex and can only be schematized as a first approximation, or for small displacements, with the classic constraints of mechanics.

Locomotion or upright posture can be studied both in the living environment, using wearable instruments, and in the laboratory, setting up walkways, or observation areas, adequately instrumented but intended for not disturbing the free expression of the motor task. In these cases, the only external forces that are considered, neglecting those of air resistance, are those exchanged between the feet and the ground, and determine the progression of the subject, or its balance. It is also possible to imagine more complex situations, such as those of a work environment, where the aim of the measurement is, for example, to determine the loads on the musculoskeletal structure during the execution of particularly demanding tasks. Once again, the study can be performed both in the field and in the laboratory; in the latter condition, the part of the environment with which the subject directly

[1]Department of Cardiovascular and Endocrine-Metabolic Diseases and Aging, Istituto Superiore di Sanità (Italian National Institute of Health), Rome, Italy
[2]Centre for Biomechanics and Rehabilitation Technologies, Staffordshire University, Stoke on Trent, UK

interacts must be reproduced, and the advantage that can be obtained is the possibility of inserting in this model of the environment a measuring instrumentation of adequate accuracy, possibly without binding the subject.

The preferable solution for measuring external forces is instrumenting the environment, a solution that besides avoiding harnessing the subject and disturbing the execution of the movement, also simplifies the measurement process. Force or pressure platforms, embedded in an ad-hoc walkway, perfectly achieve this objective in the case of level walking in the laboratory – barefoot walking when using the latter. Single sensors or sensorized insoles, to be placed in the shoes, are instead used when the intention is to leave the subject free to move in his/her living environment, or the shoes themselves are part of the system under examination. Typically, in a laboratory setting, these force or pressure devices are integrated with other biomechanical measurement devices (e.g., optical motion capture and electromyography) to allow a comprehensive assessment of movement.

In the work environment, the above instruments can be used in conjunction with tools with instrumented handles, in order to measure the forces transmitted to the hands, or with instrumented seats in order to study correct postures. Likewise, aids and prostheses for the disabled can be instrumented; a cane is for example available in the market with a built-in sensor for measuring the support force (Advanced Mechanical Technology Inc., Watertown, MA, USA), and sensors can be placed on the prostheses to improve their adaptation to the patient (e.g., pressure sensors placed in the socket of a prosthetic leg, as in [1]).

In the following, the measurement of ground reaction force will be discussed, with reference to the main types of sensors and focusing on measurement accuracy.

4.2 Force transducers

4.2.1 Introduction and properties

In this section, some basic concepts for the design of force transducers are exposed, mainly referring to devices for the measurement of the ground reaction force [2]. It is known from mechanics that any system of forces acting on a rigid body can be replaced by a single force acting in a well-determined point and the moment of force (torque). Therefore, in general, the load on a structure is defined when six components are known, the three components of the force and the three components of the torque. "Force transducers" implicitly refer to both forces and torques, and those transducers that measure all six components of the load are said to be six-component transducers.

The methods for measuring forces are based on the effects of the variation of the electrical properties of the sensor, caused by the mechanical deformation of the material and directly proportional to the magnitude of the applied forces. Pressures can then be calculated from the forces acting on a known surface area. The quality of the elastic response of the material deformation affects the quality of the transducer. Ideally, a transducer should respond linearly and without phase distortion to changes in force across the whole range of amplitude and rate of change of interest. The frequency response of a transducer depends on the type of sensitive element, its geometric dimensions and the electronic conditioning of the signal. In general, in the design of the sensitive element, a compromise must be established between flexibility and rigidity, which are directly proportional to the sensitivity and the maximum response frequency, respectively.

As for all measurement systems where the prevailing conditions of use are dynamic ones, it is important to consider the quality of the transducer response in a transient regime. With reference to systems theory, a transducer system can be described as a second-order linear system, in which the damping factor must be slightly less than one in order to maximize the velocity of the response. With such a damping factor, the step response has a small overshoot, with a short, damped oscillation at the natural frequency of the system; correspondingly, the frequency response presents a small peak at the same natural frequency, also known as the resonance frequency, which will immediately precede the range of frequencies at which the response progressively attenuates. The resonance frequency is therefore indicative of the velocity of the response and the maximum frequency at which the transducer can be used. It is a characteristic parameter normally supplied by manufacturers and can be measured with relative simplicity by subjecting the transducer to an impulse of the quantity it transduces.

Relevant static characteristics of transducers are the linearity of the response and the hysteresis. Regarding linearity, it is usually considered in terms of its opposite, non-linearity; a method for its quantification consists in determining the maximum deviation of the transducer response from the straight line that best approximates it, and in expressing it as a percentage of the full scale. If there is no hysteresis or this is negligible, the non-linearity can be corrected numerically during calibration. The hysteresis of a transducer describes the dependence of its response on the previous loading history; for a force transducer, it is defined as the maximum difference in response to two forces of the same intensity during a loading–unloading cycle and is expressed as a percentage of this difference with respect to the full scale. The hysteresis greatly affects the measurement accuracy; therefore, its value is a critical characteristic for judging the quality of a measurement device.

For the construction of force transducers, the most commonly used technologies are those based on strain gauges or piezoelectric crystals. In the following, strain gauge transducers will be described more extensively, and some basic problems addressed. The approach provides an example of how the same problems can be addressed with other technologies.

4.2.2 Strain gauge transducers

A strain gauge allows the conversion of a variation in length into an electrical signal by changing its electrical resistance. The sensitivity is maximum according to a main direction, which is that of the prevalent dimension of the resistive element. With the strain gauge strips, which are commercially available in a wide range of dimensional and electrical characteristics, used in the applications this chapter is referring to, it is possible to reduce the resistances of the transverse connections to very low values; therefore, the sensitivity in the direction perpendicular to the main one may be lowered to negligible values. The variation in resistance of a strain gauge divided by its total value is proportional to its deformation. Since the total resistance is several orders of magnitude larger than its variation, this ratio can be considered linearly proportional to the strain.

Usually, four strain gauges are used for one measurement channel, positioned in such a way that for the same increase in force, two of them have a positive and two a negative variation of resistance. The four strain gauges are connected in a Wheatstone bridge configuration, those with resistance variation of the same sign are placed on opposite branches of the bridge. Thus, in the response of the bridge, the effects of variation of the resistance on the two adjacent branches are subtracted, while those on the two opposite branches are added. This configuration is therefore the one of maximum sensitivity, and also allows minimizing thermal effects, since a temperature variation induces a resistance variation of the same sign on all strain gauges. The strain gauges must have the same nominal resistance in order to balance the bridge and obtain the same sensitivity on the four branches. The combination of the strain gauges on the bridge branches must be such as to make the transducer sensitive to only one component of the force. To clarify this aspect, the three examples which follow, and which refer to the preferably adopted geometry in the force platforms – the cylindrical geometry – illustrate three arrangements of the strain gauges suitable for measuring a single component of the force or the moment of force. Obviously, in multi-component transducers, all the strain gauges will be housed on the same mechanical structure; this operation is made possible by the commercial availability of small assembled systems with different combinations of strain gauges.

Figure 4.1 shows the arrangements of four strain gauges for measuring the axial component of the force and the torque (along and around the vertical axis, respectively), and the shear component of the force. To easily understand why these transducers are sensitive only to the labeled load component, it is helpful to refer to the connection of the strain gauges in the Wheatstone bridge, which is indicated in Figure 4.1a, and to the above reported explanations. The strain gauges arrangement in Figure 4.1b does not need any additional clarification. As regards the measurement of the torque around the vertical axis (Figure 4.1c), the strain gauges must be positioned along helices inclined at 45°, which are those which experience the maximum deformation. Finally, the transduction of the shear component of force (Figure 4.1d) is based on the difference between the bending moment measurements at two different levels along the cylinder axis, a difference that will be zero for any bending torque applied to the sensor surface along an axis normal to the shear force, and which will take place at the same time along the entire axis of the cylinder.

When several transducers are used in the same measuring instrument, as in the force platform which is supported by four instrumented pylons, it is possible to combine in the same bridge strain gauges belonging to different transducers, to

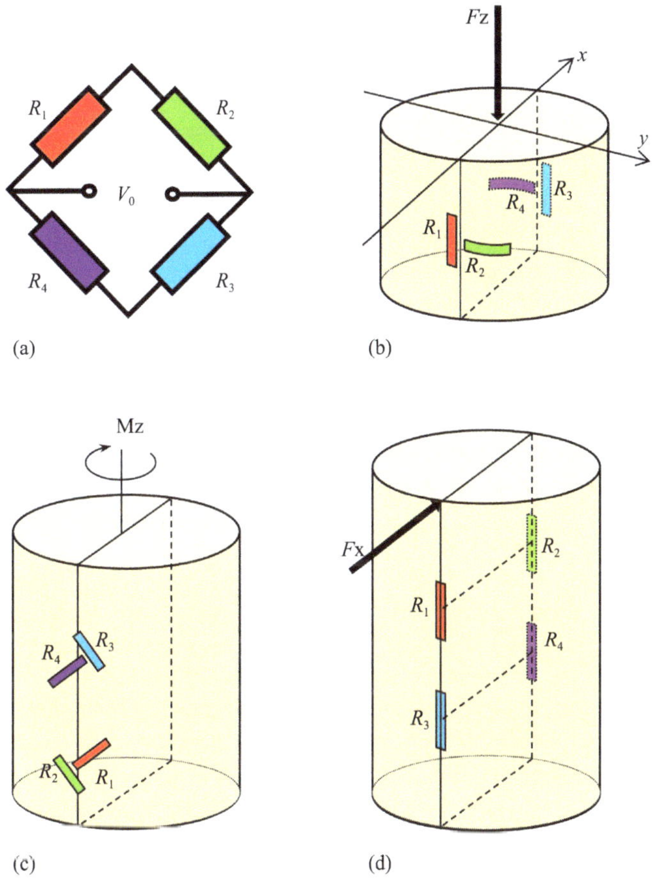

Figure 4.1 Strain gauges in the Wheatstone bridge (a) and their arrangement in a transducer for measuring: (b) the vertical force; (c) the torque around the vertical axis; (d) the cutting force normal to the surface of the transducers

have exactly the measurement channels corresponding to the components of the load to measure. In this case, the strain gauges must have the same dimensions, so as to have the same sensitivity; strain gauges subject to the same type of strain for the force component to be measured will be connected in series in the same branch of the bridge, or on the opposite branch, following the criterion of obtaining, as far as possible, branches with the same total nominal resistance and the same absolute sensitivity.

4.2.3 Piezoelectric transducers

Some crystals show a piezoelectric effect, with the emergence of electrical charges on their surface when they are mechanically stressed. Quartz crystals have this property, and their stability makes them suitable for making transducers. Piezoelectric transducers are not suitable for measuring static loads because the electric charge is not renewed under a constant load. Their response curve will extend towards low frequencies – without however ever including the null frequency – to the extent that the electric charge is preserved, that is, the impedance of the amplifiers with which the transducers are connected is high. In commercially available transducers (Kistler Instrumente AG, Winterthur, Switzerland), the quartz crystal is cut into discs; depending on the orientation of the discs with respect to the crystalline structure, elements sensitive to both the force normal to the circular surface of the discs and the tangential force can be obtained. Discs with the same axis of sensitivity are used in pairs for constructive reasons and to increase sensitivity. Three-component transducers to measure the axial force and two orthogonal shear forces are obtained by superimposing three pairs of discs. The transduction of the six components of the load is obtained in an integrated mechanical structure (platform) that includes at least three three-component transducers.

4.2.4 The six components of the load in a real transducer

The non-sensitivity of a measuring bridge to load components other than the desired one is an ideal characteristic. In general, a measurement channel for one component of the load may also show a certain sensitivity to other components; this is known as crosstalk of the latter on the desired one. Crosstalk can be caused by dimensional inaccuracies and tolerances in the orientation of the strain gauges and their sensitivity. Crosstalk does not pose particular problems if the transducer has been designed to measure all the components of the load it can transmit, as it can be compensated for during calibration through the system of equations shown in (4.1). Crosstalk cannot be compensated for if it relates to a component of the load that is not measured by the transducer.

In general, in a six-component transducer, the system of equations that relate signals to forces can be written in matrix form as:

$$F = \left[c_{ij}\right] * S \tag{4.1}$$

where F is the load vector (three components of the force and three of the moment of force) and S is the vector of the six signals. The c_{ij} elements of the 6×6 matrix are the calibration coefficients, the diagonal elements being the main ones; the other elements account for the crosstalk and would be null in its absence. The procedure would be more complex in the case of non-linearities, which however can be made negligible with an appropriate design of the transducers.

4.3 Force platforms

Force platforms (or plates) measure the resultant of the ground reaction according to an orthogonal reference system oriented as in Figure 4.2.

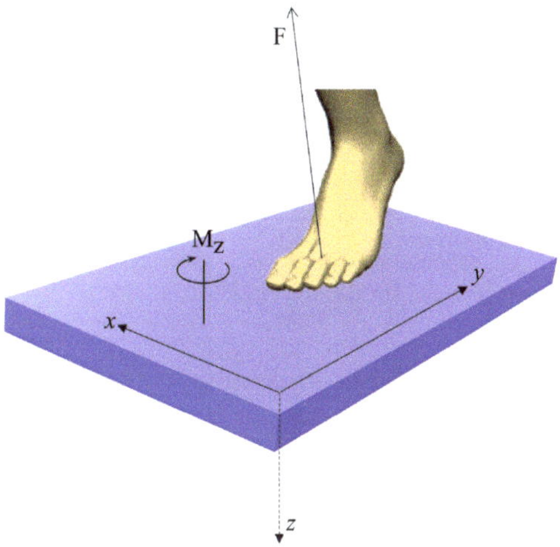

Figure 4.2 Six-component force platform: it measures the resultant F of the forces and the free moment M_z

The six-component platforms measure the three components of the resulting force and the three components of the resulting torque. In platforms with strain gauge transducers, these quantities are calculated with the system of equations in (4.1), since the strain gauges of the four pylons are combined in order to obtain an overall six-component transduction system. In platforms with piezoelectric transducers where, as described above, each of the four transducers has three components and the horizontal components are added 2 by 2, the signal vector will instead have 8 terms and the calibration coefficient matrix will have 6 × 8 terms.

Once the load vector is known, the coordinates of the point of application of the resulting force can be calculated using the following equations:

$$\begin{aligned} x &= -h * F_x - M_y/F_z \\ y &= -h * F_y + M_x/F_z \end{aligned} \tag{4.2}$$

where h is the height at which the actual support surface is located with respect to the reference plane xy, and also account for the possible thickness of a mat superimposed on the platform.

These coordinates identify the point at which the horizontal moments cancel out. There remains the vertical moment M_z, also known as the free moment. The output of the measurement process is therefore the resultant F of the force; three components and their point of application, and a torque with the vertical moment. These are the stresses that the ground transmits to the human body. The point of application of the force, also called the center of pressure (CoP) because it represents the centroid of the pressure distribution in the entire support surface of the foot, is very important and its possible inaccuracy greatly affects the calculation of the joint moments.

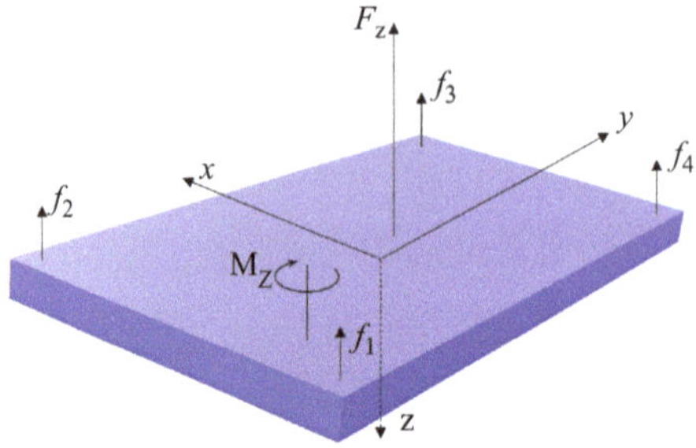

Figure 4.3 Three-component force platform

The study of the temporal progression of the CoP coordinates is the goal of posturographic examinations. For this purpose, low-cost platforms are created by placing a transducer of the vertical component of the force in correspondence with each of the four edges. These platforms, as shown in Figure 4.3, are called three-component platforms because they allow to calculate, using the system of equations in (4.3), only three components of the load, namely, the vertical component of the resulting force and the two horizontal moments:

$$\begin{Bmatrix} F_z \\ M_x \\ M_y \end{Bmatrix} = \begin{bmatrix} k_{11} k_{12} \; k_{13} \; k_{14} \\ k_{21} \; k_{22} \; k_{23} \; k_{24} \\ k_{31} \; k_{32} \; k_{33} \; k_{34} \end{bmatrix} \begin{Bmatrix} f_1 \\ f_2 \\ f_3 \\ f_4 \end{Bmatrix} \tag{4.3}$$

where k_{ij} are the calibration coefficients. Starting from these quantities, it is then possible to calculate the CoP coordinates by means of equations (4.2), in which h is considered null because those moments are reduced to the support surface of the platform. CoP coordinates can also be calculated directly starting from the measured forces, considering that this center is the point with respect to which the sum of the moments of each of these forces is canceled. What is reported here regarding the three-component platforms is intended to be only a description of their operating principle; however, different mechanical and electrical implementations are possible.

Recent advances have resulted in the availability of treadmills instrumented with force platforms; they perform measurements of the ground reaction force while the subject walks at an imposed, precisely known speed (Gaitway, Kistler Instrumente, Winterthur, Switzerland). The motor task remains almost unchanged from cycle to cycle and can be evaluated in extreme speed conditions. In this way it is easier, for example, to highlight walking abnormalities, train the subject to compensate for them, or adapt the functional parameters of a lower limb prosthesis.

These treadmills are available in different configurations; for example, side-by-side treadmill and tandem treadmill (AMTI, Waltham, MA, USA). The side-by-side treadmill consists of two belts separated longitudinally by a small gap allowing independent kinetic data for each leg during the "double support" phase of walking.

Each belt can be run at a speed that is independent of the other. The tandem treadmill, with its front and rear belts, eliminates many issues inherent in side-by-side, split-belt treadmills with no requirement to maintain medial–lateral position or straddle the belt split that runs down the center of the treadmill.

4.4 Force platform calibration

Force platforms are precision instruments and the "deterioration" of their response resulting from use means the adoption of periodic verification and recalibration protocols is required. This need, which has long been ignored, has finally received the attention of several research groups (e.g., [3]). Some methods and algorithms have been proposed, as well as some hardware prototypes; three of them are briefly described in the following.

4.4.1 An active calibration device, oriented to six-component platforms

The prototype was proposed in 2003 and is shown in Figure 4.4, which is a free adaptation from the published documentation [4]. The test device consists of a three-footed platform, only one of which is placed on the platform under test; the device applies a force that varies over time, and whose vertical and horizontal components are measured by precision load cells. The vertical component, F_V, is generated by standing on the device and varying the distance of the center of mass from the single support. The horizontal component, F_H, is supplied by a linear

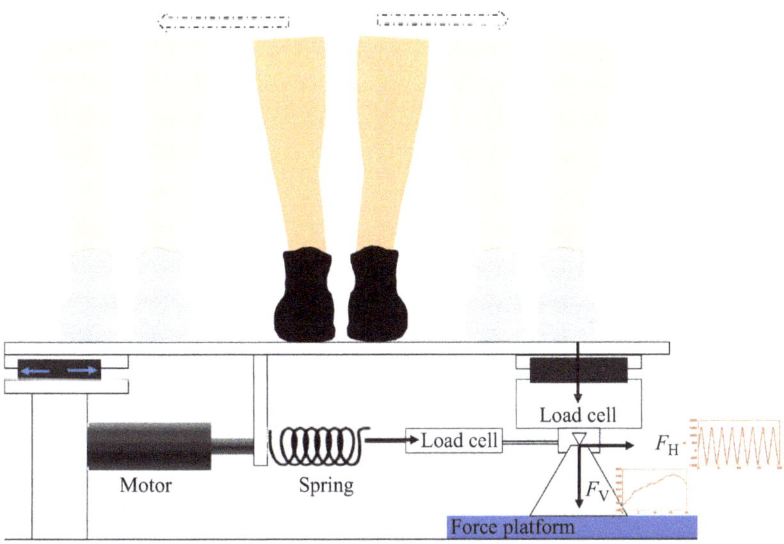

Figure 4.4 Active calibration device for six-component platforms (adapted from [4]). F_V: force vertical component; F_H: force horizontal component.

motor and transmitted to the support by means of a spring and a load cell. The force components applied are plotted in the figure. The procedure involves the accurate positioning of the support in five known points of the platform, which allows applying known moments. At each point two tests are carried out, orienting the horizontal component of the force along with the directions of the two horizontal axes, for a total of 10 tests.

The algorithm estimates the elements of the calibration matrix with the least squares approach on the data collected in the 10 tests. In its formulation, it considers and therefore allows to correct, an alignment error (azimuth and elevation) between the horizontal component of the force applied in each test and the corresponding axis of the force platform.

A simulation approach demonstrated that, in the presence of random alignment errors with standard deviation of 3°, the maximum error committed using the traditional estimation method reaches 6% on the elements of the main diagonal and 13% (of the value on the main diagonal) on the elements of crosstalk. Using the new algorithm, the error is always less than 0.4%. The proposed method was applied to a commercial platform (mod. 4080, Bertec Corporation, Columbus, OH, USA) showing a residual crosstalk of the vertical force <0.0002 [N/N] on the horizontal forces and 0.00005 [Nm/N] on the torque. The error on CoP coordinates after recalibration is <0.5 mm.

The method has therefore proved to be suitable for the periodic verification and recalibration of the force plate in the clinical setting. Compared to previous solutions, it allows applying forces that vary over time, are not point-like and result from the combination of independent horizontal and vertical components (more similar to those generated during the motor task). It allows manual and quick application of the procedure even by non-technical personnel, without moving heavy loads and without the aid of complicated tools, automatically compensating for any misalignment errors.

The same research group later proposed a simplified, easier-to-use prototype, published in 2009 [5]. Inspired to the same approach as the 2003 prototype, the new one is a passive, simple and lightweight device, basically made of a high-precision, 3-D load cell, loaded through a triangular stage, and precisely positioned on the force platform under recalibration by means of a template. After manual positioning of the load cell on the force platform, the operator is asked to exert loads on both the load cell and the force platform by his/her body movement. A dedicated algorithm then estimates the local or global six-by-six recalibration matrix of the force platform through a least squares optimization. In the experimental validation performed over four force platforms, the local calibration procedure showed a high level of performance, with a residual COP estimation inaccuracy close to 30% in the worst case.

4.4.2 A passive calibration device, oriented to three-component platforms

The calibration device consists of a ballast of mass M (adequate to simulate normal load conditions) and a smaller eccentric mass m, which rotates with respect to a vertical axis aligned with the axis of symmetry of the ballast [4,6]. The CoP position of the device depends on M, m, the instantaneous angular velocity of the rotating mass ω (t) and geometric parameters of the device. The CoP nominal trajectory, once the rotating mass has been manually set in motion, can be determined with great precision because the uncertainty regarding the position of the eccentric mass, which is of the order of 1 mm, is further reduced by the scale factor $m/(M + m)$: in this way, it is possible to obtain a resolution of the order of 0.1 mm even with rather crude construction techniques. Figure 4.5 shows the scheme of the device and the CoP trajectories before (b) and after (c) calibration.

The theory of operation is very simple: if the rotating mass is manually set in motion, the CoP performance is described by a spiral:

$$\begin{cases} \delta_x(t) = \dfrac{m}{M+m}\left[R\cdot\cos\theta(t)\left(1+\dfrac{[\omega(t)]^2 h}{g}\right)\right] \\ \delta_y(t) = \dfrac{m}{M+m}\left[R\cdot\sin\theta(t)\left(1+\dfrac{[\omega(t)]^2 h}{g}\right)\right] \end{cases} \tag{4.4}$$

where δ, θ are the polar coordinates, R is the radius of the rotating arm, and h is the height of the arm with respect to the force platform. This spiral, measured by the platform itself, constitutes a dense "training set" for the verification of the platform and its eventual recalibration. An immediate visual check consists of viewing the trend of the CoP thus recorded: any anisotropies are an obvious symptom of loss of calibration (e.g., this occurs in Figure 4.5b).

More precisely, the eigenvalues of the covariance matrix are calculated, and a scaling index is derived from their relationship. Following the same approach, a

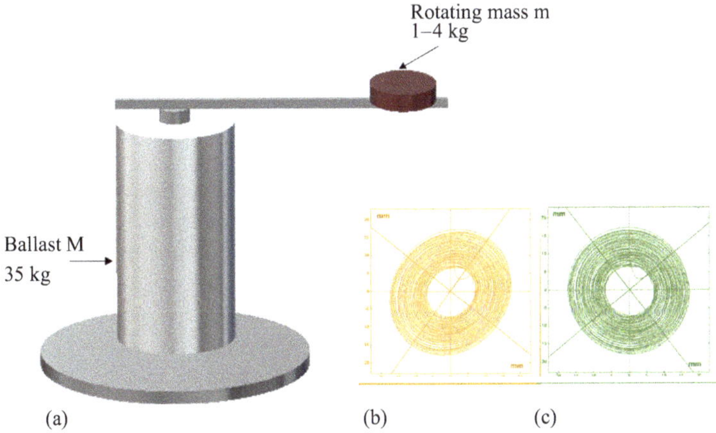

Figure 4.5 Calibration device for three-component platforms (a) and the calculated CoP trajectories before (b) and after (c) calibration (adapted from [4])

platform recalibration matrix can be calculated using a least squares algorithm which minimizes the scaling index. Figure 4.5c shows the obtainable result. The element of variability introduced by the manual start of the rotating mass is eliminated by examining, in the processing phase, only the data included between maximum and minimum set values of the angular velocity.

4.5 Foot-ground pressure measurement systems

During locomotion, the various forces between the human body and the ground are distributed under the various support structures of the foot. The measurement of the resultant force does not provide information about the load supported by these individual structures, therefore different tools are required. In relatively recent years, low-cost acquisition and calculation technologies and systems have made it possible to create measurement systems with a multiplicity of sensors of the vertical component of the force only. Since the active area of the single sensor is known, it is possible, and preferable, to express the response of these sensors in pressure rather than force units. These measuring systems, therefore, provide the distribution of pressure on a support surface, with a detail that can be established during the construction phase according to needs. Of course, a pressure measurement system can accurately assess an individual's plantar pressure pattern only during barefoot walking (conversely, when the assessment is conducted while wearing footwear, it is referred to as the shoe–ground interface). Typically, the pressure sensors are arranged according to matrices to form pressure platforms. The spatial resolution of these (distance between the centers of two adjacent sensors) is an important feature for the ability to acquire information under the smaller support structures of the foot. Another fundamental feature is the sampling frequency, which finds its limit in the large amount of data to be processed. For example, a frequency of 50 samples/s for a matrix of 2,000 elements' results in a flow of 100,000 data/s.

4.6 Pressure sensors

The most used commercial transduction technologies are based on capacitive or resistive principles. Capacitive sensors are made by placing two metal plates on the sides of an elastic dielectric material (Figure 4.6a); when a force is applied, the elastic material deforms, the distance between the plates decreases and therefore the capacity increases. Nicol and Hennig [7], whose work resulted in a commercial product (Emed system, NovelGmbH, Munich, Germany), selected the elastic material in such a way that the ratio between the dielectric constant and the thickness was a linear function of the pressure exerted on the sensor: in this way, the resulting capacity is a linear function of the pressure, thus the capacitance is a linear function of the global force exerted on the entire surface of the sensor, in the range of interest.

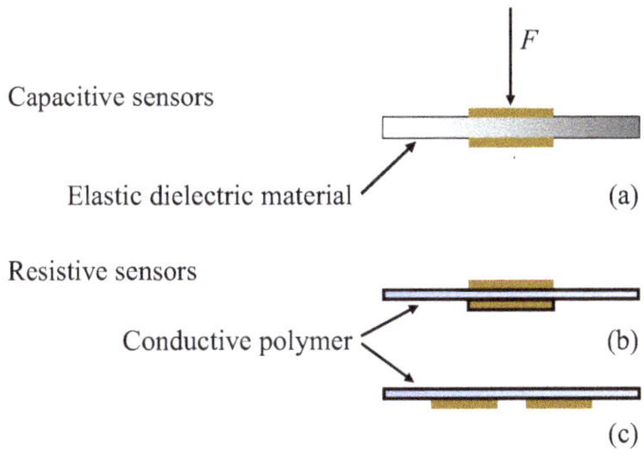

Figure 4.6 Construction diagram of the main commercial pressure sensors

Resistive sensors are made by placing two metal plates in contact with a conductive polymer. As the compression force on the surface of the polymer increases, a decrease in electrical resistance is obtained, mainly due to contact phenomena. There are two different construction schemes: the two plates are in contact with the opposite surfaces of a sheet of conductive polymer (Figure 4.6b), or they are on the same side of the conductive material (Figure 4.6c). In the second case, the plates are at a small relative distance (generally <1 mm), which is critical for the sensor response. Regarding the conductive polymer, this may be found in various constructions, most frequently with the shape of a sheet [8] or as an ink deposited on an insulating support (Interlink Electronics Inc., Santa Barbara, CA, USA; Tekscan Inc., South Boston, MA, USA).

Lastly, a third technology is available on the market of pressure platforms, which is again called capacitive, and which exploits the air as the dielectric means (AMCube, Gargas, France). Bench tests performed at the Italian National Institute of Health in 2010 showed quite a satisfactory level of performance at least in terms of precision [9].

Plantar pressure analysis is widely used in the assessment of foot function. To examine loading on different areas of the foot, a mask is applied to the footprint to sub-divide it into regions of interest (ROIs). Software generally creates the masks based on geometric features of the footprint. More recently, masking based on anatomical landmarks of the foot has been developed. This involves the integration of a 3D motion capture system, plantar pressure measurement device and a multi-segment foot model. Masking based on anatomical landmarks of the foot has been found to be as reliable and repeatable as a comparable geometrical mask, and clinically valuable especially in the presence of severe alteration of the foot-ground contact pattern [10].

4.6.1 Sensor response and main sources of error

Pressure sensors may not be very accurate, non-linearity and hysteresis mainly contribute to the inaccuracy. The most accurate sensors available in the market are those of the capacitive type, with a nominal hysteresis of 3% and an accuracy of ±5% of full scale. Some resistive sensors may reach good accuracy as well [9], even though they need accurate and continuous re-equilibration by the operator, and they likely undergo a more rapid wearing. In terms of dynamic response, it is important to know the maximum rate of change of force or pressure that can be captured with these sensors. Normally in the clinic, the need is to be able to follow phenomena with frequency content limited to about 50 Hz; most of the sensors available meet this need. In sports context, however, higher frequencies may be required to investigate much faster motor tasks.

When elastomers are subjected to a constant load, they are affected by a phenomenon normally referred to with the term creep, that is, by a gradual deformation that lasts for a long time. This involves, in sensors based on elastomers, a typical drift of the response, which makes their use not recommended when the load does not return to zero for a long time, as occurs in posturographic examinations.

To evaluate what the necessary spatial resolution should be, it is appropriate to refer to what happens under the metatarsal heads, a region of the foot very often overloaded and possible site of pain or ulcerations (in the at-risk foot). With reference to the first two metatarsal heads, the pressure profile schematically indicated in Figure 4.7 may be expected, which can be approximated with two sinusoids connected by a stretch with zero pressure. If the frequency analysis of this signal is performed, it may be reasonable to limit the response of the measurement system to the second spatial harmonic; applying the sampling theorem, according to which one must sample at a frequency at least double of the maximum frequency contained in the signal, and considering that the distance between the two metatarsal heads can go from 12 to 20 mm in the adult, it follows that the distance between the centers of adjacent sensors must not be greater than 3 mm and 5 mm, respectively. Obviously, this distance should be even less in children.

Another important aspect concerns the size of the sensor. Sampling of a pressure distribution should ideally be done with point sensors. Real sensors, on the other hand, measure an average pressure value on their sensitive surface, therefore an error is made, greater for greater surfaces, because the pressure signal is not faithfully sampled in the central point of the sensor. Unfortunately, it is not possible to construct point-like sensors because the sensitive electrical quantity, be it a capacitance or a conductance, is normally proportional to the sensor surface, and therefore it cannot go below a certain dimensional limit. However, it is useful to bear in mind that, other things being equal, arrays with small sensors are preferable.

4.7 Pressure platform types and assemblies

When first developed, the size of these platforms allowed the recording of one footstep at a time; additionally, sampling rate was an issue, especially in the case of capacitive technology which requires that each capacitor be "charged" before each sampling. More recently, sensor and wireless technology development, as well as the huge improvement in computational power have resulted in improvements in

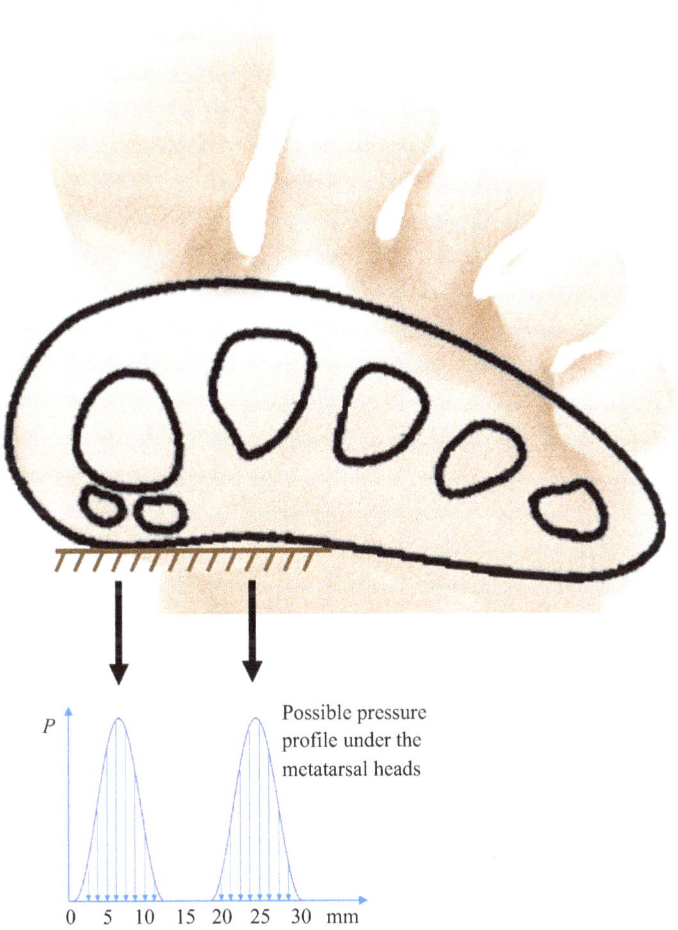

Figure 4.7 Possible pressure profile under the metatarsal heads

the performance of these systems: with much longer dimensions, with some available as modular systems to allow users to select their required size to enable capturing of multiple steps; with higher density of sensors, up to 4 sensors/cm^2 for capacitive solutions and up to 9 sensors/cm^2 for resistive solutions (1 sensor/cm^2 can also be found in automotive application, but limited to a very low sampling rate); with higher sampling rate, commonly up to 100 samples/s for clinical and biomechanical applications, up to 400 or 500 samples/s in sports experiments; with improved portability and usability, as is the case for modular, very thin and lightweight platforms, or in the case of platforms with wireless communication technology; with different assemblies, as for example in the case of instrumented treadmills which allow for plantar pressure measurement with similar spatial resolution as platform systems.

4.7.1 Comparison of force and pressure platforms performance in posturography

The responses of the two types of instruments can be compared in terms of CoP trajectory and of the main parameters that can be calculated from it, commonly used in posturography. In the case of force platforms, the CoP instantaneous coordinates are calculated according to equation (4.2). In the case of pressure platforms, the applied formula is the one used for the calculation of the instantaneous center of gravity:

$$\begin{aligned} x &= \Sigma x_i F z_i / \Sigma F z_i \\ y &= \Sigma y_i F z_i / \Sigma F z_i \end{aligned} \tag{4.5}$$

where the summation is extended to all points that were active at the considered time point, x is the coordinate along the mid-lateral axis (M-L) and y is the coordinate along the anterior–posterior axis (A-P).

The use of pressure platforms with resistive sensors is generally not recommended in the posturographic field, since the hysteresis (influenced by the creep) from which these sensors are affected, can significantly alter the estimate of the parameters of interest.

In an old, yet still valid experiment [4], a pressure plate with capacitive sensors (Emed system, $\text{Novel}_{\text{GmbH}}$, four sensors/cm^2, sampling rate of five samples/s in static conditions) was rigidly superimposed on a six-component force platform (Bertec Corporation, Columbus, OH, USA); on this integrated system, some very simple posturographic tests were carried out, lasting 52 s, in the standing position, with eyes open. The output signals from the force platform were sampled at 100 Hz, filtered at 2.5 Hz and 10 Hz and resampled at 5 Hz and 20 Hz, respectively (case A and case B). The pressure plate signals were sampled at 5 Hz (the only option allowed by the software), realigned, expressed in the reference system of the force plate. The results are reported in Table 4.1. With reference to case A (2.5 Hz cut-off frequency), the response of the pressure plate seems acceptable for some parameters, and the error, expressed as a percentage of the parameters calculated for the force platform, exceeds 10% (low estimate) only for the 95% confidence circle area, the sway area and the integrals of the frequency spectrum. In case B, the spatial parameters differ considerably (especially sway path and sway area which have errors of up to 30%) while the spectrum in the 0.5–2 Hz frequency range essentially does not change. However, it should be noted that, from the clinical point of view, such low bandwidth filtering can induce the loss of information, useful for example for the discrimination of particular pathologies such as Parkinson's disease. Furthermore, the narrower the CoP bandwidth, the more the final result resembles the trajectory of the center of mass. There is no doubt that the comparison must be made by conditioning the response signals of the two

Table 4.1 Comparison between posturographic parameters acquired simultaneously using a force and a pressure platform, at two different sampling frequencies

		Case A: 2.5 Hz			**Case B: 10 Hz**		
		Pressure platform (PP)	**Force platform (FP)**	**PP error (%FP)**	**Pressure platform (PP)**	**Force platform (FP)**	**PP error (%FP)**
Central point (cm)	M-L	−0.13	−0.13	−3	−0.21	−0.21	0
	A–P	9.39	9.39	0	9.35	9.35	0
Mean distance (cm)		0.95	1.00	−4	0.92	0.97	−5
A–P mean distance (cm)		0.47	0.49	−4	0.45	0.48	−5
A–P range (cm)		2.86	2.98	−4	2.85	3.02	−6
RMS distance (cm)		1.10	1.14	−4	1.06	1.12	−5
A–P RMS distance (cm)		0.60	0.63	−5	0.59	0.62	−6
Sway path length (cm)		68.29	73.86	−8	66.17	87.69	**−25**
Mean velocity (cm/s)		1.68	1.82	−8	1.63	2.16	**−25**
Mean frequency (Hz)		0.28	0.29	−4	0.28	0.36	**−20**
A–P path length (cm)		33.09	34.88	−5	32.78	46.92	**−30**
A–P mean velocity (cm/s)		0.82	0.86	−5	0.81	1.16	**−30**
A–P mean frequency (Hz)		0.31	0.31	−1	0.32	0.43	**−26**
95% confidence circle area (cm^2)		27.13	31.25	**−13**	25.45	30.75	**−17**
95% confidence ellipse area (cm^2)		10.10	11.05	−9	9.53	10.68	**−11**
Half long axis a (cm)		2.28	2.38	−4	2.18	2.30	−5
Half short axis b (cm)		1.41	1.48	−5	1.39	1.48	−6
Sway area (cm^2)		0.46	0.53	**−14**	0.45	0.65	**−30**
Power spectrum integral 0.02–0.2 Hz	M-L	0.006	0.007	−6	0.002	0.002	−6
	A–P	0.002	0.003	−9	0.001	0.001	−9
Power spectrum integral 0.2–2 Hz	M-L	0.001	0.001	**−13**	0.000	0.000	−1
	A–P	0.001	0.001	**−17**	0.000	0.000	−2

M-L, medio-lateral; A–P, anterior-posterior.

platforms in the same way. However, the fact that the differences are all of the same sign does not appear meaningless, but showcases their systematic nature.

4.8 Pressure insoles

The same technologies used for pressure platforms are utilized for the realization of pressure sensitive insoles to be inserted into shoes or socks. Both resistive and capacitive matrices, arranged in the form of insoles, are commercially available. The capacitive insoles are built by the same company Novel$_{\text{GmbH}}$ mentioned above and once again are those with higher precision sensors. The resistive ones are made for the most part with conductive ink technology and are characterized by a higher density of the sensor matrix (e.g., F-Scan by Tekscan, Inc., South Boston, MA, USA; WLAN by T&T medilogic Medizintechnik GmbH, Schönefeld, Germany).

A third technology is at present available for insole matrices, namely, a capacitive technology exploiting the air as the dielectric means (AMCube, Gargas, France). When first developed these systems were wired, with data capture limited by the length of the system's cables, however, in recent years these systems now utilize wireless technology. For the wireless systems, a portable data logger is usually worn around the waist or on the shoe.

Pressure insoles solve the measurement problem for long acquisition times, also leaving the subject free to move in the daily environment, but they are never to be considered as a substitute for pressure platforms, as the measurement accuracy that can be achieved with them is lower. Among the major causes of inaccuracy, it is worth mentioning the following: (a) preloads and spurious loads, that are variable even synchronously with the gait cycle, due to the unavoidable deformation of the sensor to comply with the internal surface of the shoe; (b) possible alteration of the pressure value due to the cushioning effect of the insole itself, especially when made of materials with viscoelastic characteristics; (c) lack of knowledge, and possible variation during movement, of the position of the individual sensors with respect to the anatomical sites of interest. The last cause of inaccuracy can be overcome by using insoles with a large number of small sensors. But these generally have a reduced temporal resolution and generate a large amount of data that limits the portability of the acquisition and calculation equipment.

Despite their drawbacks, measurement insoles may greatly contribute to biomechanical and clinical studies, especially when a friendlier environment than a gait lab is recommended for the assessment, or in the case of motor tasks requiring irregular surfaces and paths, or for long-lasting monitoring, or for the specific assessment of footwear or boots performance. For all those reasons, interesting solutions entered the market in recent years. Two of these are worth mentioning, even though none of them can deliver the accurate measurement of the complete plantar pressure distribution. The first solution is a patented evolution of the capacitive technology (Loadsol, NovelGmbH, Munich, Germany): based on an innovative capacitor and dedicated algorithms, the insole can resemble an in-shoe force measurement device, which can measure the instantaneous vertical force acting on the foot with a declared inaccuracy $<10\%$. Each insole can be at maximum divided into three major areas, with a sampling rate of up to 100 samples/s. The second innovative insole (Moticon, Munich, Germany) is based on a set of 16 capacitive sensors and dedicated algorithms which interpolate the discrete measurements so as to estimate the whole matrix pressure distribution. In both cases, all needed electronics are onboard and powered by the coin-cell battery, and all communications occur between the insoles and a smartphone (synchronized with a PC in the latter case).

4.9 Calibration of foot pressure measurement devices

Pressure sensors are generally subject to aging, so it is absolutely necessary to repeat the calibration procedure with a certain periodicity. Calibration objectives are different in the factory, laboratory and clinical settings. In the factory, the sensor response must be fully characterized, and this requires the application of accurate and variable pressure/time profiles. It is worth mentioning that whenever a manufacturer puts a pressure measurement device (PMD) on the market with a medical purpose, the PMD must be compliant within forced regulations on medical devices. Specifically, PMDs shall be classified as medical devices with measuring functions, which means that their technical performance must be mandatorily checked in the factory and certified. In the laboratory, on the other hand, isolated checks are often sufficient, with the application of a few pressure values in specific areas of the sensitive surface. This obviously speeds up and simplifies the assessment procedures. Further simplification is finally desired in clinical settings, where few, feasible tests shall be recommended to check the maintenance of at least the minimum quality requirements: an effective test may be the single stance body weight test (BWT), as suggested in the Technical Standard CEI-62-236 [11]. Briefly, the BWT, which can be executed by the operator before any measurement session, only asks to use the PMD to measure the body weight of a volunteer who stands still on one leg only for a few seconds on at least five different areas of the platform surface. When the root mean square error (RMSE) in one or more areas shows deviation >10% of the expected/declared accuracy in vertical force estimation, the platform performance is considered doubtful (and the manufacturer should be contacted for platform maintenance action). Figure 4.8 synthesizes the outcome of the test on a pressure platform used in laboratory and clinical settings.

Manufacturers of plantar pressure measurement instruments sometimes make pneumatic calibration devices available, which are designed to test their products and aimed at applying known pressures on the entire sensitive surface of their

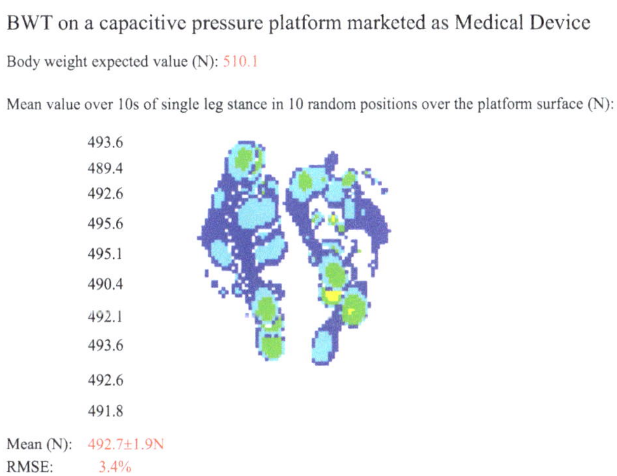

Figure 4.8 Synthesis of implementation and outcome of the BWT applied on a capacitive pressure platform

instruments. Of course, it is indeed demanding and expensive to construct and use such press machines for the calibration of large platforms (indicatively with a sensitive surface >30 × 30 cm^2) and especially for laboratory tests. This problem was addressed several years ago at the Biomedical Engineering Laboratory of the Istituto Superiore di Sanità, where a portable testing device was created that applies known pressures to limited areas of the sensitive surface [12]. Briefly, a probe is moved and then positioned on the sensitive area under test by means of a metal bridge structure, illustrated in Figure 4.9. At the end of this probe, a stainless-steel element forms the pressure chamber and is pressed against the surface of the platform by means of a lever acting on a compression spring; this compensates for the bending of the bridge structure under the load and for any differences in the thickness of the pressure platform. The pressure is imposed by filling the chamber with pressurized air that is created when the steel element is in contact with the surface of the platform. No membrane was used to make the chamber; rather, a metal tooth (0.3 mm) was created along the edge, in order to guarantee, together with the protective rubber, the hermetic closure of the area subjected to pressure.

The device has been designed to withstand a maximum pressure of 10^3 kPa, which is generally the full scale of pressure platforms. The pressure can be varied

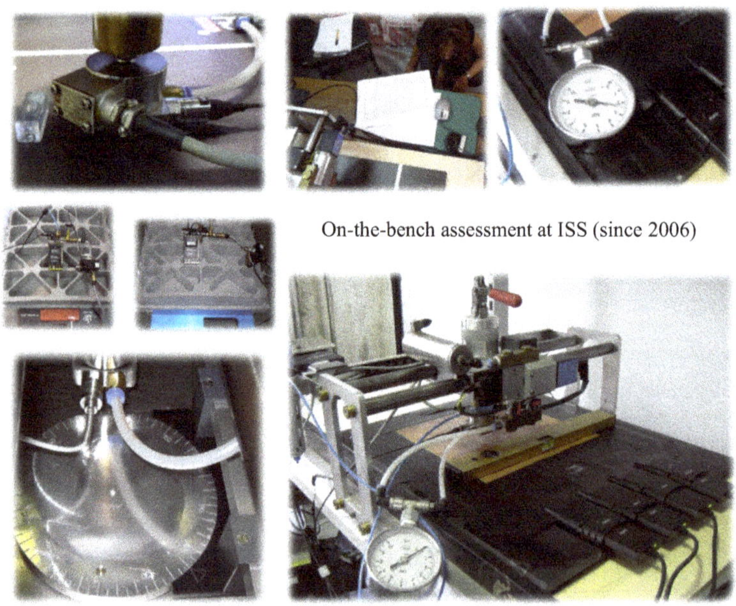

Figure 4.9 Overall on-the-bench assessment equipment at Istituto Superiore di Sanità, Rome, Italy, including the portable device more recently also equipped with a graduated table to assess CoP accuracy, and a heavy press machine for the simultaneous loading of surfaces up to 50 × 60 cm^2 (middle left)

rapidly thanks to the small volume of the chamber, whose internal base is 2.6×2.6 cm^2, and is measured by means of a commercial miniaturized pressure transducer (frequency response 1 kHz, accuracy 0.25% of full scale). Pressure pulses, with short rise times up to 1 ms (measurable limit, well below what is required), can be imposed by means of an electrically controlled pneumatic on/off valve. Alternatively, a second proportional pneumatic valve (SMC model, EIT2040-F02B) can be used, with which sinusoidal pressures up to a frequency of about 1 Hz can be achieved and delivered. Using the on/off solenoid valve, the responses to the step are investigated, and it is possible to measure the sensor response in static conditions, at different pre-defined pressure levels. This type of test is particularly useful for periodically checking whether there have been significant variations in the responses of individual sensors. The device is relatively light, transportable, adaptable to platforms of different sizes, and can be used for the complete characterization of both single sensors and matrix sensors.

4.10 Recommendations for data collection

When operating force and pressure platforms, there are a range of factors which affect measurements that need to be considered, including (1) targeting, (2) step protocol and (3) walking speed. If a person alters their gait, for example, by either shortening or lengthening stride length to contact the force/pressure platform, it is considered as "targeting" the platform. These gait alterations will result in measurements that are not representative of the person's "typical" gait and therefore will not provide useful information. To avoid this, subjects can be instructed to walk with their head up and where platforms are not embedded in the floor, the use of a walkway of mats, with the same thickness as the height of the platform, is recommended. The step protocol, i.e., the number of steps the person takes before contacting the platform, can also have a significant effect on measurements, as evidenced by research conducted by [13], and therefore should be monitored and standardized to allow comparisons between data collection sessions and/or between subjects. Different walking speeds influence ground reaction forces [14] and pressure measurements [15] and therefore should be monitored during data collection. Another important consideration when using PMDs is to understand that it is not possible to directly compare the measurements from pressure measurement systems produced by different manufactures or to compare between platform and in-shoe devices [16].

4.11 Conclusions

The operating principles of the main instruments used to measure the external forces acting on the human body have been reported, their main characteristics have been commented on and indications on their use have been given. Some issues relating to measurement accuracy were also addressed and in particular its periodic verification in the field. For this purpose, recommendations from an Italian

Technical Standard and some calibration devices made in the laboratories of some of the authors have been described.

Although the use of most of the described devices is already well established in the biomechanics of human motion, there is still a significant demand for further development in the sector. The question of their reliability, that is to say of adequate accuracy and its preservation over time, has been highlighted here as one of the most relevant issues to address. Two further main features can be identified that may be further improved: the portability of bulkier and heavier tools such as force platforms, and the wearability of devices to be fixed to the body for monitoring freely walking subjects, such as pressure insoles.

Portability is crucial to allow these devices to leave the motion laboratories and therefore find wider clinical use. This should go hand in hand with the identification of effective and efficient analysis protocols. The wearability includes the minimization of the overall dimensions, weight and encumbrance on the subject, due to the transducer hardware itself as well as to electrical connections and data collection and recording units. This question is currently stimulating, among other things, research on fabrics for garments, in order to obtain conductive or sensory capacities on selected areas [17].

References

[1] Sanders J.E., Bell D.M., Okumura R.M. and Dralle A.J. (1998). Effects of alignment changes on stance phase pressures and shear stresses on transtibial amputees: measurement from 13 transducer sites. *IEEE Trans Rehabil Eng*, 6(1):21–31.

[2] Berme N. (1990). Measurement of force, pressure, and muscle activity. In: Berme N. and Cappozzo A. (Eds.), *Biomechanics of Human Movement: Applications in Rehabilitation, Sports and Ergonomics*, Bertec Corporation, Worthington, OH, pp. 140–149.

[3] Chockalingam N., Giakas G. and Iossifidou A. (2002). Do strain gauge force platforms need in situ correction? *Gait Posture*, 16(3):233–237.

[4] Macellari V., Giacomozzi C., Cappello A. and Morasso P.G. (2003). "Misure delle forze esterne in bioingegneria della postura e del movimento" a cura di Angelo Cappello, Aurelio Cappozzo. *Pietro Enrico di Prampero Ed Patron*, 233–256. ISBN 88–555-2717-7

[5] Cedraro A., Cappello A. and Chiari L. (2009). A portable system for in-situ re-calibration of force platforms: experimental validation. *Gait Posture*, 29 (3):449–453. ISSN 0966–6362, https://doi.org/10.1016/j.gaitpost.2008.11.004.

[6] Morasso P.G., Re C. and Casadio M. (2004). *'Spot Check and Recalibration of Stabilometric Platforms'*, pp. 293–304.

[7] Nicol K. and Hennig E.M. (1973). Time-dependent method for measuring force distribution using a flexible mat as a capacitor. In: Komi P.V. (Ed.), *Biomechanics V-B,* University Park Press, Baltimore, MD, pp. 433–440.

[8] Macellari V. and Giacomozzi C. (1996). Multistep pressure platform as a standalone system for gait assessment. *Med Biol Eng Comput*, 34(4):299–304.

[9] Giacomozzi C. (2010). Appropriateness of plantar pressure measurement devices: a comparative technical assessment. *Gait Posture*, 32(1):141–144. doi:10.1016/j.gaitpost.2010.03.014. Epub 2010 Apr 15. PMID: 20399101.

[10] Giacomozzi C. and Stebbins J.A. (2017). Anatomical masking of pressure footprints based on the Oxford Foot Model: validation and clinical relevance. *Gait Posture*, 53:131–138.

[11] CEI (Italian Electrotechnical Committee) CT-62. Technical Norm CEI 62-236: 2015-01. Plantar Pressure Measurement Devices (PPMD): Measurement Performances. https://mycatalogo.ceinorme.it

[12] De Angelis G., Giacomozzi C., Maccioni G., Macellari V. and Mattia F. (1999). A pneumatic device for the calibration of pressure platforms. Abstracts of the 5th Conference of the European Society for Engineering and Medicine (Barcelona, Spain, May 30–June 2 1999), pp. 133–134.

[13] Naemi R., Chevalier T.L., Healy A. and Chockalingam N. (2012). The effect of the use of a walkway and the choice of the foot on plantar pressure assessment when using pressure platforms. *Foot (Edinburg)*, 22(2):100–104.

[14] Alexander R.M. and Jayes A.S. (1980). Fourier analysis of forces exerted in walking and running. *J Biomech*, 13(4):383–390.

[15] Chung M.J. and Wang M.J. (2012). Gender and walking speed effects on plantar pressure distribution for adults aged 20–60 years. *Ergonomics*, 55 (2):194–200.

[16] Chevalier T.L., Hodgins H. and Chockalingam N. (2010). Plantar pressure measurements using an in-shoe system and a pressure platform: a comparison. *Gait Posture*, 31(3):397–399.

[17] De Rossi D., Lorussi F., Mazzoldi A., Orsini P. and Scilingo E.P. (2003). Active dressware: wearable kinesthetic system. In: Barth, F.G., Humphrey J. A.C. and Secomb, T.W. (Eds), *Sensors and Sensing in Biology and Engineering,* Springer-Verlag, New York, NY, pp. 379–392.

Chapter 5
Assessment of muscle function

Steven Lindley[1,2], *Dan Robbins*[1] *and Amy Robinson*[2]

5.1 Introduction

While it is rarely something that is consciously thought about, walking requires careful coordinated organisation of multiple groups of muscles. Exactly how muscles work has been an active area of research for decades.

Researchers and clinicians consider many questions about muscle function in human movement, for example:

- When do muscles start/stop working?
- Which muscles work together?
- Which muscles work in opposition?
- What intensity do muscles contract at?
- How much energy does it require for the muscles to function?
- Where does the energy required come from?
- How does footwear, orthotics or prostheses affect muscle function?

Many other questions have stimulated past and present research activities, and there is an exciting future for new and innovative lines of enquiry into the function of muscles but with firm foundations from the previous decades.

This chapter aims to provide readers an overview of important aspects of muscle function and measurement through electromyography (EMG), however is far from an exhaustive exploration of EMG, and interested readers will be directed to further reading after we hope this chapter piques your interest.

5.1.1 Terminology

One approach to assess how muscles are working is to measure the electrical signals that control the muscles. The process which is used to record the electrical activity of muscles is referred to as 'Electromyography', a relatively complex term that can be divided into three constituent parts. The prefix 'electro' indicates

[1]School of Allied Health, Biomechanics Lab Lead, Faculty of Health, Education, Medicine & Social Care, ARU, Chelmsford, UK
[2]Delsys Europe, Stamford House, Sale, Greater Manchester, UK

electrical activity, the root 'myo' indicates related to muscle tissue and the suffix 'graphy' indicates 'to record' or 'take a picture'. An alternative version of the word is 'electromyogram', where the suffix 'gram' represents the visual output of the electrical activity of the muscle tissue. It can therefore be stated that electromyography produces an electromyogram. Typically, the abbreviation EMG is used to represent both electromyography and the electromyogram.

When reviewing or analysing EMG signals, the terms 'muscle activation' or 'myoelectrical activity' are commonly used to refer to electrical activity within a specific time period.

5.1.2 History of EMG

The earliest records of muscle-generated electricity were documented in 1666 by Francesco Redi who deduced that the source of electrical ray fish's energy was from muscle tissue. It was not until 1791 when Luiggi Galvani demonstrated that electricity can be used to initiate a muscle contraction by depolarising the muscles of a frog's legs by touching them with metal rods [1]. The first evidence of an electrical activity being recorded in human muscles was presented by Du Bois-Reymond in the mid-1800s. Interestingly, the signal magnitude was enhanced when a portion of the subject's skin was removed, potentially offering the first evidence of signal attenuation of tissues between the recording electrode and the muscle [2]. The first recording of the electrical activity of muscles was reported in 1890 by Marey, who coined the term 'electromyography' [3]. In 1922, Gasser and Erlanger used an oscilloscope to study action potentials being conducted along a nerve axon [4] further advancing knowledge of the interaction between the electrical activity of nerves and muscles.

Throughout the first half of the twentieth century, technology developed further, providing the opportunity for research to investigate dynamic movements and conditions associated with pain. During the 1960s, Basmajian developed a fine wire technique to record activity from single MUs, furthered by Carlo De Luca in 1972 with the Quadrifilar needle electrode [5], with research that would ultimately lead to the cutting-edge decomposition of surface EMG (sEMG) recordings available today. In 1965, Basmajian was one of the founders of the International Society of Electrophysiological Kinesiology (ISEK), an organisation still actively furthering the field of EMG today. Work in the study of muscle function has continued over the decades with the same scientific tenacity and energy, but now coupled with the inclusion of innovative and disruptive technologies.

5.2 Muscle physiology

To better contextualise the use and application of EMG, and its associated technologies, one should familiarise themselves with the underlying physiology and anatomy. We would encourage readers to immerse themselves within the many excellent existing neurophysiology texts available, where here we endeavour to simply present an overview.

5.2.1 Neuromuscular anatomy

The structural unit of a muscular contraction is the muscle cells or muscle fibres, which are supplied by the terminal branches of one motor nerve fibre or axon. An axon of a nerve end in multiple terminal branches provides neural control or *innervation* to multiple muscle fibres. The term used to describe the single smallest controllable muscular unit is a motor unit (MU). A MU consists of a single alpha-motoneuron, its neuromuscular junction and the muscle fibres it innervates; as few as 3 and as many as 2000 [6].

5.2.2 Signal propagation

Neural action potential signals originate in the anterior, ventral horn of the spinal grey matter where they propagate along motoneurons, via terminal branches to the muscle fibres that are innervated within the motor unit (Figure 5.1). A single contraction of the muscle fibres is referred to as a '*twitch*', the characteristics of the muscle fibre type (slow oxidative, fast oxidative or fast glycolytic) will determine the velocity, tension and duration of the twitch (Figure 5.2).

Neural action potentials work on an 'all or nothing' principle, therefore the amplitude of an action potential is always the same. To control the overall rate, tension and duration of a *whole muscle contraction*, a complex interaction between the frequencies of the action potentials a MU receives and the number of MUs that are receiving action potentials is required.

5.3 Measuring neuromuscular function

Understanding the functional characteristics of MUs is vital in the pursuit of knowledge on how force is produced and controlled, yet the measurement brings

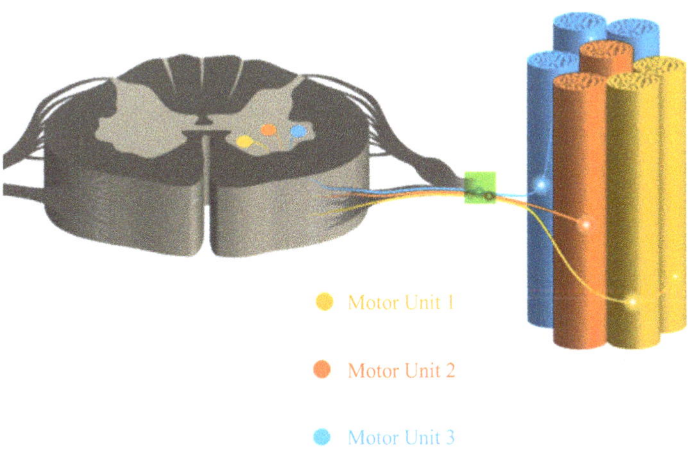

Figure 5.1 MUs originating in the ventral horn of the spinal cord, innervating a group of muscle fibres

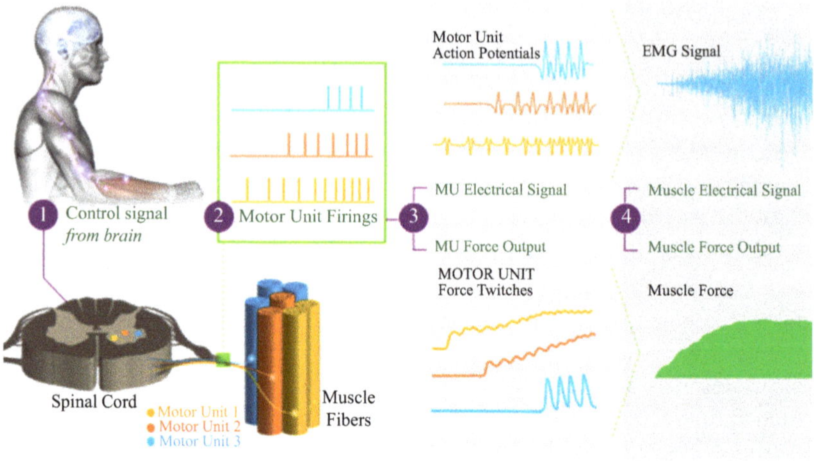

Figure 5.2 How electrical signals originating in the brain, ultimately manifest at the muscle

with it a series of physiologically technical challenges that require meaningful consideration.

5.3.1 Innervation zone (IZ)

When an action potential reaches muscle fibres, the signal is propagated along the muscle fibres in both directions away from the point of innervation. While it is currently only possible to visualise the exact location of innervation by taking biopsies and using histochemical staining, there are some unique physiological characteristics of the pattern of signal propagation that allow identification of *innervation zones* non-invasively using sEMG.

When recording myoelectrical signals on and around the location of MU innervation, it can be observed that directly above the innervation point there is a zone in which the amplitude on the signal is attenuated compared to recordings of the signal which have been propagated along the muscle fibres. This area of reduced amplitude is referred to as the *innervation zone*. If an array electrode is positioned directly above the IZ, aligned so the IZ is approximately in the middle of the array and the ends of the array follow the direction of the muscle fibre orientation, the resulting recordings will display propagated signals at simultaneous sensors along the array moving away from the IZ. The signals will also have an amplitude that is reflected in opposite directions (Figure 5.3). To identify these unique characteristics, monopolar sensors are typically used in EMG array sensors.

The reduced amplitude of the signal over the IZ is the primary reason for avoiding electrode placement over the location of IZs. To facilitate the avoidance of IZs and promote the acquisition of quality EMG signals, an atlas of innervations zones has been produced by Barbero, Merletti and Rainoldi [7].

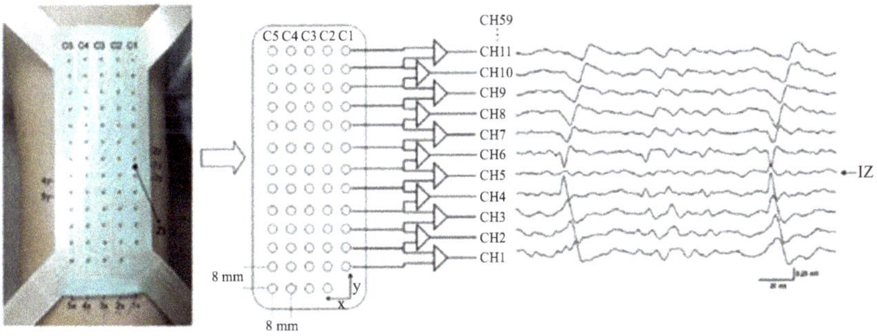

Figure 5.3 An array sensor highlighting the effect of IZs on the amplitude of the EMG signal

5.3.2 Sensor placement

The location and positioning of EMG sensors is a much discussed topic with various recommendations either based upon surveys (SENIAM Recommendations) [8] or through empirical evidence [9]. It has been shown that the location of the sensor is one of the most important factors in obtaining a high-fidelity EMG signal, with differing locations rendering dramatically different signal characteristics [10]. Sensor placement has also been shown that it may have a considerable effect on the temporal relationship of the vasti muscles [11]. Roy *et al.* [9] demonstrated the effects of positioning the sEMG sensor across varying areas of the muscle with respect to sEMG signal characteristics. With the electrode being positioned in proximity to the tendinous insertion of the muscle and the IZs, it was found to be detrimental to signal fidelity. The differing placements and consequent differing signal characteristics are represented in Figure 5.4. The placement of sEMG sensors has therefore been recommended to avoid IZs, which often reside near the edge of the muscles, tendinous origin and insertions with the optimal placement being the middle of the muscle belly [10].

While the concept of sensor placement to avoid IZs is practical and effective in larger muscles, when assessing smaller muscles, such as those on the foot, avoiding IZs is a more challenging task. Particular care with sensor placement and selection of sensor type and size is paramount. While avoiding an IZ during a static contraction is feasible, movement of IZs under the skin's surface and hence relative to a surface recording electrode cannot be avoided and must be considered when interpreting results.

5.3.3 Tissue properties

In addition to the location of the IZ along the length and width of a muscle, the depth of the MU innervation point is an additional physiological factor that can influence EMG signals. EMG recordings from the surface of the skin above muscles with deep innervation points will result in signals with smaller amplitude than those of muscles where the MU innervation point is relatively superficial (Figure 5.5) [12].

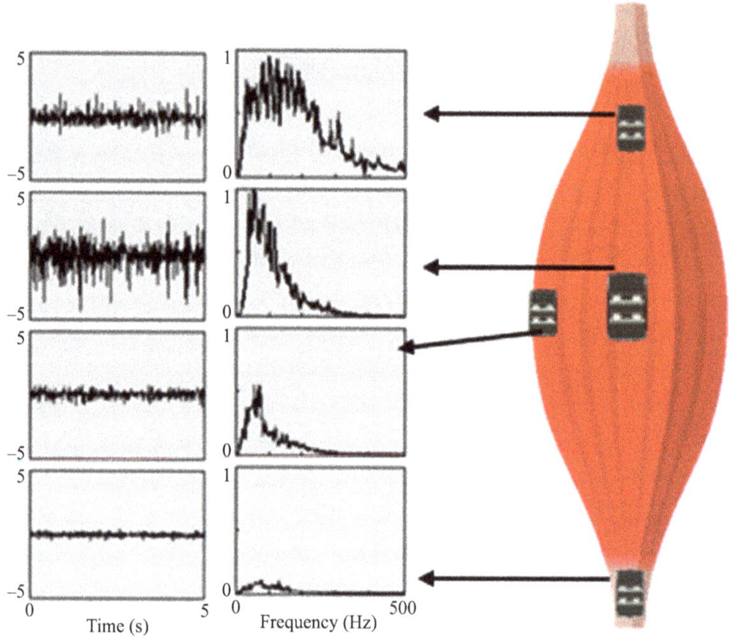

Figure 5.4 *Example of how sensor placement can affect the temporal and frequency spectra of the signal*

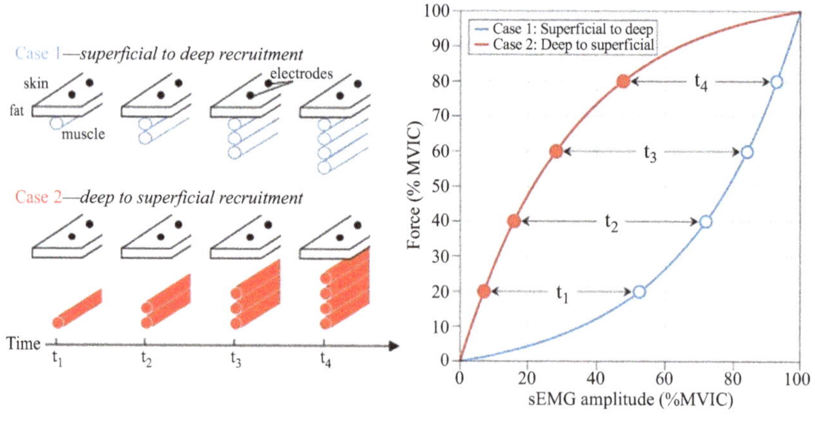

Figure 5.5 *Effects of tissue properties of the recording muscle on the resulting signal*

Other than the location and depth of the innervation of a MU, the primary physiological aspect influencing the acquisition of good quality signals is the volume of non-muscular tissues and fluids between the recording electrode and the active muscle tissue. Obviously, this is a much greater issue when recording via surface electrodes compared to electrodes that are inserted directly into the muscle tissue and therefore bypass subcutaneous barriers created by non-muscular tissues and fluids.

The influence of the subcutaneous non-muscular tissues and fluids is analogous to that of a low pass filter; therefore, high-frequency components of the acquired signals will be filtered to a greater degree than low-middle frequency components of the signal. Research has indicated that even displacements of 100 μm from the active muscle tissue will result in an attenuation of 80% of the signal strength [13]. The influence of subcutaneous fat has been investigated via the application of electrical stimulation to the surface of the skin and recording of the signal via needle electrodes inserted into muscle tissue; the results indicated a linear relationship between the thickness of subcutaneous fat and signal rise time for square wave signals [14]. Logic would suggest that signals travelling in the opposite direction would also be filtered, though we cannot assume the resulting filtering characteristics would be equivalent. However, it should be noted that the influence of subcutaneous fat only appears to attenuate electrical activity from non-voluntary contractions, i.e. electrically stimulated muscles [15].

It has also been suggested that changes in blood volume in exerting muscles change the quantity and characteristics of the barriers between active muscle fibres and surface electrodes, again further influencing the filtering effect on the signal prior to reaching the recording electrode [16].

While the exact characteristics of the filtering effect of physiology have yet to be conclusively defined, normalising signals using maximal voluntary or electrically stimulated muscle contractions provides a relative signal value that can be used within a single data collection session.

One final aspect which must be considered in relation to electrode placement is the potential influence of crosstalk, or the acquisition of electrical activity from muscles closely located to the muscle from which signals are being acquired from. In addition to the physical distance between the muscles, increased subcutaneous fat levels have been shown to influence the amount of crosstalk acquired within a signal [17].

5.3.4 Skin preparation

Once the appropriate location for an electrode has been identified, the application site should be prepared. While it is obvious any visible dirt must be removed, non-visible barriers must also be considered. Cleaning the application area to remove the naturally occurring oils, any skin creams or lotions, dead skin cells or excessive body hair are all important aspects of site preparation and can directly influence the quality of the signal acquired. Typically, the minimal skin preparation applied would be the use of 70% isopropyl alcohol wipe over the application site. However, depending on your research question, subject group and quality of the EMG technology, the user may need to also shave the application site to remove excessive hair and hydrate the skin if excessively dry. Occasionally, researchers use abrasive scratch pads, tape or gel to gain a high-fidelity signal, though the vast majority of publications do not include this aspect of skin preparation.

5.3.5 Signal quality check

It is possible to assess the quality of skin preparation using an impedance meter before and after skin preparation, though this process is rarely undertaken, or at least reported in scientific studies. It has been suggested that ensuring that skin impedance is below 55 kΩ by good skin preparation practices will ensure that the frequency spectrum of the EMG will not be affected [18]. However, skin impedance being susceptible to environmental conditions (hydration, temperature, etc.) and ignorant of the other elements important in accessing a high-fidelity EMG signal therefore would not be recommended as a reliable use of assessing signal quality. Instead, it would be recommended to look at various elements of the EMG signal, including baseline noise, line interference, signal clipping and signal-to-noise ratio (SNR).

The SNR is arguably the most accurate single measure in assessing the quality of an EMG signal; ideally, the user would review the various characteristics of the signal but in absence of this, the SNR will provide the user a qualified metric of signal quality.

SNR can be defined by the following equation:

$$SNR = \frac{\text{sEMG signal amplitude (rms)}}{\text{Baseline signal amplitude (rms)}}$$

SNR values greater than 10 would be deemed acceptable but with optimal conditions, users should aim for an SNR of about 30. For more considered signal quality approaches, then the users should consider the constituent components of both the signal and the noise.

The baseline noise is the sum of the inherent sensor electronic noise and the skin-electrode interface; the lower baseline noise component allows enhanced discrimination of the EMG signal and specifically so the onset of muscle contraction. Baseline noise values <4.8 μV root mean square (rms) would be suggested as a recommendation. If the baseline noise component exceeds <4.8 μV rms, then the skin preparation procedure should be repeated and perhaps reviewing the sensor location.

Line interference or electrical interference from power lines and other sources can adversely affect the EMG signal. The frequency of the line noise is typically found at 50 Hz or 60 Hz and is therefore within the same interested frequency bandwidth as EMG signals. Frequency analysis of the signal will expose any line interference with spikes in the 50 Hz or 60 Hz domain. Alternatively, visual inspection of a small-time window can also prove to be a quick and effective method to identify line interference, for example, looking at a signal with a window of 0.1 s would reveal 5 or 6 peaks of a sinusoidal wave created by line interference.

Signal clipping or signal spikes can be found in an EMG signal and are frustrating as they can mask the EMG signal or be wrongly interpreted as an EMG signal. These signal artefacts can often be described as motion artefacts, where there is a disturbance of the skin–electrode interface caused by motion of wires (in wired EMG systems) or through a mechanical event that disturbs the electrode contact on the skin such as a heel strike. By following good skin preparation and adequate fixation on the skin, you can reduce the likelihood of introducing these artefacts. For highly dynamic activities, it is often advised to apply additional fixation of the EMG sensor on the subject.

5.4 EMG electrode and system design

Arguably the most influential element in obtaining and utilising a high-fidelity EMG signal is the electrical apparatus being used, namely, the EMG sensor. EMG sensors and EMG electrodes are terms that are often used interchangeably; however, they describe different aspects of the EMG system. The EMG electrode refers to the part of the hardware that makes direct electrical contact with the tissue and detects electrical activity. The EMG sensor encompasses the whole sensing unit, consisting of the EMG electrodes, the pre-amplifier, signal conditioning electronics, the sensor enclosure and often hundred more components. The engineering methods, the quality of the electrical components and the electrode geometry largely determine the EMG signal presented to the user and its consequent interpretation to the underlying physiological processes. Therefore, understanding design features and considerations of EMG technology is crucial in making informed decisions about the type of electrode to be used for your application.

5.4.1 Electrode design

EMG electrodes can be found in all different shapes and sizes, where some are designed to be placed within the muscle belly (intramuscular EMG electrodes) others are designed to be placed on the surface of the skin over the muscle belly (sEMG electrodes). Although the construction and type of EMG electrode can differ, the application is the same; placed close enough to the muscle they can detect ionic movement and therefore represent the electrical signal from the muscle.

There are predominately two methods for recording from an EMG electrode; either through a single electrode (monopolar) or pair of electrodes (bipolar), placed within or over the top of the muscle belly. The monopolar EMG electrode consists of a single detection surface, recording the electrical potential from within the capture area of that electrode relative to a reference electrode. The bipolar EMG electrode consists of a pair of detection surfaces and records a differential signal between the two detection surfaces, with respect to a reference electrode. Each method has its merits and preferred applications. The monopolar EMG electrode presents a simple design and application, however, often lacks the ability to adequately remove unwanted electrical signals from the EMG signal. The design of a bipolar EMG electrode differs as it utilises a differential measurement and therefore the ability to remove any common signals to both electrode sites, such as power line noise, and therefore improving the signal accuracy and fidelity.

5.4.1.1 Intramuscular EMG

Intramuscular needle electrodes are typically used in clinical assessment and explorations. Depending on local legislation, there may be limitations on what qualifications and training are required to insert intramuscular electrodes.

Fine-wire EMG electrodes, popularised in the early 1960s, are the most common approach within human movement sciences consisting of a single or pair of fine nylon-coated wires (<50 μm in diameter) that have barbed tips and a sharp gauged hypodermic needle. The fine-wire EMG electrode is inserted into the muscle, through the skin and fat layers, and the hypodermic needle is withdrawn leaving the wires within the muscle due to the barbed tips of the wires. The exposed wires are then connected to an EMG recording system for data recording.

Intramuscular EMG electrodes are generally very selective; the small size of the electrode's detection area results in an EMG signal generated by only a few MUs that are located in close proximity to the electrodes. Due to the close proximity to the muscle fibres, the distance between the source of the electrical signal (muscle fibres) and the detection area (fine-wire) is relatively short (0.5–3 mm). As a result, the frequency spectrum of the intramuscular EMG signal contains more energy at the higher frequencies than that of sEMG, which is low pass filtered by subcutaneous fat, tissues and fluids.

These electrodes do have potential complications including patient discomfort [19] and wire fracture [19]. However, the incidence of these problems is extremely low [19] and is not considered a significant threat to subjects by most experienced electromyographers.

Within the field of human movement sciences, intramuscular modes of EMG signal acquisition are predominately used in applications where access to deeper muscles is required, whereby sEMG cannot accurately detect, such as the Tibialis Posterior. Due to the relative complexity, patient's concern of needles, required skill and qualification and associated costs, then sEMG is the more commonly applied technique.

5.4.1.2 sEMG

Over the past hundred years, sEMG electrodes have become more sophisticated in detecting the EMG signal, and they are now presented in a wide variety of constructions whilst the fundamental concept remains the same since 1912 where Piper used a metal plate to detect human muscle activity. It wasn't until 1979 when De Luca *et al.* [20] developed a dry active EMG electrode (Figure 5.6a) and as such a novel alternative to existing electrodes that required conductive gels or pastes.

Constructed as silver-silver chloride electrode contacts on the sensor, either as parallel bars or pins, the dry EMG electrode offers superior EMG fidelity whilst removing the requirement for any conductive interface and instead requiring inherent moisture to be present on the skin to form the electrolytic interface. The

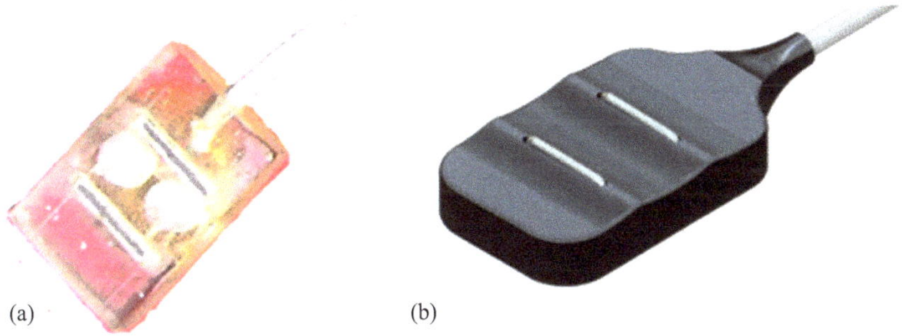

Figure 5.6 Examples of active dry electrodes, where (a) represents the first example of its type and (b) the later Delsys Bagnoli EMG electrode

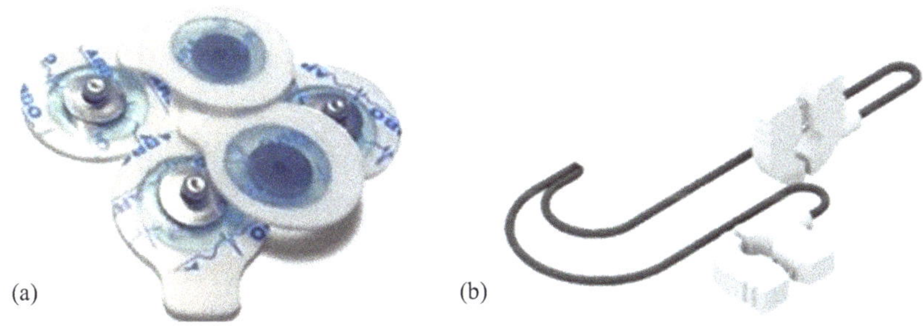

Figure 5.7 Examples of alternative EMG acquisition methods where (a) shows pre-gelled disk contacts and (b) shows the connective snap leads

EMG electrodes presented by De Luca *et al.* [20] evolved over the years to become a commercially available and patented product design with the Delsys Bagnoli EMG electrode (Figure 5.6b) that maintained the original parallel bar dry EMG electrode design.

Alternative EMG electrodes are constructed of pre-gelled disk contacts (Figure 5.7a) that are connected to snap or pinch connectors (Figure 5.7b) and insulated cables to the main sensor body or EMG amplifier. Although a popular design choice amongst commercially available technologies, the requirement of the conductive gel as the interface means it is susceptible to movement on the skin and it requires a short, albeit insulated, cable between the EMG contacts and amplification is thus more sensitive to cable movement artefacts or noise in the EMG signal.

The aforementioned EMG electrode and sensor designs all provide a single channel of EMG through their differential detection; however, in recent years, new innovative approaches to detect multiple channels of EMG from within one sensor design have been developed. Sensors seen in Figure 5.8 offer four channels of EMG through one sensor whilst maintaining the dry electrode differential EMG

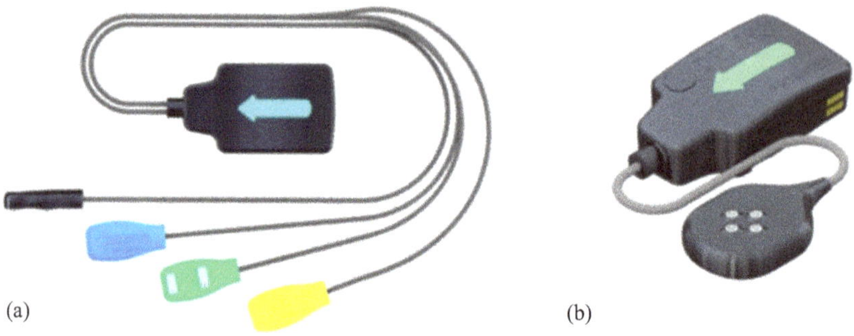

Figure 5.8 *Examples of ways of procuring four channels of EMG data via the (a) Delsys Trigno Quattro sensors and (b) Delsys Trigno Galileo sensor*

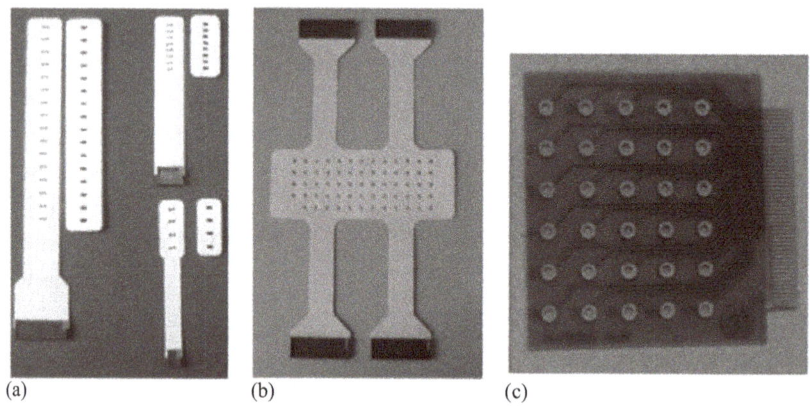

Figure 5.9 *Examples of high-channel count EMG showing (a) linear and (b and c) array configurations*

measurement, therefore providing the user the ability to study more muscles or spatial resolution of individual muscles.

Alternative approaches for studying single muscles through high-channel count have also been considered since the 1970s through the use of grid or high-density EMG electrodes (Figure 5.9). These EMG electrodes typically contain a large number of EMG detection points and are installed on a flexible interface, with generally small (4–10 mm) inter-electrode spacing. These electrodes provide the user with access to EMG information that is suited to explorations of spatial properties, rather than temporal, of the EMG signal such as conduction velocity or IZ mapping. However, new applications of similar technologies have been applied in recent years to study MU control and other spatial metrics. The requirement for application of conductive pastes, exposed electrodes and cables causing noise and artefact have all deterred wide-spread adoption of these types of applications in human movement sciences.

5.4.2 Sensor and system design

As discussed previously, EMG electrodes can come in various forms as can the EMG sensors. Historically, EMG sensors have been tethered to a desktop amplifier and then connected via a data acquisition device to a computer (Figure 5.10). These 'wired' EMG sensors and systems offer high-fidelity EMG data from multiple sensors (typically up to 32), providing the user with various configuration options and flexibility in acquiring other physiological data types too. The wired EMG sensors contain active electronics that are typically shielded with a nickel-silver layer and encased within a polyurethane enclosure to protect from sources of noise and connected to amplifiers through shielded cables. Wired EMG systems are inherently limited with the requirement of being tethered to a desktop system and therefore recording data during more dynamic applications such as gait analysis may be challenging.

The late 2000s brought around the introduction of 'wireless' EMG sensors and systems, where small non-intrusive wireless sensing units transmit data over wireless communication networks to a receiving unit, and thus removing the main limitation of tethered technologies. Wireless sensors bring with them the freedom of movement and seemingly unlimited application and specificity towards capturing the full range of human movement. With the advent of wireless and wearable sensor technologies in the last 15 years, there has been a dramatic increase in the measurement of human movement through the use of EMG. Although the fundamental components of a wireless sensor are readily available, the design and implementation of these components in creating a high precision technology to

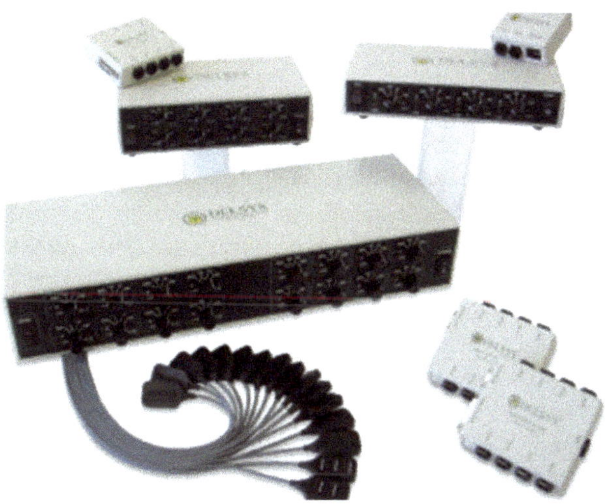

Figure 5.10 A desktop amplifier

measure a high-fidelity EMG signal is an engineering art form. Aside from the EMG circuitry that is often the result of decades of electrical engineering design and finesse, there are other critical components in the successful execution of a research-grade EMG sensor, including the wireless communication protocol. Although this may seem a transactional component of the design, without proper consideration the sensor will not be able to robustly transmit data, provide adequate sampling rates or offer inter-sensor synchronisation. Users would be advised to ensure a good understanding of the implemented design of the wireless communication so that confidence in the data collected aligns with the physiological interpretations of the data.

5.4.3 Sampling rate

Selecting the sampling rate or sampling frequency is dependent on various factors, two of the main factors being Nyquist Theorem and the type of activity being measured. Nyquist Theorem [21] states that the sampling frequency should be at least twice the highest frequency component being measured in order to accurately reconstruct the signal. Therefore, it is important to consider the type of activities being captured to establish the highest frequency component and the consequent minimum frequency. The frequency content of sEMG signals can be up to 400 Hz [6]. In order to abide by Nyquist Theorem, sampling at no less than twice its frequency, the sEMG signal should be sampled at a minimum of 900 Hz to avoid aliasing of the signals. Typically, EMG systems are sampled at a minimum of 1,000 Hz, although more advanced technologies and research topics that require it can sample at rates of up to 4,000 Hz.

5.4.4 Reference electrode

Reference electrodes, also known as ground or earth electrodes, are necessary components of any EMG recording method in order to accurately remove common noise sources (line interference, motion artefacts and static field disturbances) and improve the common-mode rejection ratio. When a mechanical disturbance occurs, such as a heel strike, the electrolytic interface between the electrode contacts and the skin is electrically destabilised. The disturbance can saturate the EMG signal and be challenging to remove post-processing, therefore a method to remove or reduce the introduction of these disturbances is sought after. The most commonly applied technique is to use a separate reference electrode(s) that utilises a triangular configuration of a differential EMG recording, and thereby decreases and stabilises the impedance of the skin–electrode interface. For wired EMG systems, the approach is to reference each EMG electrode to a single conductive hydrophilic gelled electrode. For wireless EMG systems, which are commonly applied in more dynamic applications, a single reference electrode wired to the wireless sensors is not convenient and would not perform well at removing noise sources. Instead, wireless EMG systems either provide a reference ground electrode contact on each sensor or attempt to electrically ‘float’ and provide a ‘reference free’ solution. The problems of artefact and sensitivity to electro-static fields are especially severe for

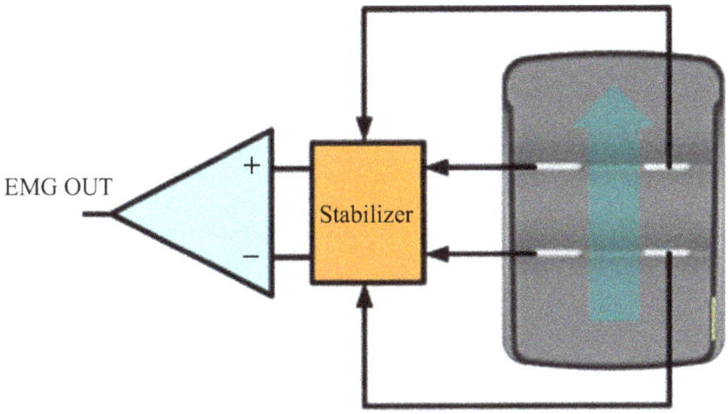

Figure 5.11 Example of a combined EMG and reference contact unit

wireless “reference free” sEMG sensor designs, instead wireless solutions that utilise differential recording from a triangular configuration of three or more electrodes are much less affected by disturbances. As an example, the Delsys Trigno EMG sensor (Figure 5.11) employs a single differential configuration with two EMG electrode contacts and two reference ground contacts to provide a high-fidelity EMG signal whilst reducing the introduction of noise sources through the dual stabilising reference electrode contacts.

5.4.5 Inter-electrode spacing

The distance between EMG electrode contacts is described as the inter-electrode spacing. These distances vary depending on equipment selection and user’s applications, but typically vary between 5 mm and 40 mm. The goal of the inter-electrode spacing is to maximise the EMG signal amplitude from the muscle of interest (target signal) and minimise the signal amplitude from muscles in proximity to the target muscle (crosstalk signal). Although the distance between electrode contacts could be seen as a nuanced product design consideration, it has often dramatic effects on the resultant EMG data due to the influence of muscle crosstalk. De Luca *et al.* [22] demonstrated the influence of inter-electrode spacing on the degree of crosstalk contamination in the Tibialis Anterior and Triceps Surae muscles during walking gait, whereby increasing the inter-electrode distance also increased the likelihood of signal suffering from crosstalk. In addition to the chosen inter-electrode distance of the sEMG sensor, users should consider the consistency across muscle groups, subjects and testing sessions. A difference in the inter-electrode spacing consistency leads to an inconsistency in the pickup volume, which would cause differences in the amplitude and frequency of the captured signals and therefore cause concern for data interpretation. Users should consider methods of keeping a fixed inter-electrode spacing and the chosen spacing based on peer-reviewed evidence for their application. It was found [22] that a 10 mm inter-electrode spacing was optimal in amplifying the target muscle whilst offering minimal crosstalk.

5.5 EMG decomposition

Typical sEMG measurements provide information of the gross muscle behaviour, such as the muscle activation, muscle amplitude or muscle fatigue. These metrics are explored through the gross EMG signal, in that the represented signal is a composition of all the motor unit action potentials (MUAPs) within the measured area. An alternative and supportive method of understanding the muscle is to study the individual MUAPs. Through studying the muscle behaviour at the MU level, it can provide desirable information on how MUs are controlled by the central nervous system (CNS) in generating force or for assessing the degree of dysfunction in upper motoneuron diseases [23]. In the past decade, developments have allowed the exploration of MUs through sEMG technology [23], where previous attempts of the MU study were through invasive methods, and as such provide users access to a high yield of MUs and consequently greater representation of the MU pool [24] through a process called sEMG decomposition.

5.5.1 Decomposition techniques

As first described by LeFever and De Luca [25] (Figure 5.12), the process of decomposing an EMG signal into its constituent components can present information on the firing characteristics and morphology of the MUs, consequently detailed information on the behaviour of the muscle. The original process was achieved by manually identifying individual MUAPs and allocating them, by their shape and amplitudes, to a specific motor unit action potential train (MUAPT).

Although this process could provide a unique insight into the behaviour of the muscle, its application was limited. Due to the selective nature of the intramuscular electrode, only a small number of MUs could be detected, with 2–5 MUs typical for most measurements, which significantly limits the representative information that can be gained from a muscle contraction where hundreds of MUs are likely to be active.

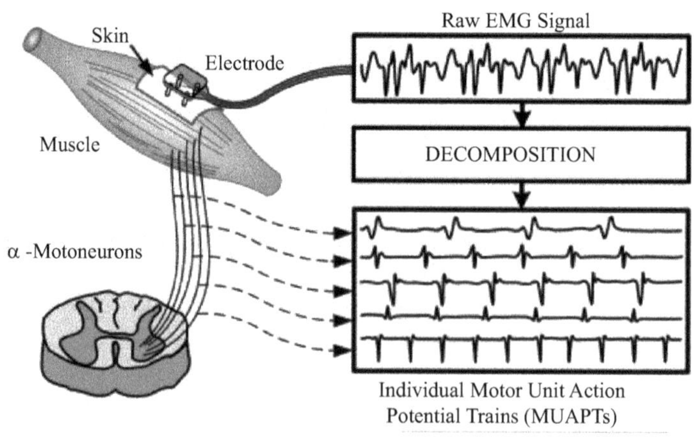

Figure 5.12 Decomposing the raw EMG signal into the MUAPTs

A further limiting factor with invasive EMG decomposition is that the signal becomes redundantly complex when the force contraction increases, this is because the additional force the muscle requires to contract requires additional MUs and their firing behaviour changes thus the superposition of the MUs becomes near impossible to resolve into the MUAPTs at all or with much confidence. Therefore, the invasive measurements have typically been conducted with low force levels (1–10% maximal voluntary contraction (MVC)), thus limiting the application to small and low force level activities.

Decomposition of an sEMG signal is an attractive notion that has been considered for many years [26], however, the complexity of an sEMG signal with multiple superimposed MUAPs and noise contamination make it a difficult task to resolve. In the early 2000s, development towards a non-invasive sEMG decomposition solution was shown to have impressive results [23,24] in providing a solution for investigating MU properties from the surface of the skin. Artificial intelligence-based algorithms were developed [23,24] to automatically decompose sEMG signals from isometric contractions at force levels up to 100% MVC through the application of a specialised four-channel EMG sensor (dEMG, Delsys Inc., Natick, USA).

The advent of sEMG decomposition technology provided ground-breaking new insights into the hierarchical control of the neuromuscular system, whereby it was documented that, at any time and force during voluntary constant-force contractions, earlier-recruited MUs maintain higher firing rates than later-recruited ones, providing an inverse orderly hierarchy of nested firing rate curves resembling the layers of the skin of an onion [23]. The developments provided users the ability to decompose an sEMG signal up to a maximum contraction and accurately observe high yields of MUs (typically 30–50 MUs) and new information on the nervous systems' hierarchical control of MUs. These insights provided significant advancements compared to the ability to explore a normal sEMG signal for its amplitude, timing or fatigue. The advances have led to a greater understanding in various population groups and activities; during voluntary isometric contractions in non-symptomatic individuals [27], in elderly subjects [28], during muscular fatigue [29,30], strength training [31] and stroke sufferers [32]. The advantages of the ability to decompose an sEMG signal are vast, but specifically the ability to measure up to 80 MUs from a single contraction, with force contractions up to 100% MVC and all from the surface of the skin far outweighs the aforementioned limitations of invasive EMG decomposition methods.

However, the methods were still limited to constant force and isometric muscle contractions until the evolution of the algorithms in 2015 [33]. De Luca and colleagues clearly demonstrated the successful approach of decomposing the sEMG signal collected from cyclic dynamic muscle contractions and introducing the concept of seeing MUs in-action during concentric and eccentric muscle actions for the first time (Figure 5.13).

The application of sEMG in cyclical dynamic activities can be applied to common activities such as gait analysis to provide unique insight into the behaviour of a muscle, and groups of muscles, through the detailed inspection of the MU as opposed to traditional amplitude or timing metrics of gross EMG signals (Figure 5.14). These novel approaches to measure the muscle offer never-seen-before control mechanisms, and open the clinical communities to not only

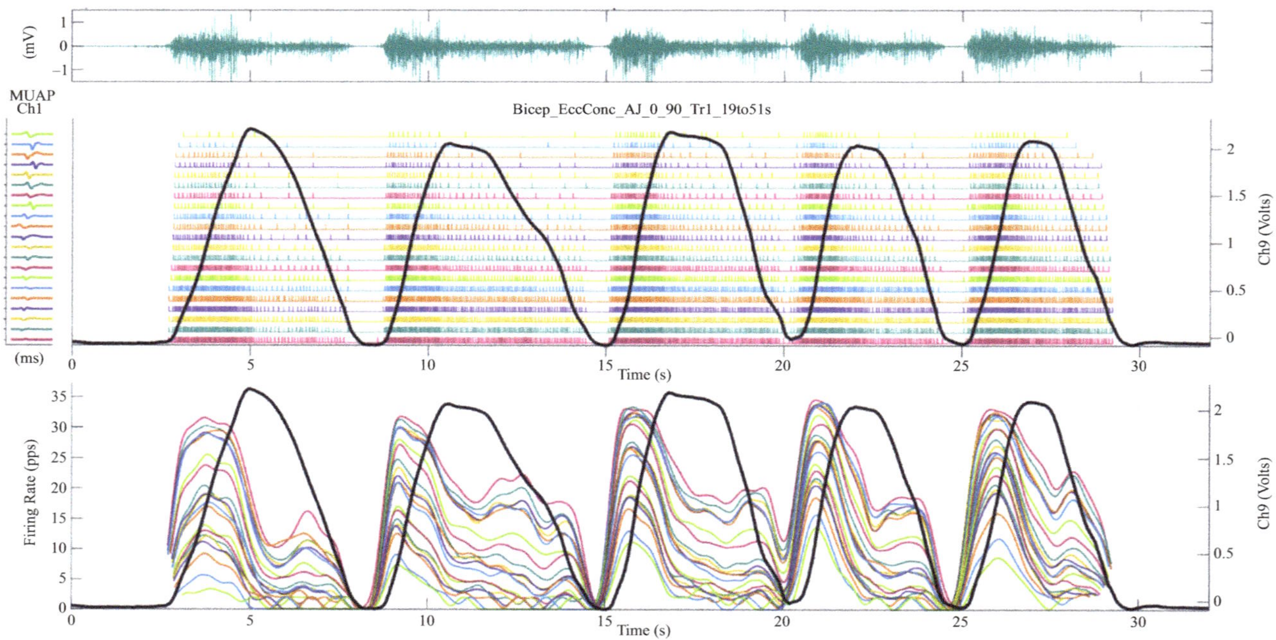

Figure 5.13   Highlighting the decomposition of raw EMG (top row) into the individual MU firings (middle row) and MU firing rates (bottom row) during a cyclic contraction

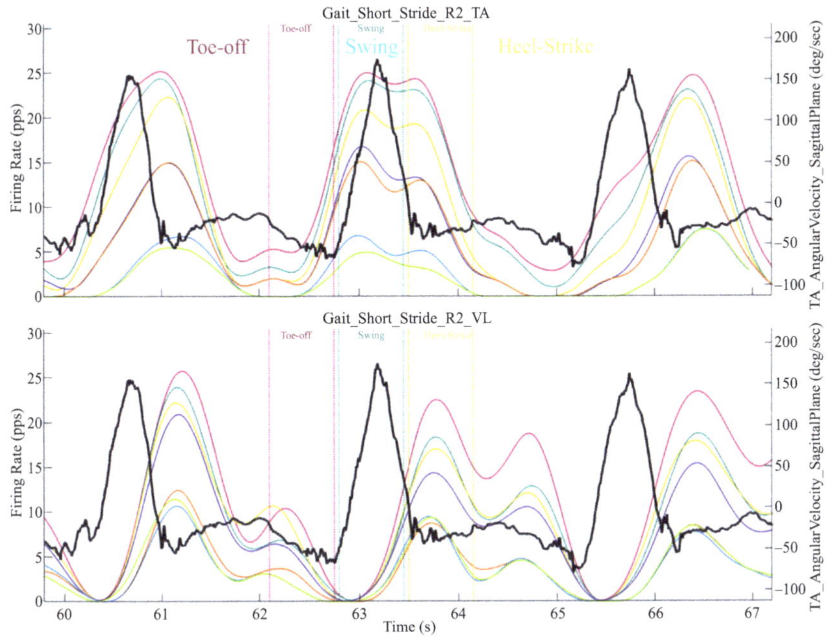

Figure 5.14 The employment of EMG decomposition into its constituent MU firing rates when applied to gait analysis

understanding more about the neural control during gait but how common interventions interact with MU behaviour.

5.5.2 EMG decomposition metrics

sEMG decomposition outputs the firing instances and morphology of MU, which can be processed to extract information about the firing rate, recruitment and derecruitment, morphological changes in action potentials, correlations in MU firing, firing rate variability within and across MUs all of which can be applied to a wide range of fields in exercise physiology, kinesiology, motor control or rehabilitation.

The complex inter-relationship between the MU firing rate, MU recruitment threshold and the common drive offers unique strategies at achieving the same objective in the generation of force and its control within the muscle.

The firing rate, or discharge rate, of a MU is one of the determining factors in producing force within the muscle [34]. This is the frequency at which the muscle fibres within a MU are excited by their innervating axon, whereby in general the MU firing rate increases with increasing muscular effort until a maximum firing rate is achieved. The firing rate is closely aligned with the recruitment of MUs and should not necessarily be viewed alone, but instead along with other MU variables.

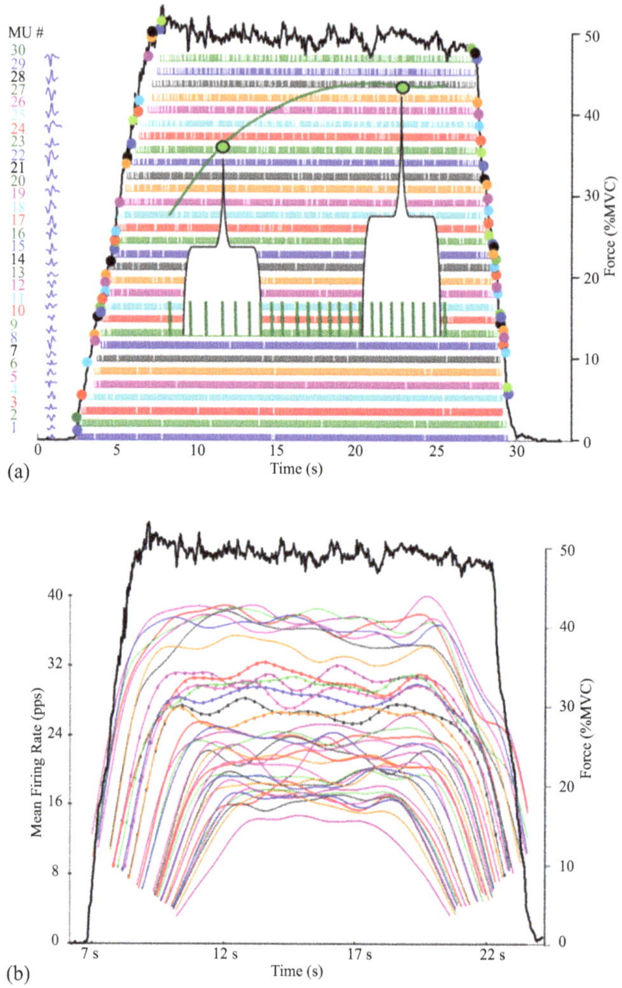

Figure 5.15 Representation of (a) the timing information associated with each of the identified MUAPs detected in the signal and (b) the average firing behaviour of each of the MUs

Another mechanism of changing the force level of a muscle in a voluntary contraction is the recruitment of MUs [29] and the strategy used to achieve this (Figures 5.15 and 5.16). It is well documented that the recruitment of the MUs and the modulation of firing rates is structured and inversely proportional to the mean firing rate of MUs [27,28,35–37]. The recruitment threshold provides information on the recruitment of the MU pool and its strategy to generate force within the muscle through either an altered MU firing rate and/or the level of force required to recruit the MUs. Tucker and Hodges [38] showed that the recruitment strategy of

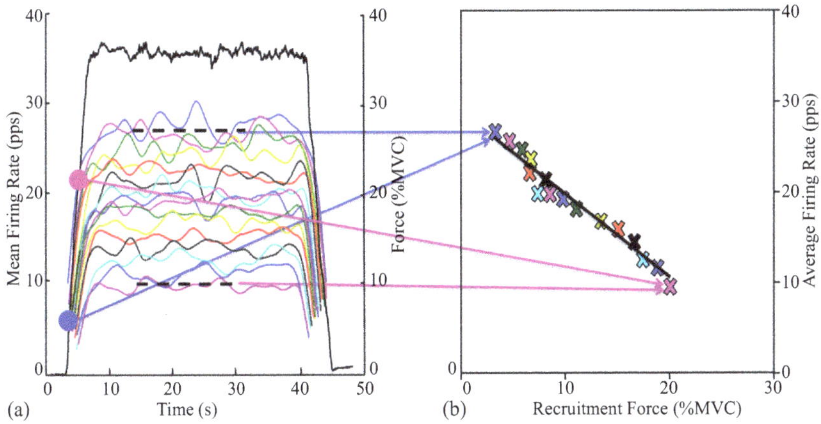

Figure 5.16 Correspondence of the mean firing rate (a) to the recruitment threshold (b)

the vasti MUs changed during pain which then altered the direction of muscle force, however, the study was limited as they were only able to measure a small number of MUs at a low level of force and in a non-functional or challenging task. Stock *et al.* [37] using sEMG decomposition methods found that the recruitment threshold of the Vastus Lateralis was altered in a fatigued muscle, demonstrating that the fatiguing protocol either recruited higher threshold MUs or increased the firing rate of the existing higher threshold MUs.

It has been shown that MUs rather than being controlled individually are in fact receiving a 'common drive' and are being controlled in unison [35], most likely mediated centrally [39]. Erim *et al.* [28] eloquently presented the effects of ageing on MU properties by showing that there is a significantly decreased commonality in the firing of the MUs in the elderly, compared to young subjects, suggesting a less efficient strategy from the CNS. Complimentary to this work, De Luca and Erim [35] demonstrated that a common drive exists between MUs belonging to two synergistic muscles and in agreement with earlier work that a common drive exists between antagonist muscles [40]. The notion of a disrupted commonality in the firing rates and the response to an intervention is clinically a tantalising prospect, potentially providing information on the control and efficiency of the muscle.

5.6 Real-time EMG decomposition

Current approaches for EMG decomposition involve collecting raw sEMG data from multiple electrodes, and perhaps sensors, and then storing offline for processing. The processing utilises advanced algorithms to resolve the complexity of superpositioned MU data, which often involves 100,000s lines of code in software programming, and even with the most advanced approaches and high specification computers, the processing can be 10+ min for each contraction. Although it could be considered a simple inconvenience to wait for data processing, it, in fact, excludes the possibility of accessing more novel approaches to use and interact with MU data. The current approaches limit the number of muscles you can work with, the types of contractions, the activities you want to measure, provision of biofeedback, interacting with robotic and computer devices. Therefore, it is exciting and perhaps daring to consider the possibility of decomposing an sEMG signal from dynamic and ballistic contractions, from multiple muscle groups and in real-time, but in the past 5 years, there has been promising progress towards an approach to achieve this.

Offering an algorithm and software solution to provide users with MU characteristics in real-time is now a very real prospect. Early work from researchers at Altec Inc. (Natick, USA) has demonstrated that through the use of high-fidelity sEMG technology, you can perform a short training phase and immediately afterwards decompose the sEMG signal into the firing instances of the muscles MUs. These advancements in the algorithms to resolve the complexity of MU superposition are supported with advancements in computer processing and coding ingenuity, thus allowing real-time processing and of MU data to provide users with metric dashboards and visualisations for more clinical application. In addition to the research and clinical gains of a real-time EMG decomposition solution, it offers a unique opportunity for the field of robotics and brain–machine interfacing. The possibility of controlling prosthesis, machinery and computers through electrical signals is not a new one, but the possibility of such a refined control input signal, MU control, is one that is unrivalled.

5.7 High-density surface EMG

Whilst traditional methods of sEMG can provide valuable insight when seeking to record the activation–deactivation dynamics of muscles and estimating muscle force [12], there are some applications, where singular bipolar electrodes may not encapsulate the entire picture of what is taking place in the muscle. Using both intramuscular recordings and single-channel sEMG methods of examination, electrical events originating in the MU are exclusively measured as a time-varying signal [41]. Consequently, spatial MU characteristics such as position, spatial extent, MU innervation point (see Section 5.2 above), and direction and lengths of the muscle fibres and action potential propagation, which are essential for the MU force-generating capacity are not taken into consideration. Muscles are notoriously non-homogenous in their activation [42], and therefore, amplitude variations can be observed in both parallel and perpendicular directions owing to the proximity of innervation zones and tendons, as well the propagation properties of the conductor volume. For investigations where a more holistic view of muscle behaviour is deemed important, then taking into account the spatial distribution of muscle excitation can provide valuable insight.

To overcome some concerns pertaining to missing electrophysiological activity when recording the EMG signal, the implementation of high-density EMG (HDEMG) offers an attractive alternative, providing a recorded signal with two spatial and one temporal dimension through a grid design in order to generate a greater representation of muscle excitation behaviour [41]. The use of a grid-like electrode covering the muscle surface allows the user to observe the non-homogenous activation of skeletal muscle [42] and consequently, HDEMG has the potential to improve force estimation accuracy compared with conventional bipolar electrodes due to their ability to capture more detail about neuromuscular activity [43]. Neuromuscular subdivisions (or compartments; [44]) of MUs within a muscle enable portions of the muscles to become active selectively whilst other areas remain passive. Intramuscular subdivisions have been observed in animals during *in vivo* and during locomotion [45–47]. Voluntary selective activation in compartmentalised human skeletal muscle has been observed in the trapezius [49] and gastrocnemius [50] and suggests that this is an important consideration when trying to understand mechanical force outputs of the active muscle as a single bipolar measurement may not be representative of complete neuromuscular behaviour.

5.7.1 Instrumentation

HDEMG uses a two-dimensional grid comprising of many closely spaced electrodes and is designed to measure the distribution of activity over a confined area. There is, theoretically, no limit to the number of surface electrodes over the muscle, the number of channels or configuration one chooses to arrange complex montages from different electrodes for each channel. Of utmost importance in grid geometry is the speed of application, signal quality and subject comfort. Like sEMG, HDEMG electrodes are categorised as being either dry active or wet passive, in that the latter provides an interface with the skin by using a conductive gel in conjunction with a skin preparation. These electrodes can be presented in linear formation (Figure 5.9a), which can detect the measurement of sarcolemma propagation by comparing time delays of the consecutive signal, or as arrays (Figure 5.17a) and are adept at handling skin contours. However, there are certain

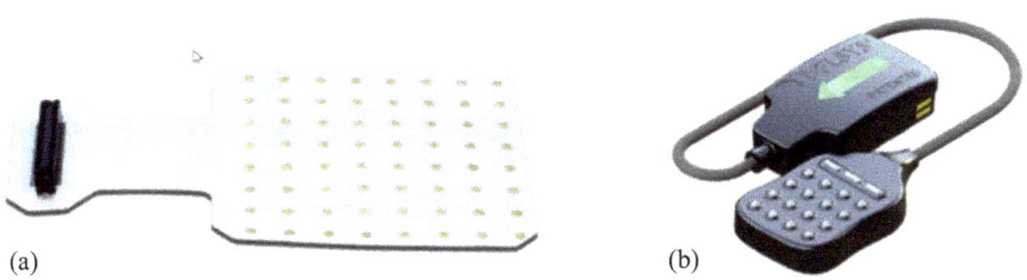

(a) (b)

Figure 5.17 Examples of (a) a passive wet HDEMG electrode and (b) an active dry HDEMG sensor limitations associated with this approach as their fixation can be time consuming and the electrodes can easily shift and lessen classification accuracy [50]. In contrast, active dry electrodes allow for a fast application as well as ease of replacing the grid to find an optimal recording position (Figure 5.17b), whilst the direct contact of a silver–silver chloride electrode on the skin can improve the quality of the signal without the use of conductive gels [20].

5.7.2 What can be found in the signal

During voluntary contractions, the action potentials from several MUs superimpose, forming an intricate inference pattern. The level of muscle excitation can then be understood in terms of the characteristics of the HDEMG interference pattern via a topographical EMG representation, mapping out the electrical potential [51]. The topographical distribution of muscle activity can provide insight into the spatial distribution of the myoelectric intensity of each muscle, allowing users to observe the regions of greater and lesser excitation which can occur following changes in contraction level, contraction duration and joint positions. The topographical activation map associated with HDEMG modes of recording data represents a matrix where each pixel in the map corresponds to an RMS value associated with the corresponding channel on the electrode grid (Figure 5.18). Individual features can then be extracted from the map such as the centroid or barycentre, which employ methods of detecting the central point of concentrated excitation. Additionally, skin surface topography relates directly to often pertinent anatomical and physiological properties of the underlying MU [52], making HDEMG an interesting tool when exploring muscle function.

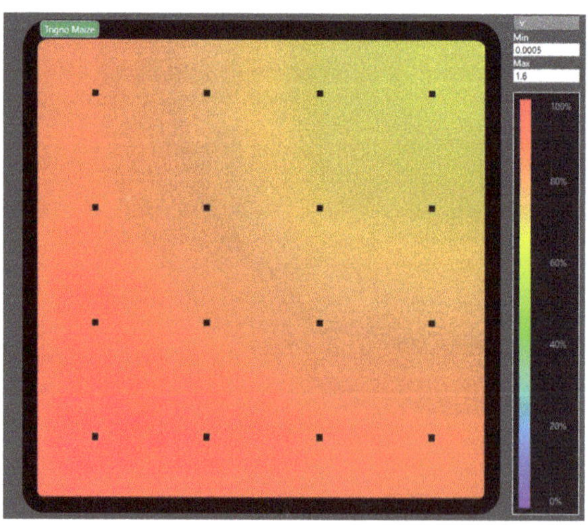

Figure 5.18 Example of a topographical representation of spatial excitation across the muscle

5.7.3 A look to the future

Due to the complexity of understanding the *in vivo* behaviour of the human neuromuscular system during locomotion, much of the existing information is derived from isometric muscle contractions using reductionist techniques, where muscle geometry is kept constant during the collection period. Data derived under such conditions has been influential in our understandings of the foundations of muscle function, however, their direct application to voluntary movement is flawed as a result of the highly controlled conditions the results derive from [53]. During locomotion, the contraction of the muscle is typically dynamic, the active muscle moves independently of the skin, and therefore the active MU changes as a function of time, causing significant changes in the MUAP shape and impacting the information which can be extracted about the MU. In lieu of this, there is a demand for technology that can provide a window into the complexities of human movement.

sEMG has been one such tool capable of providing quantitative descriptions of muscle behaviour, and HDEMG has taken this further by allowing key spatial parameters to be detected during dynamic activities [54–56]. However, dynamic movement introduces additional complications into EMG signal acquisition and many potential sources of contamination. Many HDEMG systems contain wires, which not only limit communication distances and tethers the participant to an amplifying unit but can also introduce electromagnetic interference into the signal [57]. Additionally, cumbersome cables associated with current HDEMG technology together with increments in locomotive activity can cause the frequency of motion artefacts to amplify and overlap with the low-frequency content of the muscle activity [58]. These issues imply that choice of equipment is an important factor worthy of careful consideration with emphasis on the data processing techniques that might be employed when filtering the signals which can contribute to the quality of the result. However, as with any technology, the engineering of HDEMG has come a long way and continues to evolve. Delsys Trigno Maize (Figure 5.17b) and upcoming Trigno Tiber sensors (Delsys Inc, Boston, MA, USA) offer a wireless range of HDEMG sensors, capable of delivering high-fidelity signals without the antiquated and bulky hardware often employed. Such solutions emerging on the market suggest that more progressive steps can now be taken to better understand muscle behaviour which would greatly benefit our understanding of human locomotion.

5.8 Summary

It is clear that the field of EMG has profited from the rapid technological advancements in the past decade, providing users with high-precision recording tools, intelligent algorithms and new perspectives on neural control. The possibilities of real-time EMG decomposition, multi-channel EMG sensor technologies and advanced processing abilities will revolutionise our approach in understanding the neural control of human movement.

References

[1] Kazamel, M. and Warren, P. P., 'History of Electromyography and Nerve Conduction Studies: A Tribute to the Founding Fathers', *Journal of Clinical Neuroscience*, (2017), 43: 54–60. https://doi.org/10.1016/J.JOCN.2017.05.018

[2] Du Bois Reymond, E., Untersuchungen über thierische Elektikstät. Vol. 2: 425–30 (1849). (G Reimer, Berlin). Cited in: Clarke E, O'Malley CD. *The Human Brain and Spinal Cord: A Historical Study*, 2nd ed. San Francisco: Norman (1996): 192–203.

[3] Reaz, M. B. I., Hussain, M. S. and Mohd-Yasin, F., 'Techniques of EMG Signal Analysis: Detection, Processing, Classification and Applications', *Biological Procedures Online 2006 8:1*, (2006), 8: 11–35. https://doi.org/10.1251/BPO115

[4] Richards, J., *The Comprehensive Textbook of Clinical Biomechanics*, 2nd ed. New York, NY: Elsevier (2018).

[5] De Luca, C. J. and Forrest, W. J., 'An Electrode for Recording Single Motor Unit Activity During Strong Muscle Contractions', *IEEE Transactions on Biomedical Engineering*, (1972), BME-19: 367–372. https://doi.org/10.1109/TBME.1972.324140

[6] Basmajian, J. V. and De Luca, C. J., *Muscles Alive: Their Functions Revealed by Electromyography*, 5th ed. Baltimore, MA: Williams & Wilkins (1985).

[7] Barbero, M., Merletti, R. and Rainoldi, A. (2012). *Atlas of Muscle Innervation Zones: Understanding Surface Electromyography and its Applications* – Marco Barbero, Roberto Merletti, Alberto Rainoldi – Google Books. https://books.google.co.uk/books?hl=en&lr=&id=rWF-8HE9kf8C&oi=fnd&pg=PR3&dq=barbero-merletti+and+rainoldi&ots=FDPpj_sR6J&sig=wkOGdYG0rc6Cw247svTB5rb8zbE&redir_esc=y#v=onepage&q=barbero-merlettiand rainoldi&f=false

[8] Hermens, H. J., Freriks, B., Disselhorst-Klug, C. and Rau, G., 'Development of Recommendations for SEMG Sensors and Sensor Placement Procedures', *Journal of Electromyography and Kinesiology*, (2000), 10: 361–374. https://doi.org/10.1016/S1050-6411(00)00027-4

[9] Roy, S. H., De Luca, C. J. and Schneider, J., 'Effects of Electrode Location on Myoelectric Conduction Velocity and Median Frequency Estimates', *Journal of Applied Physiology*, (1986), 61: 1510–1517. https://doi.org/10.1152/JAPPL.1986.61.4.1510

[10] De Luca, C. J., 'The Use of Surface Electromyography in Biomechanics', *Journal of Applied Biomechanics*, (1997), 13: 135–163. https://doi.org/10.1123/JAB.13.2.135

[11] Wong, Y. M. and Ng, G. Y. F., 'Surface Electrode Placement Affects the EMG Recordings of the Quadriceps Muscles', *Physical Therapy in Sport*, (2006), 7: 122–127. https://doi.org/10.1016/J.PTSP.2006.03.006

[12] Vigotsky, A. D., Halperin, I., Lehman, G. J., Trajano, G. S. and Vieira, T. M., 'Interpreting Signal Amplitudes in Surface Electromyography Studies in

Sport and Rehabilitation Sciences', *Frontiers in Physiology*, (2018), 8: 985. https://doi.org/10.3389/FPHYS.2017.00985/BIBTEX

[13] Andreassen, S. and Rosenfalck, A., 'Impaired Regulation of the Firing Pattern of Single Motor Units', *Muscle & Nerve*, (1978), 1: 416–418. https://doi.org/10.1002/MUS.880010514

[14] Petrofsky, J., 'The Effect of the Subcutaneous Fat on the Transfer of Current through Skin and into Muscle', *Medical Engineering & Physics*, (2008), 30: 1168–1176. https://doi.org/10.1016/J.MEDENGPHY.2008.02.009

[15] Cooper, M. A., Trent J. H., Vardiman, J. P., Gallagher, P. M. and Fry, A. C., 'Relationships between Skinfold Thickness and Electromyographic and Mechanomyographic Amplitude Recorded during Voluntary and Non-Voluntary Muscle Actions', *Journal of Electromyography and Kinesiology*, (2014), 24: 207–213. https://doi.org/10.1016/J.JELEKIN.2013.12.005

[16] Kamen, G., 'Electromyographic Kinesiology', in: Robertson, D.G.E., Caldwell, G.E., Hamill, J., Kamen, G. and Whittlesey, S.N. (ed.) *Research Methods in Biomechanics*, 2nd ed. Champaign, IL: Human Kinematics.

[17] Solomonow, M., Baratta, R., Bernardi, M., *et al.*, 'Surface and Wire EMG Crosstalk in Neighbouring Muscles', *Journal of Electromyography and Kinesiology*, (1994), 4: 131–142. https://doi.org/10.1016/1050-6411(94)90014-0

[18] Hewson, D. J., Hogrel, J. Y., Langeron, Y. and Duchêne, J., 'Evolution in Impedance at the Electrode-Skin Interface of Two Types of Surface EMG Electrodes during Long-Term Recordings', *Journal of Electromyography and Kinesiology*, (2003), 13: 273–279. https://doi.org/10.1016/S1050-6411(02)00097-4

[19] Jonsson, B., Omfeldt, M. and Rundgren, A., 'Discomfort from the use of wire electrodes for electromyography', *Electromyography*, (1968), 8: 5–17. PMID: 5675553

[20] De Luca, C. J., Le Fever, R. S. and Stulen, F. B., 'Pasteless electrode for clinical use', *Medical & Biological Engineering & Computing*, (1979), 17: 387–390.

[21] Shannon, C. E., 'A mathematical theory of communication', *The Bell System Technical Journal*, (1948), 27: 379–423, 623–656.

[22] De Luca, C. J., Kuznetsov, M., Gilmore, L. D. and Roy, S. H., 'Inter-Electrode Spacing of Surface EMG Sensors: Reduction of Crosstalk Contamination during Voluntary Contractions', *Journal of Biomechanics*, (2012), 45: 555–561. https://doi.org/10.1016/J.JBIOMECH.2011.11.010

[23] De Luca, C. J., Adam, A., Wotiz, R., Gilmore, L. D. and Nawab, S. H., 'Decomposition of Surface EMG Signals', *Journal of Neurophysiology*, (2006), 96: 1646–1657. https://doi.org/10.1152/JN.00009.2006/ASSET/IMAGES/LARGE/Z9K0090676200010.JPEG

[24] Nawab, S. H., Chang, S. S. and De Luca, C. J., 'High-Yield Decomposition of Surface EMG Signals', *Clinical Neurophysiology*, (2010), 121: 1602–1615. https://doi.org/10.1016/J.CLINPH.2009.11.092

[25] LeFever, R. S. and De Luca, C. J., 'A Procedure for Decomposing the Myoelectric Signal Into Its Constituent Action Potentials—Part I:

Technique, Theory, and Implementation', *IEEE Transactions on Biomedical Engineering*, (1982), BME-29: 149–157. https://doi.org/10.1109/TBME.1982.324881

[26] Masuda, T., Miyano, H. and Sadoyama, T., 'A Surface Electrode Array for Detecting Action Potential Trains of Single Motor Units', *Electroencephalography and Clinical Neurophysiology*, (1985), 60: 435–443. https://doi.org/10.1016/0013-4694(85)91018-1

[27] De Luca, C. J. and Contessa, P., 'Hierarchical Control of Motor Units in Voluntary Contractions', *Journal of Neurophysiology*, (2012), 107: 178–195. https://doi.org/10.1152/JN.00961.2010/ASSET/IMAGES/LARGE/Z9K0011211060011.JPEG

[28] Erim, Z., Beg, M. F., Burke, D. T. and De Luca, C. J., 'Effects of Aging on Motor-Unit Control Properties', *Journal of Neurophysiology*, (1999), 82: 2081–2091. https://doi.org/10.1152/JN.1999.82.5.2081/ASSET/IMAGES/LARGE/9K1090514010.JPEG

[29] Adam, A. and De Luca, C. J., 'Recruitment Order of Motor Units in Human Vastus Lateralis Muscle is Maintained during Fatiguing Contractions', *Journal of Neurophysiology*, (2003), 90: 2919–2927. https://doi.org/10.1152/JN.00179.2003/ASSET/IMAGES/LARGE/9K1133444007.JPEG

[30] Contessa, P., Adam, A. and De Luca, C. J., 'Motor Unit Control and Force Fluctuation during Fatigue', *Journal of Applied Physiology*, (2009), 107: 235–2343. https://doi.org/10.1152/JAPPLPHYSIOL.00035.2009/ASSET/IMAGES/LARGE/ZDG0070985980005.JPEG

[31] Beck, T. W., DeFreitas, J. M. and Stock, M. S. 'The Effects of a Resistance Training Program on Average Motor Unit Firing Rates', *Clinical Kinesiology (Online); San Diego*, (Spring 2011), 65(1): 1–8.

[32] Suresh N., Li X., Zhou P. and Rymer W. Z., 'Examination of motor unit control properties in stroke survivors using surface EMG decomposition: a preliminary report', *Annual International Conference of the IEEE Engineering in Medicine and Biology Society*, (2011), 2011: 8243–8246.

[33] De Luca, C. J., Chang, S. S., Roy, S. H., Kline, J. C. and Hamid, N. S., 'Decomposition of Surface EMG Signals from Cyclic Dynamic Contractions', *Journal of Neurophysiology*, (2015), 113: 1941–1951. https://doi.org/10.1152/JN.00555.2014/ASSET/IMAGES/LARGE/Z9K0061529020008.JPEG

[34] Burke, R. E., 'Motor Units: Anatomy, Physiology, and Functional Organization', *Undefined*, (1981), 345–422. https://doi.org/10.1002/CPHY.CP010210

[35] De Luca, C. J. and Erim, Z., 'Common Drive of Motor Units in Regulation of Muscle Force', *Trends in Neurosciences*, (1994), 17: 299–305. https://doi.org/10.1016/0166-2236(94)90064-7

[36] De Luca, C. J. and Hostage, E. C., 'Relationship between Firing Rate and Recruitment Threshold of Motoneurons in Voluntary Isometric Contractions', *Journal of Neurophysiology*, (2010), 104: 1034–1046. https://

doi.org/10.1152/JN.01018.2009/ASSET/IMAGES/LARGE/Z9K00810028 10010.JPEG

[37] Stock, M. S., Beck, T. W. and Defreitas, J. M., 'Effects of Fatigue on Motor Unit Firing Rate versus Recruitment Threshold Relationships', *Muscle & Nerve*, (2012), 45: 100–109. https://doi.org/10.1002/MUS.22266

[38] Tucker, K. J. and Hodges, P. W., 'Changes in Motor Unit Recruitment Strategy during Pain Alters Force Direction', *European Journal of Pain*, (2010), 14: 932–938. https://doi.org/10.1016/J.EJPAIN.2010.03.006

[39] De Luca, C. J., Gonzalez-Cueto, J. A., Bonato, P. and Adam, A., 'Motor Unit Recruitment and Proprioceptive Feedback Decrease the Common Drive', *Journal of Neurophysiology*, (2009), 101: 1620–1628. https://doi.org/10.1152/JN.90245.2008/ASSET/IMAGES/LARGE/Z9K0080890020004.JPEG

[40] De Luca, C. J. and Mambrito, B., 'Voluntary Control of Motor Units in Human Antagonist Muscles: Coactivation and Reciprocal Activation', *Journal of Neurophysiology*, (1987), 58: 525–542. https://doi.org/10.1152/JN.1987.58.3.525

[41] Zwarts, M. J. and Stegeman, D. F., 'Multichannel Surface EMG: Basic Aspects and Clinical Utility', *Muscle & Nerve*, (2003), 28: 1–17. https://doi.org/10.1002/MUS.10358

[42] Eng, J. J. and Hoffer, J. A., 'Regional Variability of Stretch Reflex Amplitude in the Cat Medial Gastrocnemius Muscle during a Postural Task', *Journal of Neurophysiology*, (1997), 78: 1150–1154. https://doi.org/10.1152/JN.1997.78.2.1150/ASSET/IMAGES/LARGE/JNP.AU51F3.JPEG

[43] Staudenmann, D., Kingma, I., Daffertshofer, A., Stegeman, D. F. and Van Dieën, J. H., 'Improving EMG-Based Muscle Force Estimation by Using a High-Density EMG Grid and Principal Component Analysis', *IEEE Transactions on Biomedical Engineering*, (2006), 53: 712–719. https://doi.org/10.1109/TBME.2006.870246

[44] English, A. W., Wolf, S. L. and Segal, R. L., 'Compartmentalization of Muscles and Their Motor Nuclei: The Partitioning Hypothesis', *Physical Therapy*, (1993), 73: 857–867. https://doi.org/10.1093/PTJ/73.12.857

[45] De Ruiter, C. J., De Haan, A. and Sargeant, A. J., 'Physiological Characteristics of Two Extreme Muscle Compartments in Gastrocnemius Medialis of the Anaesthetized Rat', *Acta Physiologica Scandinavica*, (1995), 153: 313–324. https://doi.org/10.1111/J.1748-1716.1995.TB09869.X

[46] English, A. W., 'An Electromyographic Analysis of Compartments in Cat Lateral Gastrocnemius Muscle during Unrestrained Locomotion', *Journal of Neurophysiology*, (1984), 52: 114–125. https://doi.org/10.1152/JN.1984.52.1.114

[47] Wakeling, J. M., Tijs, C., Konow, N. and Biewener, A. A., 'Modeling Muscle Function Using Experimentally Determined Subject-Specific Muscle Properties', *Journal of Biomechanics*, (2021), 117: 110242. https://doi.org/10.1016/J.JBIOMECH.2021.110242

[48] Holtermann, A., K. Roeleveld, P. J. Mork, *et al.*, 'Selective Activation of Neuromuscular Compartments within the Human Trapezius Muscle',

Journal of Electromyography and Kinesiology, (2009), 19: 896–902. https://doi.org/10.1016/J.JELEKIN.2008.04.016
[49] Vieira, T. M. M., Windhorst, U. and Merletti, R., 'Is the Stabilization of Quiet Upright Stance in Humans Driven by Synchronized Modulations of the Activity of Medial and Lateral Gastrocnemius Muscles?', *Journal of Applied Physiology*, (2010), 108: 85–97. https://doi.org/10.1152/JAPPLPHYSIOL.00070.2009/ASSET/IMAGES/LARGE/ZDG0011088790009.JPEG
[50] Pan, L., Zhang, D., Jiang, N., Sheng, X. and Zhu, X., 'Improving robustness against electrode shift of high density EMG for myoelectric control through common spatial patterns', *Journal of NeuroEngineering and Rehabilitation*, (2015), 12: 110.
[51] Kleine, B. U., Peter Schumann, N., Stegeman, D. F. and Scholle, H. C., 'Surface EMG Mapping of the Human Trapezius Muscle: The Topography of Monopolar and Bipolar Surface EMG Amplitude and Spectrum Parameters at Varied Forces and in Fatigue', *Clinical Neurophysiology*, (2000), 111: 686–693. https://doi.org/10.1016/S1388-2457(99)00306-5
[52] Roeleveld, K., Sandberg, A., Stålberg, E. V., Stegeman, D. F. and Stegeman, D. F., 'Motor unit size estimation of enlarged motor units with surface electromyography', *Muscle and Nerve*, (1998), 21: 878–886. https://doi.org/10.1002/(SICI)1097-4598(199807)21:7
[53] Martin, J., 'Muscle power: the interaction of cycle frequency and shortening velocity', *Exercise and Sports Sciences Reviews*, (2007), 35: 74–81.
[54] Cronin, N. J., Kumpulainen, S., Joutjärvi, T., Finni, T. and Piitulainen, H., 'Spatial Variability of Muscle Activity during Human Walking: The Effects of Different EMG Normalization Approaches', *Neuroscience*, (2015), 300: 19–28. https://doi.org/10.1016/J.NEUROSCIENCE.2015.05.003
[55] Hegyi, A., Dániel Csala, A. P., Finni, T. and Cronin, N. J., 'High-Density Electromyography Activity in Various Hamstring Exercises', *Scandinavian Journal of Medicine & Science in Sports*, (2019), 29: 34–43. https://doi.org/10.1111/SMS.13303
[56] Schmitz, J. P. J., Van Dijk, J. P., Hilbers, P. A. J., Nicolay, K., Jeneson, J. A. L. and Stegeman, D. F., 'Unchanged Muscle Fiber Conduction Velocity Relates to Mild Acidosis during Exhaustive Bicycling', *European Journal of Applied Physiology*, (2012), 112: 1593–1602. https://doi.org/10.1007/S00421-011-2119-5/FIGURES/6
[57] Chowdhury, R. H., Reaz, M. B. I., Mohd Alauddin Bin Mohd Ali, Bakar, A. A. A., Chellappan, K. and Chang, T. G., 'Surface Electromyography Signal Processing and Classification Techniques', *Sensors*, (2013), 13: 12431–12466. https://doi.org/10.3390/S130912431
[58] De Luca, C. J., Gilmore, L. D., Kuznetsov, M. and Roy, S. H., 'Filtering the surface EMG signal: movement artifact and baseline noise contamination', *Journal of Biomechanics*, (2010), 43: 1573–1579.

Chapter 6

Considerations for data analysis

Brian L. Davis[1], Alexandria Mallinos[1] and Brittany Sommers[1]

6.1 Introduction

The idea of gait analysis being utilized as a clinical tool has been a topic of much debate [1]. In their editorial comments, Brand and Crowninshield considered various gait analysis approaches and concluded that they had little impact on clinical practice. They based their arguments on the facts that measured gait parameters:

- Did not correlate well with a patient's functional capacity (often due to the selection of measured variables).
- Were often directly observable and semi-quantifiable by a physician or therapist. (Detecting Trendelenberg gait, for instance, can easily be done by a clinician without the use of sophisticated gait equipment).
- Did not clearly distinguish between normal and abnormal.
- Were collected in a manner that significantly altered the performance of the evaluated activity.
- Were not accurate or reproducible.
- Were communicated in a form which is not readily identifiable in a physical or physiological setting.

Due to advancements in technology and gait analysis methodologies, many of the issues discussed by Brand and Crowninshield nearly 30 years ago have become relatively obsolete. One example is in the use of pressure measurements to determine diabetic patients who are at risk for foot ulceration, or in monitoring the benefits of various insoles that reduce ulcer risk. In these cases, the data are not directly observable, can distinguish normal from abnormal, are reproducible, and can be communicated in clinician-friendly graphics. For other patients, technology advancements have made it possible to understand how amputee socket alignment not only impacts their gait but also the internal moments that are subsequently produced. Through the implementation of three-dimensional motion capture and force plates, clinicians can obtain a better understanding of the amputee's gait that

[1]College of Engineering, Cleveland State University, Cleveland, OH, USA

will allow for optimal socket alignment to be achieved. In doing so, the use of gait analysis technology not only allows the clinician to pinpoint gait abnormalities but also ensures that the patient is achieving ideal gait patterns. This ultimately results in a reduced static and dynamic alignment time.

In this chapter, various considerations for data analysis will be discussed. For each topic, a representative clinical condition will be examined. For instance, time domain analyses will highlight applications to amputee gait, factor analyses will focus on EMG synergies, wavelet analyses will showcase applications to commonly encountered movements, and normalization issues will be applied to pediatric gait.

6.2 Time domain analyses

Technology for analyzing human gait has been in use for well over a century. However, it was not until after World War II that these measurement systems were then applied to amputee gait [2]. Since then, gait patterns and deviations of amputees have become a focal point in many clinical and engineering-based research. As a result, not only have distinct amputee gait patterns been identified, but the gathered data has been instrumental in improving prosthetic componentry and clinical practice guidelines.

Optimizing amputee gait is necessary in order to minimize the complications that arise due to prolonged abnormal gait patterns. Gait abnormalities, if experienced for an extended period of time, can result in asymmetrical muscle firing and weakness, contracture development, lower back pain, and increased energy expenditure. Nevertheless, many non-pathological amputee gait deviations can be corrected by a prosthetist who can modify prosthetic componentry and/or correct improper alignment. Ultimately, achieving a "normal" amputee gait is the goal of the clinical and engineering team. Technologies such as load transducers, force plates, and three-dimensional (3D) cameras have made it possible to quantitatively analyze amputee gait. These tools allow for the visualization of ground reaction forces (GRF), moments, and frame-by-frame joint angle analysis which help establish the fundamental building blocks of amputee gait.

In this section, we shall discuss data analysis techniques that have been utilized to understand the biomechanics behind common amputee gait deviations.

6.2.1 Strain gauge force transducers and socket reaction moments

The alignment of the socket with respect to the prosthetic foot is crucial in order for the patient to have proper function of their prosthesis. Performed by the prosthetist, socket alignment is initially set during bench alignment and is then followed by static and dynamic alignment which is based upon observed gait trends and patient feedback. Alignment of the prosthesis has been shown to significantly impact the loading on the prosthesis [3–5]. Abnormal loading of the prosthesis due to malalignment of the socket can lead to skin breakdown [6,7] and increased energy expenditure [8]. Computerized gait analyses have made it possible to completely analyze the 3D kinetics and kinematics of gait. This has allowed researchers to visualize the GRF patterns and subsequent socket reaction moments.

By relying on custom-made force transducers with strain gauges at the base of the socket, researchers have been able to analyze the socket reaction moments with respect to alignment [9,10].

Coronal translation and angulation affect socket reaction moments in the following manner. When a valgus moment occurs, the load is located along the proximal–lateral and distal–medial aspects of the prosthesis. Furthermore, medial translation of the socket results in an overall increased valgus moment when compared to that of an aligned socket. In contrast, a laterally translated socket results in a significantly increased varus moment during the stance phase. These moments are heightened when the socket is angularly displaced in the coronal plane.

6.2.2 Force plates and GRF

Factors such as prosthetic foot alignment, stiffness, and bumper settings greatly impact the gait dynamics of an amputee. Gait deviations such as foot slap and drop-off are common abnormalities seen in amputee gait that are due to a combination of the alignment and forefoot flexibility of the prosthetic foot. Foot slap occurs when the foot plantar flexes too rapidly at heel strike creating a slapping sound upon impact. This is a result of either the plantar-flexion bumper being too soft (and thereby not providing enough resistance to the foot during weight transfer) or the foot being too anterior on the prosthesis relative to the socket. In contrast, drop-off arises when heel rise occurs too quickly, thereby causing early knee flexion. This gait deviation is a result of a too soft heel bumper or posteriorly aligned foot.

Ultimately, these ankle-related deviations can be observed through analyzing the GRF throughout the gait cycle (see Figure 6.4). Many researchers have analyzed the effect of ankle prosthetic alignment and componentry with respect to their subsequent GRF [11–13]. It has been observed that the flexibility of the forefoot plays a significant role in the progression of the center of pressure for the GRF. Prosthetic feet with highly flexible forefoot components do not allow a forward progression of the GRF, yielding similar results to those found with feet with shorted keels (or moment arms). This creates a drop-off effect.

Through analyzing data gathered from force plates, drop-off results in a decreased GRF on the prosthetic side during the late stance phase [14,15]. For instances of foot slap, increased GRF on the prosthetic side has been observed due to the increased lever arm at initial contact. Similar to socket reaction moments, understanding and visualizing GRF during gait are crucial in order to recognize why certain prosthetic gait deviations occur.

6.2.3 Motion capturing systems (3D)

3D motion capturing systems have been utilized to analyze joint moments and relative angles of amputees [16–20]. This has proven to be a tremendous tool to visualize and understand the biomechanical properties of normal and abnormal amputee gait patterns. Similar to the techniques discussed previously, the data obtained from 3D motion capturing systems has allowed investigators to see how factors such as alignment and prosthetic componentry can impact an amputee's gait. Furthermore, data collection in amputee gait analysis has made it possible to make kinematic and kinetic measurements clinically comprehensible [21].

Common kinematic properties that have been obtained using this technology consist of knee, hip, and ankle joint rotations and moments, in addition to other components such as pelvic tilt obliquity, and rotation.

These parameters have made it possible to analyze how different prosthetic componentry can influence the gait of amputees of various K-levels. Activity levels or K-levels are used by Medicare to rate an amputee based on their functional and rehabilitation potential. K-levels are determined based on the outcome measure, AMPRO, which measures the patient on their ambulation and balance. The K-levels include the following [22]:

- K0: The patient does not have the ability or potential to ambulate or transfer safely with or without assistance and a prosthesis does not enhance the quality of life.
- K1: The patient has the ability or potential to use a prosthesis for transfers or ambulation on level surfaces at fixed cadence. Typical of the limited and unlimited household ambulator.
- K2: The patient has the ability or potential for ambulation with the ability to traverse low-level environmental barriers such as curbs, stairs, or uneven surfaces. Typical of the limited community ambulator.
- K3: The patient has the ability or potential for ambulation with variable cadence. Typical of the community ambulator who has the ability to traverse most environmental barriers and may have vocational, therapeutic, or exercise activity that demands prosthetic utilization beyond simple locomotion.
- K4: The patient has the ability or potential for prosthetic ambulation that exceeds basic ambulation skills, exhibiting high impact, stress, or energy levels. Typical of the prosthetic demands of the child, active adult, or athlete.

In addition, 3D motion capturing systems have also been used to better understand the gait of transfemoral amputees as it differs greatly from non-amputees. The absence of the anatomical knee joint results in the patient adapting a coping response that is present throughout the entire body. As a result, the interaction between the thorax and the pelvis is of utmost interest. In their study, Goujon-Pillet *et al.* [23] identified specific 3D motion patterns of the upper body during the gait of transfemoral amputees. The data indicated that there was a huge difference between the gait patterns of transfemoral amputees and "healthy" individuals which suggests that the motion of the upper body is used to enhance gait.

The results from gait analyses allow investigators to break down the fundamental kinematic building blocks and develop biomechanical models that help standardize typical amputee walking styles. These data can then be compared to that of "normal" gait and changes caused by alignment and componentry can be observed. Furthermore, 3D motion capture systems also provide researchers with the opportunity to understand the components of pathological amputee gait and assess the effectiveness of gait treatment programs. In doing so, understanding and visualizing the joint moments and relative angles during gait are necessary to recognize how amputee gait patterns differ from non-amputees and how deviations impact the overall biomechanics of their gait.

6.3 Frequency domain

Fast Fourier transforms (FFTs) are widely used for applications in engineering, medical imaging, music, science, and mathematics. The FFT has been considered "the most important numerical algorithm of our lifetime," and it was included in the Top 10 Algorithms of the twentieth century by the IEEE magazine Computing in Science & Engineering [24]. In the field of gait analysis, FFT is an established method that has been utilized in studies on patella-femoral pain [25], abnormal gait in children [26], individuals at risk of falling [27], GRF during walking and running [28], to name a few. The use of FFT analyses, in general, has a number of distinct advantages [29]. These include (i) facilitating efficient data storage and reconstruction of the original graphic representation, (ii) simplifying the interface between mathematical models and measurements collected on human subjects, and (iii) delineating differences in gait pathologies.

6.3.1 Accelerometry studies

Studies of the impact shock wave experienced during running have been conducted using skin-mounted accelerometers. However, in such experiments, the soft tissue between the bone and the transducer can modify the acceleration signal [30]. The implications of skin-mounted devices for measurement of the transmission and attenuation of heel strike accelerations during running have been reviewed by Valiant [31]. Ideally, accelerometers would be rigidly attached to the bone in order to maximize the frequency response of the attachment, but this is not practical in most cases.

In the case of accelerometry data, studies focusing on impact forces in running have to transform signals from the time domain to the frequency domain using FFTs [30,32–34]. The resulting spectrum allows the separation of peaks due to limb motion or the natural frequency of the accelerometer attachment from the shock wave due to the impact [30]. In addition, this transformation permits the calculation of the power spectral densities of the accelerometer data, and this measure has been shown to be sensitive to changes in running speed (i.e., to changes in impact load and loading rate) [30].

6.3.2 Application of FFT to quantify neural control

Besides examining gait data and the transmission of impacts through the human body, other researchers have used FFT approaches for studying the "connectivity" between brain (electroencephalography (EEG) signals) and muscle activity. Synchronization between EEG and electromyography (EMG) signals at a particular frequency band (coherence) has been reported by many studies [35,36]. Coherence can be calculated between multiple EEG channels and specific muscles as follows [37].

Assume that there are two EMG signals I ($I = 1, 2$), and $F_{i,l}(f)$ are 256 points of the FFT derived from the signal I over the lth 256-point section; then the cross-spectrum is:

$$X_{12}(f) = \frac{1}{L}\sum_{l=1}^{L} F_{1,l}(f)F_{2,l}^{*}(f)$$

where * denotes the complex conjugate.

The autospectrum of one signal is:

$$A_i(f) = \frac{1}{L}\sum_{l=1}^{L} F_{i,l}(f)F_{i,l}^{*}(f)$$

Coherence is given by:

$$C_{12}(f) = \frac{|X_{12}(f)|^2}{A_1(f)A_2(f)}$$

An application of this to motor control can be seen in a study by Yue and colleagues [38]. In this case, coherence measurements were obtained from EEG and EMG data in young and elderly cohorts. Using this application of FFT methodology, they showed significantly weakened corticomuscular connections in the elderly group.

Cautions when using FFTs include the issues pertaining to both the confounding factors and the algorithm itself. An example of the former is to be seen in the study by Schneider and Chao [29]. While they found differences in FFT results for patients with and without knee pain, what was not considered is the fact that patients with knee pain walked slower than control subjects – and FFTs are sensitive to gait speed. As for the algorithm itself, Voloshin *et al.* [34] performed a Fourier analysis of the vertical component of the GRF and tibial acceleration recorded during the first 60 ms of foot contact during walking. They reported that impact shock waves in 10 healthy young male subjects occurred at frequencies between 15 Hz and 25 Hz. However, the use of a 60-ms sample time in this study would have precluded the measurement of frequency components below 16.67 Hz.

6.4 Wavelet analyses

While FFT analyses provide valuable information about the frequency content of a signal, what is lost is the timing of different frequencies. For instance, an FFT analysis of ground contact during running may show high frequencies, what is not known is when these frequencies occur. Wavelet analyses provide some of this information.

The first known wavelet can be traced to Alfréd Haar in 1910 [39] however, wavelet analyses truly began in the early 1980s with the work of Meyer, Mallet, and others [40]. Wavelets became an influential tool in data analysis in the late 1980s through the work of Ingrid Daubechies [41]. Daubechies was determined to simplify and implement existing wavelets by Meyer and Mallet. Daubechies' work in defining new and practical wavelet transforms and their applications led her to be the first woman to receive the National Academy of Sciences (NAS) award in mathematics. Since then, wavelets have become a prominent signal processing tool used by mathematicians, statisticians, and engineers. This leads to the question, what is a wavelet?

6.4.1 *Wavelets and wavelet transforms*

Wavelets are an extension of the Fourier transform, however, unlike the Fourier transform, wavelets produce a multiresolution analysis of the signal, allowing for time-frequency localization [42]. This makes wavelets an extremely powerful tool when analyzing non-stationary, transient signals. The wavelet function, also known as the "mother wavelet," consists of a scaling parameter, a, and translational parameter, "*b*" [43,44]:

$$\psi^{a,b}(x) = \frac{1}{\sqrt{|a|}}\psi\left(\frac{x-b}{a}\right) \tag{6.1}$$

These parameters of $\psi^{a,b}$ allow the wavelet to expand and compress around a signal at various frequencies [40,42]. The scaling parameter, a, changes with the frequency range. Smaller values of "a" correspond to higher frequencies, and larger values of a correspond to lower frequencies. The translational parameters, "b," determines the point where time localization is occurring. such that each wavelet transform $\psi^{a,b}(x)$ is centered around $x = b$. This is a crucial difference between the wavelet transform and Fourier transform. For the Fourier transform, $g^{\omega,t}$, width of the transform will never change regardless of the angular frequency, ω [42]. A wavelet transform applied to a transient signal can detect changes in a way that the Fourier transform cannot due to the unique scaling and translation parameters.

6.4.2 *Wavelet families*

Another important distinction between wavelet transforms and the Fourier transform is that the basis functions for wavelets are not specified before analysis [42,44]. This allows the user to specify the mother wavelet (or wavelet "family") that will be applied to the signal. This is advantageous because not all wavelets are created equal, and it is important to choose the proper mother wavelet depending on the type of signal (this will be discussed in more detail later). Examples of mother wavelet families include Haar, Daubechies, Symlets, Meyer, Biorthogonal, Morlet, and Mexican Hat wavelets [40,42]. Describing the characteristics of the wavelet families would be extensive and will not be covered in this brief overview of wavelets. However, a useful tool for learning more about specific wavelet family characteristics is MATLAB® "waveinfo" command.

6.4.3 Continuous and discrete wavelet transforms

Continuous wavelet transforms (CWT) are defined by:

$$W[h](a,b) = \frac{1}{\sqrt{a}} \int_{-\infty}^{\infty} \psi_{a;b}^{*}(x) h(x) dx \tag{6.2}$$

The complex conjugate of $\psi(x)$ is represented by $\psi^{*}(x)$ and $h(x)$ is the signal or function being analyzed [42]. CWTs are best used for detecting discontinuities within signals and functions [45]. CWTs produce a wavelet coefficient, which responds to the local behavior of the function by revealing the local smoothness of a signal at increasing resolutions [46]. The amplitude of wavelet coefficients will remain relatively small so long as the local behavior of the function is smooth and will alternatively increase amplitude if the function is locally unstable or contains singularities. Although useful for detecting singularities, CWTs are redundant. They operate by shifting the scaling function continuously across the signal to determine the correlation between the wavelet and the signal, transforming a one-dimensional signal into a two-dimensional space [43]. Thus, a more practical application often used in literature and research is the discrete wavelet transform (DWT).

DWTs are an extension of the continuous wavelet transform, however, now both the scaling and translation parameters are discretized as $a = 2^j$ and $b = 2^j k$. This reduces the redundancy seen in a CWT due to the DWTs orthonormal structure [39,45]. Our wavelet from Eq. (6.1) then becomes [47]

$$\psi^{j,k}(x) = \frac{1}{\sqrt{|2^j|}} \psi\left(\frac{x}{\sqrt{2^j}} - k\right) \tag{6.3}$$

And the DWT is represented as

$$d_{j,k} = \int_{-\infty}^{\infty} \psi_{j;k}^{*}(x) h(x) dx \tag{6.4}$$

Discrete wavelet decompositions

In order to apply a DWT to a signal, it is important to understand the decomposition that is occurring at each level. A decomposition "level" identifies the number of frequency bands that the signal is set to be compressed into [39]. Similarly, to the CWT, the DWT will result in a set of wavelet coefficients; however, the number of wavelet coefficients in a DWT equals the number of levels. The number of levels should be chosen based on the sampling frequency and desired end frequency range of the signal.

Wavelets integrate a high pass and low pass filter on the signal at every time scale resulting in a set of detail coefficients and approximation coefficients [43,48]. Detail coefficients contain high-frequency content from the signal, whereas approximation coefficients contain the low-frequency content. Each subsequent decomposition level is equal to half the frequency of the previous (i.e., twice the resolution), and the approximation coefficients are used in each subsequent decomposition to produce new coefficients (Figures 6.1 and 6.2).

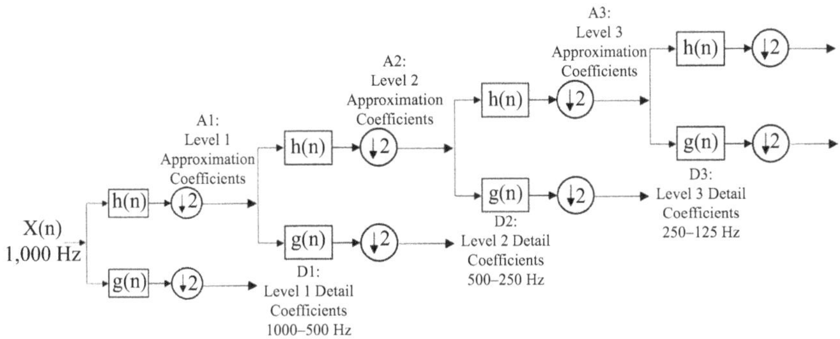

Figure 6.1 Discrete wavelet decomposition process

6.4.4 Choosing the proper mother wavelet

To reduce data analysis error, the following four considerations should be made when choosing the proper mother wavelet application for a signal. If not selected correctly, the wavelet analysis can yield inaccurate results due to not being properly fit for the signal.

1. Wavelet type
2. Wavelet length or support
3. Vanishing moments
4. Level of decomposition

6.4.4.1 Wavelet type

Wavelet type refers to the wavelet family mentioned earlier. The wavelet type should be chosen such that the shape is similar to the shape of the signal to reduce inaccurate results [49,50]. Wavelet type should also be chosen based on characteristics such as orthogonal or biorthogonal and what would be most appropriate for the signal.

6.4.4.2 Wavelet length of support

Since wavelet smoothness increases with wavelet length (also called support), the chosen length should not be too short that the roughness impairs the signal, but also not too long that it overlooks important signal characteristics [50]. Wavelets with smaller support are best for signals with more frequent transient changes. Likewise, larger support wavelets are more suitable for more sparse signals. If a large support wavelet is used on a rapidly changing signal, the wavelet coefficients may not capture all of the detailed information of the signal.

6.4.4.3 Vanishing moments

Vanishing moments influence the support of a wavelet [49]. Higher-order vanishing moments must have longer wavelet lengths. Wavelets increase smoothness with increased vanishing moments [50]. Ingrid Daubechies demonstrated that for a

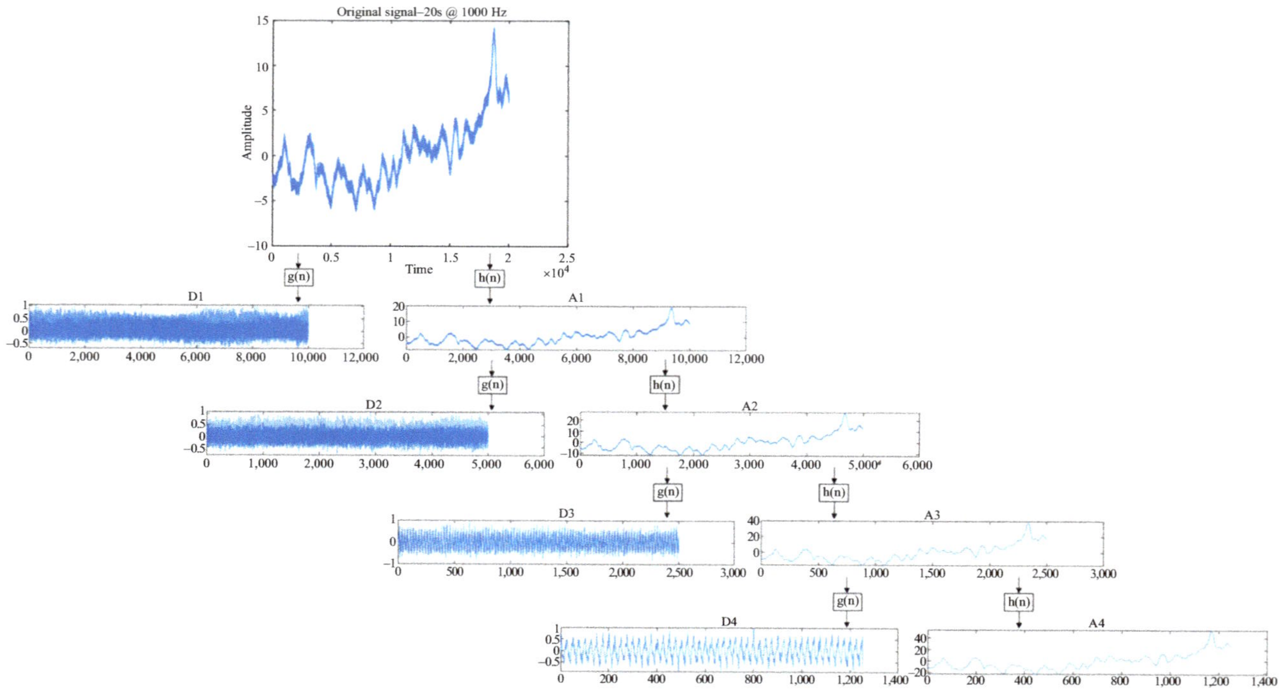

Figure 6.2 Representation of approximation and detail coefficients as a result of the wavelet decomposition

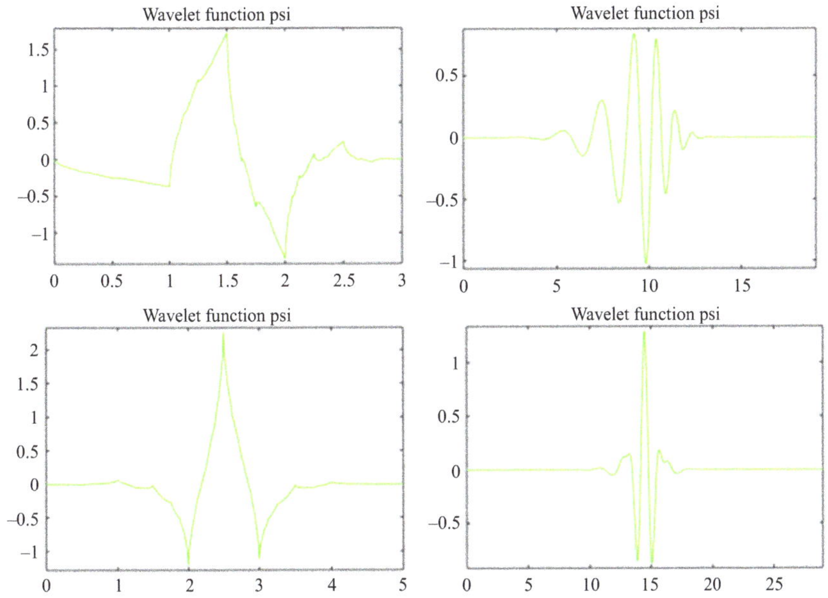

Figure 6.3 Examples of mother wavelets and their relationship to vanishing moments and length of support. Top: db2 (left) and db10 (right). Bottom: coif1 (left) and coif5 (right).

wavelet with *p* vanishing moments, the support must be *2p-1* [40]. Wavelet oscillation loosely increases with the number of vanishing moments; however, the integral of a wavelet will always remain zero. Wavelets are usually named for the number of vanishing moments they have. For example, a dbN or symN are Daubechie and Symlet wavelets with N vanishing moments. Coiflet wavelets labeled coifN contain 2*N* vanishing moments. Choosing an amount of vanishing moments depends on the signal and is similar to that of choosing the length of support. More rapidly changing, transient signals (Figure 6.3) should use fewer vanishing moments (and shorter length of support, examples: db2, sym2, etc.) as this will result in less smoothing and removing of detail [51]. Likewise, more regular signals should use more vanishing moments (and longer lengths of support).

6.4.4.4 Level of decomposition

The number of decomposition levels should be chosen from the sampling frequency and the desired end frequency range of the signal [52,53]. Consider a simple example: studying the postural control from a center-of-pressure signal obtained during quiet standing. The sampling frequency is 1,000 Hz, you are interested in frequency information <0.5 Hz to understand the postural feedback mechanisms. This requires 12 decomposition levels to reach the desired frequency band. L_1 = 500–1,000 Hz, L_2 = 250–500 Hz, L_3 = 125–250 Hz, L_4 = 62.5–125 Hz, and so on until L_{12} = 0.24–0.48 Hz. If you only chose 10 decomposition levels, your final frequency band would be $L_{10} = 0.98–1.95$ Hz. This would not decompose the signal enough to reveal important signal information, and the interpretations of the results would be lacking.

6.4.5 Applications of wavelets

The average wavelet coefficient (AWC) method was defined by Simonsen and Nes [44] and is an application of the DWT. The AWC method consists of performing the DWT and determining the mean of the absolute value of the detail coefficients (the AWC values). By creating a log–log plot of the AWC values versus the discretized scaling parameter, *a*, a linear regression is produced from which signal characteristics can be calculated [46]. The Hurst exponent (H), a statistical parameter defining correlations in any moving system, can be extracted from this linear regression [44] and used in biomechanics as a measure of stability.

The slope of the log–log of AWC vs. *a* is ½ + H.

The applications of wavelets are extensive, especially in signal and image processing for speech processing and image compression. Wavelets are also used in facial and fingerprint recognition, biotechnology and bioinformatics, analysis of EEG and other biosignals, data analysis, disturbance quantification, and many other applications [54].

Let us consider two scenarios and their respective continuous wavelet transform plots in MATLAB. CWT plots produce a scalogram of time on the x-axis, and frequency on the y-axis. If the sampling frequency is specified, the y-axis will represent Hertz. If not, the y-axis will represent samples/cycle. The y-axis of the CWT plot has a scale of log10.

The scalogram includes the cone of influence, the gray area from the frequency, and time axes to the white dashed line [55]. The cone of influence represents areas potentially impacted by edge effects. Edge effects are most prevalent during the beginning and end of the time series data, where the time series is limited. We have already discussed that wavelets are not only stretched and compressed along the signal but also translated over time. At the beginning and end of the signal, the wavelet may be "seeing" information outside of the interval of analysis [55,56]. These regions become the areas in the cone of influence that is suspect to misinterpretation due to edge effects.

The first example is GRF data in the z-direction (F_z) of normal, over-ground gait for a pediatric patient (Figure 6.4).

Plotting a CWT for the left and right GRF data results in the following (Figure 6.5).

Time series data for the left GRF show heel strike and toe-off occurring between approximately 2.3 s and 3.2 s. Time series data for the right GRF show heel strike and toe-off occurring between approximately 3 s and 3.8 s. This is reflected in the CWT plots, demonstrating time-frequency localization of the GRF time series data.

A second example consists of a simplified simulation of jumping on a trampoline for 15 s. In the first 5 s (0–5 s), you are instructed to jump softly. You are

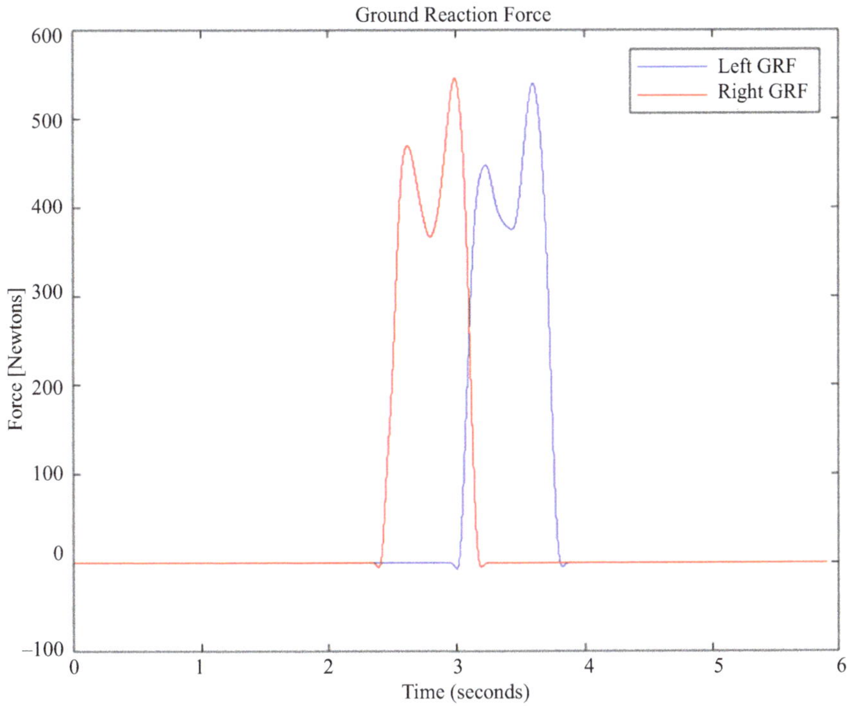

Figure 6.4 Time series GRF data

then instructed to not jump for the next 5 s (5–10 s). Finally, you are instructed to jump hard for five more seconds (10–15 s). This results in the following time series data and CWT plot (Figure 6.6).

The time localization properties of wavelets are highlighted here again, as the change in magnitude reflects the change in oscillation amplitude over time.

6.5 Factor analyses

One of the goals of biomechanical research is to uncover some of the control processes that are involved in regulating muscle activity during gait (see Section 3.2). In this regard, one of the questions frequently of interest to gait analysts concerns the phasic action of muscles. For this reason, researchers [50] have often relied on EMG signals to infer what "programming" is taking place in the central nervous system (CNS).

Measurement of the coactivation of muscles can help in quantifying the manner in which movement is controlled. Co-contraction can be regarded as a specific case of coactivity that refers to an agonist/antagonist pair being simultaneously active. In this situation, one may suspect a need to stiffen a joint (e.g., during alternating knee flexion and extension, or at the ankle during

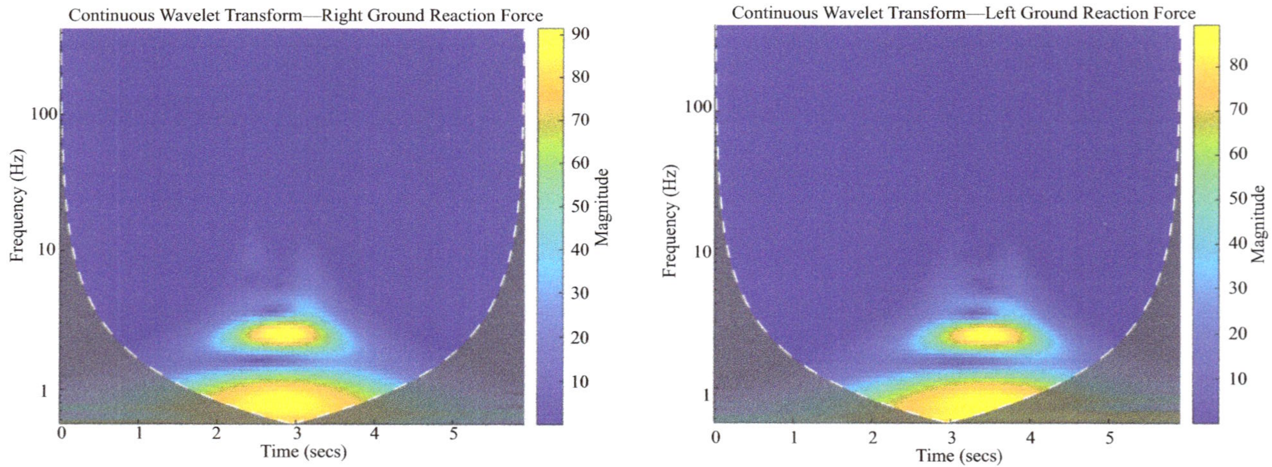

Figure 6.5 CWT plots for right and left GRF during over-ground gait

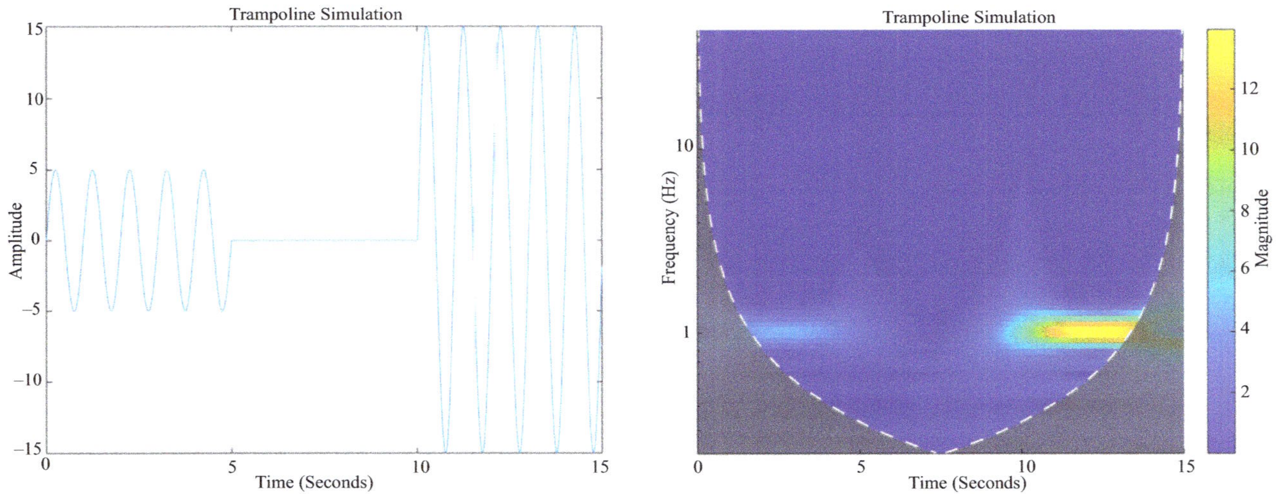

Figure 6.6 Time series data (left) and CWT plot (right) of simulated jumping on a trampoline

walking). A different situation is found when muscles spanning different joints (e.g., hamstrings and tibialis anterior) are coactive. In this case, one might anticipate a need to achieve a particular kind of movement (such as during the swing phase).

One approach for assessing these muscle activation patterns is to conduct a "factor analysis" [58]. This statistical procedure permits the basic dimensions (vectors) of a matrix to be determined. In brief, the method entails the following points:

- Determining principal components for a matrix. In the case of EMG data, this could consist of "n" rows of data, each one at a different time point, and "m" columns of EMG data for various muscles of interest. If the linear envelopes of EMG data for 16 muscles were obtained for 1 s at 100 Hz, $n = 100$, and $m = 16$. The "principal components" represent the number of independent vectors needed to represent the trends in the original matrix.
- Performing a "factor rotation" (e.g., using the "varimax" criterion [59]).

The biomechanical relevance of a factor analysis is that it permits muscles with similar EMG linear envelopes to be grouped together in a relatively objective manner. There are two "subjective" aspects of the analyses. First, a decision has to be made as to how much of the variance in the data set needs to be accounted for. This can be likened to a Fourier analysis (see Section 6.3) where one uses only a few sine and cosine terms to describe the main features of cyclical data. To perfectly describe all the fluctuations in the data set may require many Fourier coefficients, and investigators usually determine how much of the variance they want to account for. In the case of a factor analysis, if the main features (dimensions) of a data set can be described by a few components (hence the term "principal components"), then there will be a few eigenvectors that are associated with large eigenvalues. Smaller amounts of variance (e.g., superimposed noise) will result in other eigenvectors being associated with small eigenvalues. Eigenvalues less than unity are usually considered to describe noise [58,60].

The second "subjective" aspect of factor analysis relates to interpreting the factors that are found. This requires some insight into what each factor could be describing, and can be based on temporal events during the gait cycle. Davis and Vaughan [58] determined that, for a normal walking sequence, there are four factors that account for most of the variation in EMG patterns in 16 muscles of the lower extremity. They could be divided into those that act at the times of (a) heel strike, (b) single limb loading response, (c) propulsion phase, or else (d) acted in a biphasic manner.

6.6 Data visualization approaches

With recent advancements in data analytics, clinicians and gait researchers have a powerful arsenal when it comes to understanding human performance. It is not uncommon for data analytics to uncover, (i) roles of combinations of uni- and biarticular muscles in determining a patient's ability to perform a task, (ii) contributions of individual moments and forces to the accelerations of adjacent and non-adjacent limb segments (induced acceleration analysis), (iii) predictions of simultaneous de-rotation osteotomies and tendon transfers on gait, and (iv) relative effects of multiple risk factors on musculoskeletal abnormalities. While the mathematical and statistical underpinnings of these analytical approaches may be academically impressive, what is often lost is the ability to obtain a succinct "take-home" message that can be translated to clinical practice. This is where elegant representations of data can have a profound effect. Creating meaningful graphical representations can help immeasurably when it comes to conveying key information.

Over a century ago, the San Antonio Light advertised its coverage of the First World War with the claim, "One picture is worth a thousand words." In the 1970s, Bread – an American soft rock band – sang, "If a picture paints a thousand words, then why can't I paint you?" Numerous other publications have reinforced the power of images [61]. Edward Tufte's bestseller, "The Visual Display of Quantitative Information," formalizes approaches that distinguish between "good" and "bad" representations of data. His recommendations [62] are to:

- induce readers to think about the data (rather than use "flashy patterns").
  Figure 6.6 is an example of a graph that, while simple at first glance, requires a reader to think about the concepts underlying human movement. It incorporates elements of force magnitudes, timing, and frequency content, all in one diagram.
- show all the data (i.e., not just bar graphs showing means for different conditions).
  Figure 6.7 is one such example – note the waves of EMG activations for most of the leg muscles used during gait.
- avoid data distortion.
- present many numbers in a small space.
  Examine, for instance, the combination of EMG and kinematic information (Figure 6.8). These data, coupled with different phases of the gait cycle, could be presented in multiple graphs of angles and EMG activity versus time – or presented as an elegant image as shown here.
- make large data sets coherent (once again, Figure 6.8 epitomizes this recommendation).
- encourage the eye to make comparisons. The well-known force-length graph that all students studying physiology are well aware of shows how degrees of actin-myosin overlap affect a muscle's length-tension relationship. In this example, students are encouraged to visualize both the microscopic (protein) levels, as well as the overall muscle's force-generating capability.
- reveal both a broad overview and fine detail (as with the actin–myosin example just mentioned).
- serve a clear purpose, and integrate statistical and verbal descriptions of the data.

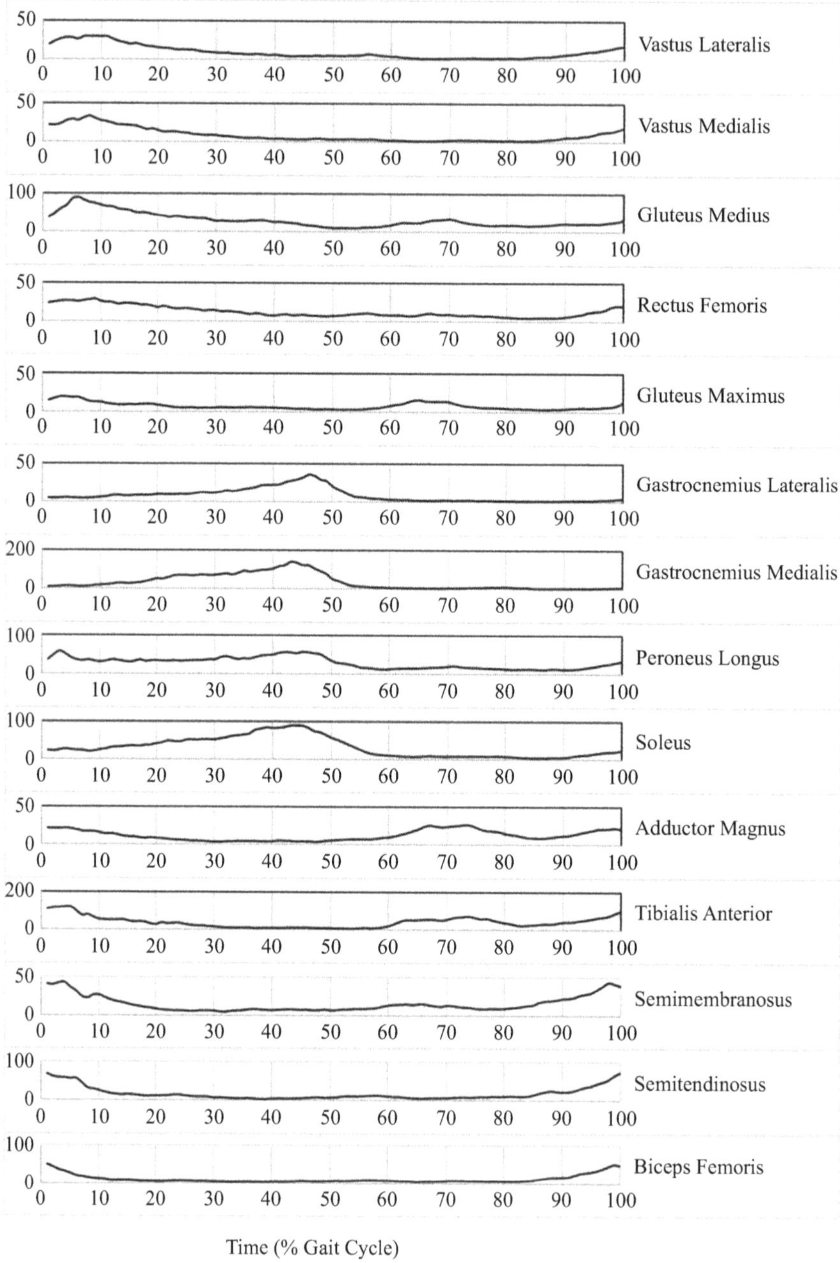

Figure 6.7 Normal EMG patterns for 14 of the major muscles in the lower extremities are plotted as a function of the gait cycle. Note. Based on data in Hof et al. [63], using the data visualization in Bechtol [64]

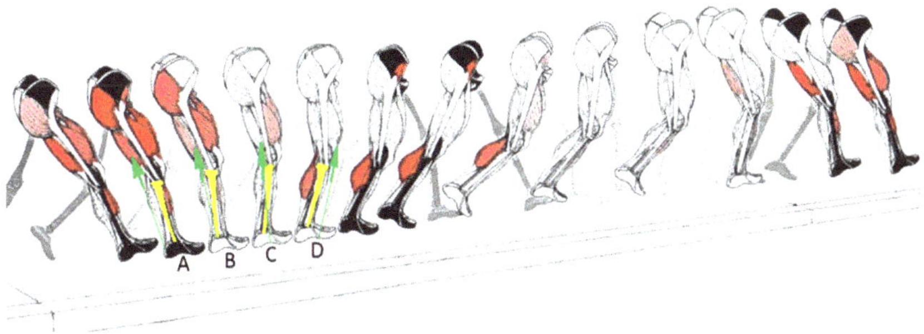

Figure 6.8 *An example of the combining multiple sets of data in a single image. In this graphic, the representation from Inman Ralston and Todd [65] has been modified to include a depiction of ground reaction forces (green arrows) and their relationship to the foot and ankle. Pauwels [66] maintained that these force vectors when combined with muscle activity, led to the tibia having relatively constant bending stresses in various phases of the gait cycle (shown in A through D). This theory is dependent on the anatomical locations of muscles, how vigorously they act during each part of the walking cycle, as well as ground reaction forces and kinematic information. Besides the effect on bone bending stresses, with this information, one can "look deeper" and get some idea of concentric and eccentric muscle activity.*

One of the early workers in biomechanics – Etienne Jules Marey – was high on Tufte's list of outstanding contributors to the development of graphical methods. Marey's data detailing the movement patterns of horses, lizards, and starfish as well as his display of a Paris–Lyon train schedule are examples. Marey, in turn, was impressed by the work of Charles Minard (1781–1870) who depicted the fate of Napoleon's army in Russia (Figure 6.9). Beginning at the left, the thick band shows the size of the army at each step of the journey. Note how thin the band is when the army returned – only 10,000 of the original 422,000 soldiers survived! Six variables are plotted: the size of the army, its location (two dimensions), the direction of movement, the temperature, and dates corresponding to certain locations.

In general, images that are geographical in nature can convey thousands of data points. One such example [67] shows the pressure distribution beneath the foot of a diabetic patient. In the same way, contour plots showing elevation above sea level are based on many measurements; these data representations show hills and valleys, frequency content, and can connect with anatomical abnormalities (such as pes planus, hallux valgus, and missing digits).

In summary, it is possible for a graphical display to be accurate and visually appealing. While it might not be possible for these displays to be generated at the click of a mouse button, it may be well worth the extra effort.

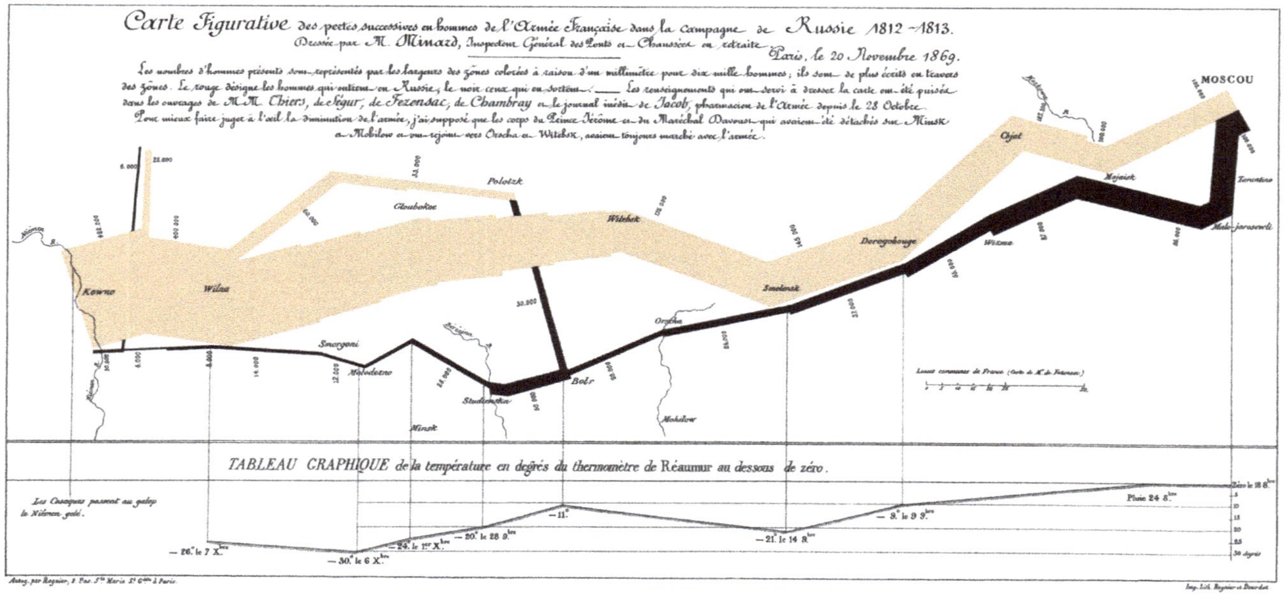

Figure 6.9 A depiction of the fate of Napoleon's army in Russia. It has been described as perhaps the best statistical graphic ever drawn

6.7 Summary

This chapter has covered a range of topics – from the early days of gait analysis in the time of Marey, to modern examples of the use of wavelet analyses and coherence estimates of brain-muscle coupling. Whatever aspect of gait may be of interest to a student, researcher, physician, or allied health professional, it is important to heed the concerns expressed by Brand and Crowninshield almost 40 years ago. Data that are collected should not be directly observable (why buy expensive equipment when you can look at a patient and make a decision on pathology or recovery status?). Moreover, the data should allow one to distinguish normal from abnormal, be reproducible, and in a form that can be communicated in clinician-friendly graphics.

References

[1] Brand, R. A., and Crowninshield, R. D. (1981). Comment on criteria for patient evaluation tools. *Journal of Biomechanics*, 14(9), 655–655.

[2] Gage, R. J., and Hicks, R. (1989). Gait analysis in prosthetics. *Clinical Prosthetics & Orthotics*, 9(3), 17–21.

[3] Fang, L., Jia, X., and Wang. R. (2007). Modeling and simulation of muscle forces of trans-tibial amputee to study effect of prosthetic alignment. *Clinical Biomechanics (Bristol Avon),* 22, 1125–1131.

[4] Sanders J, Daly C. H., Burgess E. M., and Boone D. A. (1992). An important socket design consideration determined from interface shear stress measurement. In: *Proceedings of 7th World Congress of Prosthetics and Orthotics International*, Chicago, IL, 392.

[5] Zahedi, M. S., Spence, W. D., Solomonidis, S. E., and Paul, J. P. (1986). Alignment of lower-limb prostheses. *Journal of Rehabilitation Research and Development*. 23, 2–19.

[6] Pinzur, M. S., Cox, W., Kaiser, J., Morris, T., Patwardhan, A., and Vrbos, L. (1995). The effect of prosthetic alignment on relative limb loading in persons with trans-tibial amputation: a preliminary report. *Journal of Rehabilitation Research and Development*, 32, 373–377.

[7] Seelen, H. A., Anemaat, S., Janssen, H. M., and Deckers, J. H. (2003). Effects of prosthesis alignment on pressure distribution at the stump/socket interface in transtibial amputees during unsupported stance and gait. *Clinical Rehabilitation*, 17, 787–796.

[8] Schmalz, T., Blumentritt, S., and Jarasch, R. (2002). Energy expenditure and biomechanical characteristics of lower limb amputee gait: the influence of prosthetic alignment and different prosthetic components. *Gait and Posture*, 16, 255–263.

[9] Boone, D. A., Kobayashi, T., Chou, T. G., *et al.* (2013). Influence of malalignment on socket reaction moments during gait in amputees with transtibial prostheses. *Gait & Posture*, 37(4), 620–626.

[10] Kobayashi, T., Orendurff, M. S., and Boone, D. A. (2015). Dynamic alignment of transtibial prostheses through visualization of socket reaction moments. *Prosthetics and Orthotics International*, 39(6), 512–516.

[11] Lloyd, C. H., Stanhope, S. J., Davis, I. S., and Royer, T. D. (2010). Strength asymmetry and osteoarthritis risk factors in unilateral trans-tibial, amputee gait. *Gait & Posture*, 32(3), 296–300.

[12] Silverman, A. K., Fey, N. P., Portillo, A., Walden, J. G., Bosker, G., and Neptune, R. R. (2008). Compensatory mechanisms in below-knee amputee gait in response to increasing steady-state walking speeds. *Gait & Posture*, 28(4), 602–609.

[13] Zmitrewicz, R. J., Neptune, R. R., Walden, J. G., Rogers, W. E., and Bosker, G. W. (2006). The effect of foot and ankle prosthetic components on braking and propulsive impulses during transtibial amputee gait. *Archives of Physical Medicine and Rehabilitation*, 87(10), 1334–1339.

[14] Hansen, A. H., Meier, M. R., Sessoms, P. H., and Childress, D. S. (2006). The effects of prosthetic foot roll-over shape arc length on the gait of trans-tibial prosthesis users. *Prosthetics and Orthotics International*, 30(3): 286–299.

[15] Klodd, E., Hansen, A., Fatone, S., and Edwards, M. (2010). Effects of prosthetic foot forefoot flexibility on gait of unilateral transtibial prosthesis users. *Journal of Rehabilitation Research and Development*, 47(9), 899–910.

[16] Bateni, H., and Olney, S. J. (2002). Kinematic and kinetic variations of below-knee amputee gait. *Journal of Prosthetics and Orthotics*, 14(1), 2–10.

[17] Buckley, J. G. (2000). Biomechanical adaptations of transtibial amputee sprinting in athletes using dedicated prostheses. *Clinical Biomechanics*, 15(5), 352–358.

[18] Frossard, L., Cheze, L., and Dumas, R. (2011). Dynamic input to determine hip joint moments, power and work on the prosthetic limb of transfemoral amputees: ground reaction vs knee reaction. *Prosthetics and Orthotics International*, 35(2), 140–149.

[19] Perry, J., Burnfield, J. M., Newsam, C. J., and Conley, P. (2004). Energy expenditure and gait characteristics of a bilateral amputee walking with C-leg prostheses compared with stubby and conventional articulating prostheses. *Archives of Physical Medicine and Rehabilitation*, 85(10), 1711–1717.

[20] Sanderson, D. J., and Martin, P. E. (1996). Joint kinetics in unilateral below-knee amputee patients during running. *Archives of Physical Medicine and Rehabilitation*, 77(12), 1279–1285.

[21] Ferrari, A., Benedetti, M. G., Pavan, E., *et al.* (2008). Quantitative comparison of five current protocols in gait analysis. *Gait & Posture*, 28(2), 207–216.

[22] Nielsen, C., Jorge, M., and Lusardi, M. M. (2013), *Orthotics and prosthetics in rehabilitation* (3rd ed.). St. Louis, MO: Elsevier.

[23] Goujon-Pillet, H., Sapin, E., Fodé, P., and Lavaste, F. (2008). Three-dimensional motions of trunk and pelvis during transfemoral amputee gait. *Archives of Physical Medicine and Rehabilitation*, 89(1), 87–94.

[24] Dongarra, J., and Sullivan, F. (2000). Guest Editors Introduction to the top 10 algorithms. *Computing in Science & Engineering*, 2 (1), 22–23.

[25] Morgan, K. D., and Noehren, B. (2018). Identification of knee gait waveform pattern alterations in individuals with patellofemoral pain using fast Fourier transform. *PLoS One,* 13(12), e0209015. doi: 10.1371/journal.pone.0209015. eCollection 2018. PMID: 30550589

[26] Tingley, M., Wilson, C., Biden, E., and Knight, W. R. (2002). An index to quantify normality of gait in young children. *Gait & Posture*; 16(2), 149–158.

[27] Howcroft, J., Lemaire, E. D., Kofman, J., and McIlroy, W. E. (2018). Dual-task elderly gait of prospective fallers and non-fallers: a wearable-sensor based analysis. *Sensors (Basel)*, 18(4), 1275. doi: 10.3390/s18041275

[28] Alexander R. M., and Jayes A. S. (1980). Fourier analysis of forces exerted in walking and running. *Journal of Biomechanics*, 13(4), 383–390. doi:10.1016/0021-9290(80)90019-6.

[29] Schneider, E., and Chao, E. Y. (1983). Fourier analysis of ground reaction forces in normals and patients with knee joint disease. *Journal of Biomechanics,* 16(8), 591–601. doi:10.1016/0021-9290(83)90109-4. PMID: 6643531 DOI: 10.1016/0021-9290(83)90109-4

[30] Shorten, M. R., and Winslow, D. S. (1992). Spectral analysis of impact shock during running. *International Journal of Sport Biomechanics*, 8, 288–304.

[31] Valiant, G. A. (1989). Transmission and attenuation of heel strike accelerations. In P. R. Cavanagh (Ed.), *The Biomechanics of Distance Running*. Champaign, IL: Human Kinetics, 225–247.

[32] Nigg, B. M., Denoth, J., and Neukomm, P. A. (1981). Quantifying the load on the human body: problems and some possible solutions. In: A. Morecki, K. Fidelus, K. Kedzior, & A. Wit (Eds.), *Biomechanics VII*. Baltimore, MD: University Park Press, 88–89.

[33] Shorten, M. R., Valiant, G. A., and Cooper, L. B. (1986). Frequency analysis of effects of shoe cushioning on dynamic shock in running. *Medicine and Science in Sports and Exercise*, 18, 2, S80–81.

[34] Voloshin, A. S., Burger, C. P., Wosk, J., and Arcan, M. (1985). An in vivo evaluation of the leg's shock absorbing capacity. In: D. Winter, R. Norman, R. Wells, K. Hayes, and A. Patla (Eds.), *Biomechanics IX-B*, Champaign, IL: Human Kinetics, 112–116.

[35] Baker, S. N., Olivier, E., and Lemon, R. N. (1997). Coherent oscillations in monkey motor cortex and hand muscle EMG show task-dependent modulation. *The Journal of Physiology*, 501(Pt 1), 225–241. doi: 10.1111/j.1469-7793.1997.225bo.x.

[36] Conway, B. A., Halliday, D. M., Farmer, S. F., *et al.* (1995). Synchronization between motor cortex and spinal motoneuronal pool during the performance of a maintained motor task in man. *The Journal of Physiology*, 489(Pt 3), 917–924. doi:10.1113/jphysiol.1995.sp021104.
[37] Rosenberg, J. R., Amjad, A. M., Breeze, P., Brillinger, D. R., and Halliday, D. M. (1989). The Fourier approach to the identification of functional coupling between neuronal spike trains. *Progress in Biophysics and Molecular Biology*, 53(1), 1–31.
[38] Bayram, M. B., Siemionow, V., and Yue, G. H. (2015). Weakening of corticomuscular signal coupling during voluntary motor action in aging. *Journal of Gerontology, Series A: Biological Sciences,* 70(8), 1037–1043. doi:10.1093/gerona/glv014.
[39] Nanavati, S., and Panigrahi, P. (2004). Wavelet transform a new mathematical microscope. *Resonance.* 9(3), 50–64.
[40] Hubbard, B. B. (1998). *The World According to Wavelets: The Story of a Mathematical Technique in the Making*. Natick, MA: A K Peters, Ltd.
[41] Burrus, C. S., Gopinath, R. A., and Guo, H. (1998). *Introduction to Wavelets and Wavelet Transforms: A Primer*. Upper Saddle River, NJ: Prentice-Hall, Inc.
[42] Daubechies, I. (1992). *Ten Lectures on Wavelets*. Montpelier, VT. Capital City Press.
[43] Chagdes, J., Rietdyk, S., Haddad, J., *et al.* (2009). Multiple timescales in postural dynamics associated with vision and a secondary task are revealed by wavelet analysis. *Experimental Brain Research*, 197, 297–310.
[44] Simonsen, I., Hansen, A., and Nes, O. (1998). Determination of the Hurst exponent by use of wavelet transform. *Physical Review E,* 58, 2779.
[45] Farge, M. (1992). Wavelet transforms and their applications to turbulence. *Annual Review of Fluid Mechanics,* 24, 395–457.
[46] Schaefer, A., Brach, J., Perea, S., and Sejdic, E. (2014). A comparative analysis of spectral exponent estimation techniques for $1/f^B$ processes with applications to the analysis of stride interval time series. *Journal of Neuroscience Methods,* 222, 118–130.
[47] Chen, W., and Bao, S. (2017). Epileptic focus localization using discrete wavelet transform based on Interictal Intercranial EEG. *IEEE Transactions on Neural Systems and Rehabilitation Engineering,* 25, 413–425.
[48] Zhang, Z., VanSwearingen, J., Brach, J., Perera, S., and Sejdic, E. (2017). Most suitable mother wavelet for the analysis of fractal properties of stride interval time series via the average wavelet coefficient method. *Computers in Biology and Medicine,* 80, 175–184.
[49] Liu, C. L. (2010). A Tutorial of the Wavelet Transform.
[50] Valens, C. A Really Friendly Guide to Wavelets. October 16, 2000.
[51] Salem, M. A., Ghamry, N., and Meffert, B. (2009). Daubechies versus biorthogonal wavelets for moving object detection in traffic monitoring systems. *Informatik-Berichte*, 229, 8–9.

[52] Ngui, W., Leong, M., Hee, L., and Abdelrhman, A. (2013). Wavelet analysis: mother wavelet selection methods. *Applied Mechanics and Materials*. 393, 953–958.

[53] Zhang, Z., Telesford, Q., Giusti, C., Lim, K., and Bassett, D. (2016). Choosing wavelet methods, filters, and lengths for functional brain network construction. *PLoS One*. 11(6), e0157243.

[54] Leavline, J. (2017). On applications of wavelets in engineering and technology. *International Journal of Engineering Studies and Technical Approach*. 01(8), 35–40.

[55] Nobach, H., Tropea, C., Cordier, L., *et al.* (2007). Review of some fundamentals of data processing. In: C. Tropea, A. L. Yarin, & J. F. Foss (Eds.), *Springer Handbook of Experimental Fluid Mechanics*. Berlin, Heidelberg: Springer, 1337–1398.

[56] Lilly, J. M., and Olhede, S. C. (2012). Generalized Morse wavelets as a superfamily of analytic wavelets. *IEEE Transactions on Signal Processing*. 60(11), 6036–6041.

[57] Patla, A. E. (1985). Some characteristics of EMG patterns during locomotion: implications for the locomotor control process. *Journal of Motor Behavior*, 17, 443–461.

[58] Davis B. L., and Vaughan, C. L. (1993). Phasic behavior of EMG signals during gait: use of multivariate statistics. *Journal of Electromyography & Kinesiology* 3(1), 51–60.

[59] Kaiser, H. F. (1958). The varimax criterion for analytic rotation in factor analysis. *Psychometrika*. 23, 187–200.

[60] Kaiser, H. F. (1961). A note on Guttman's lower bound for the number of common factors. *British Journal of Mathematical and Statistical Psychology*, 14, 1–2.

[61] Delan, D. (2019). The Unspin Room: When a picture is worth a thousand spins. The Berkshire Eagle. https://www.berkshireeagle.com/stories/dalton-delan-the-unspin-room-when-a-picture-is-worth-a-thousand-spins,563266

[62] Tufte, E. R. (1983). *The Visual Display of Quantitative Information*. Cheshire, CT: Graphics Press, p. 06410.

[63] Hof, A. L., Elzinga, H., Grimmius, W., and Halbertsma, J. P. K. (2002). Speed dependence of averaged EMG profiles in walking. *Gait and Posture*, 16(1), 78–86. doi:10.1016/s0966-6362(01)00206-5.

[64] Bechtol, C. O. (1975). Normal human gait. In J. H. Bowker and C. B. Hall (Eds.), *Atlas of Orthotics: American Academy of Orthopaedic Surgeons*. St. Louis, MO: Mosby, 133–143.

[65] Inman, V. T., Ralston, H. J., and Todd, F. (1981). *Human Walking*. Baltimore/London: Williams and Wilkins.

[66] Pauwels, F. (1980). *Biomechanics of the Locomotor Apparatus*. Berlin: Springer-Verlag.

[67] Cavanagh, P. R., Simoneau, G. G., and Ulbrecht, J. S. (1993). Ulceration, unsteadiness, and uncertainty: the biomechanical consequences of diabetes mellitus. *Journal of Biomechanics*, 26(Suppl 1), 23–40.

Chapter 7

Novel technologies for gait analysis

Tom Shannon[1]

7.1 Introduction

The widespread development and implementation of routine clinical gait analysis across the world have had a secondary benefit with many new applications being discovered for the technologies in disparate but potentially commensurate environments including basic research, rehabilitation, ergonomics, virtual reality, engineering and animation to name but a few. Conversely, the unique demands of these environments have forged a symbiotic relationship where new requirements for technological improvements have directly benefited the original clinical and life sciences applications.

The measurement of athletic performance demanded the development of cameras capable of operating in the outdoors with higher resolution and capture rates to track fast and subtle motions with reliability and consistency. Clinical gait analysis benefitted through the improved acquisition of faster motions such as heel strike and in the better description of the motion of anatomical structures such as the foot (Figure 7.1).

Since 1997, the wide applications of the use of motion capture technologies to enhance films, games, virtual reality environments, music videos and television have brought many stories to life, thought almost impossible prior to their introduction into the entertainment sector. With it has brought many challenges including the need to capture the motion of multiple characters simultaneously, in large volumes and with high camera counts. The sector has also demanded improved subtlety in the measurement of the motion of hands, feet and facial expressions with a need to provide real-time reconstruction of the animated characters who may or may not be humanoid (Figures 7.2 and 7.3).

The demands of life sciences including clinical gait analysis and sport, engineering, ergonomics and the entertainment sector have resulted in a synergy that has driven and improved the core technologies for the benefit of all. In just one example, the life sciences' need to continuously improve skeletal modelling has benefited animators to create ever-more realistic characters, whilst their

[1]Centre for Biomechanics and Rehabilitation Technologies, Staffordshire University, Stoke on Trent, UK

Figure 7.1 Animation (Vicon Motion Systems Ltd)

Figure 7.2 Clinical gait analysis laboratory (Vicon Motion Systems Ltd)

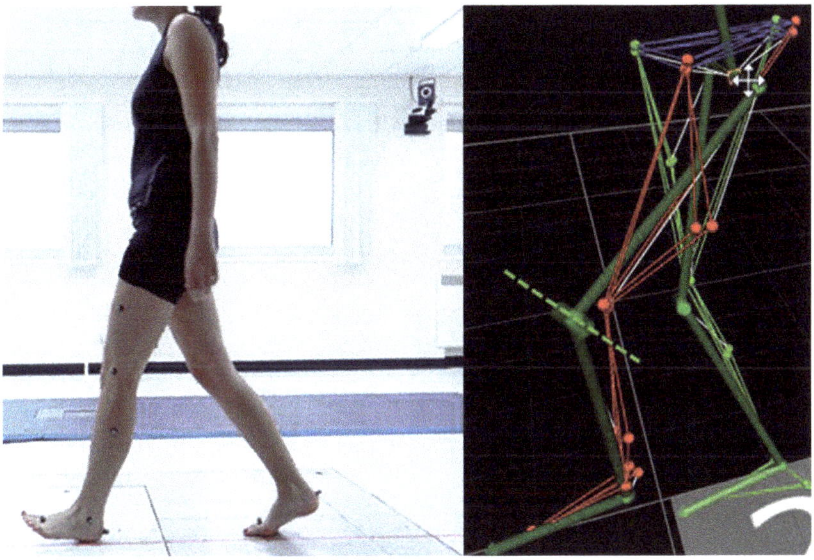

Figure 7.3 Clinical gait analysis data (Vicon Motion Systems Ltd)

requirements have directly improved how the technologies are applied and data are presented within the clinical gait and biomechanics communities.

Motion capture can be described as the process of applying technology to record the movement of objects, animals, human subjects, actors or patients. In the entertainment sector, the technique is often referred to as 'performance capture' in recognition that the subtlety of motion of actors, including the famous, are captured in a digital format and are a full physical transcription of their acting skills.

The joy of animation was first expressed in the early twentieth century when in 1911 the cartoonist Winsor McCay drew a character on multiple sheets of paper with slight differences in anatomical position from page to page. Flipping through the sheets of paper at a constant rate then created the illusion of motion [1]. Animators and movie directors quickly learned that the manual drawing process was very time consuming with every artist forced to construct each pose within each frame.

Contemporary motion capture technology is a progression over the older technique of 'rotoscoping', a labour-intensive methodology wholly dependent on manually tracing over the outline of an actor undertaking a pre-defined movement onto sheets of acetate. An artist then fills in the traced outlines, frame by frame, drawing the animated character such that the result would exactly match the movement and actions of the actor within the resulting live-action footage.

Little innovation to improve and simplify the method was undertaken until computers started to have an impact within the major studios. The introduction of 'key framing' allowed the artist to specify the start and end drawings or 'frames' of an animation sequence with the character motions within the intermediate drawings

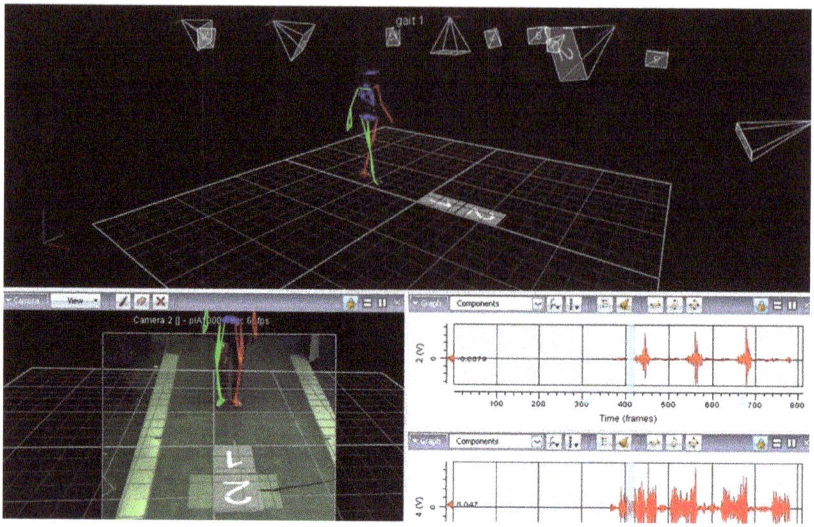

Figure 7.4 EMG (Vicon Motion Systems Ltd)

called 'in-betweens' automatically generated [2]. The technique relied on a smooth transition between the boundary frames and was found unsuitable in those cases where the motion was inherently complex such as in human walking with multiple articulations.

This chapter presents a review of the methods applied in the recognition, analysis and impact of human gait from four different but commensurate approaches: image processing via passive and active optical motion capture kinematic acquisition technologies, floor-based kinematic and kinetic technologies, energy and oxygen consumption measurement and the use of wearable sensors including for electromyography placed upon palpated anatomical landmarks (Figure 7.4).

The analysis of gait continues to be the focus of many research activities across the world centred on describing quantitative objective measures of the characteristics of human locomotion. The common goal remains to apply the measures to various fields such as sports, identification of individuals for security purposes and to clinically collect information to aid in the understanding of the aetiology of presenting gait abnormalities so as to assist in treatment decision-making. Changes in gait reveal much about a patient's quality of life in particular when attempting to quantify the impacts of the progression of different pathologies including neurological diseases such as multiple sclerosis and diabetic peripheral neuropathies; Duchenne's muscular dystrophy, cerebral palsy; systemic diseases such as cardiopathies; and alterations due to sequelae from a stroke.

Scales used to analyse clinical gait parameters have traditionally been semi-subjective with a reliance on the observation of specialists to assess the quality of a patient's gait by encouraging them to walk for multiple cycles during a clinic presentation. Surveying patient opinions to provide a personal and subjective evaluation of their gait can add to the overall assessment. The disadvantage of this method is that the error margins of the accuracy and precision of subjective acquired measures will likely have an un-predictable impact on the diagnosis and treatment regimens of presenting pathologies. As a counter, significant progress in the last three decades has given rise to the commercialisation of devices and techniques to facilitate an objective evaluation of different parameters resulting in more efficient and reliable measurement available to gait analysis specialists.

Conventional and well-established clinical gait laboratories utilising well understood passive optical movement analysis solutions have become the norm across the world since the early 1980s. The technique is based on the acquisition and re-construction in three-dimensional space of the centres of small passive retro-reflective markers using multiple, synchronous video cameras controlled by a computational device. In clinical applications, the markers are placed directly onto the skin through palpating bony landmarks to derive quantitative information about the motion of anatomical points, segments and joint angles. The reaction forces and torques produced between the patient and the ground can be simultaneously acquired to derive data including joint forces, moments and powers, as can an appreciation of commensurate muscle electrical activity using specialised electrodes.

The intended purpose of clinical gait analysis remains to provide information that is used to take decisions for therapeutic purposes and to aid in the understating of the aetiology of gait abnormalities resulting from and including a diagnosis of cerebral palsy or Duchenne muscular dystrophy among presenting children and adolescent patients. The interpretation of resulting kinematic and optional kinetic and electromyographic data remains in the hands of interdisciplinary teams of heath care professionals and bio-mechanists with substantial knowledge of normal and pathologic human gait where characteristics are measured, abnormalities are identified, causes are postulated and treatments are proposed. The same technology is routinely applied to establish the efficacy of orthoses and prostheses, and during rehabilitation among the wider population. The common goal of the therapies and interventions is to improve the efficiency of patient ambulatory locomotion.

Gait analysis presentations are usually less than one hour in duration with no restrictions on re-application. There are no identified contraindications or required precautions for normal laboratory use. In most cases, the equipment will not have been designed for use in an operating theatre, anaesthetic gas, oxygen-rich environments or where there is a risk of compromising the general safety and performance of medical electrical equipment. Most equipment is unsuitable for use in high magnetic flux (e.g. magnetic resonance imaging), ionising radiation, sterile or life- or safety-critical environments (Figure 7.5).

Human gait research remains dependent on the qualitative and quantitative evaluation of common factors that characterise the motion. The differences lie in the focus of factors within each field of study. For example, interest in subject identification may centre on classifying individual motion based on their silhouette and the relative movements of different body segments during ambulatory motion whereas the clinician is looking for changes over time that may indicate disease evolution when compared to the gait of typically developed adults (TDA).

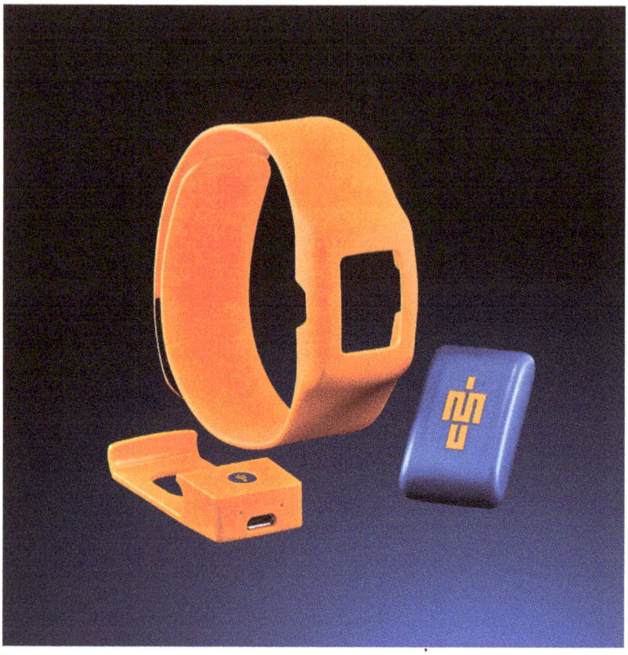

Figure 7.5 IMU (IMeasureU Ltd)

To amplify the point, consider cases of Parkinson's disease that damages neurons in the substantia nigra portion of the brain and nerves of the body as well as causing accumulations of Lewy bodies in the form of the protein alpha-synuclein [3]. In healthy subjects, the neurons produce dopamine, a chemical messenger that transmits signals from the substantia nigra to produce smooth purposeful movement. The onset of the disease results in a reduction of dopamine and a manifestation of motor symptoms commonly observed in diagnosed patients.

Hausdorff [4] characterised disturbances to gait among diagnosed patients into two types, episodic and continuous, with the former cases occurring occasionally and intermittently surfacing in an apparently random inexplicable manner in the form of festination (a quickening and commensurate shortening of stride length), start hesitation and freezing. Conversely, continuous gait disturbances are usually more commonly experienced by patients in advanced cases of disease progression and are manifested in persistent alterations in walking patterns. Patient gait [5] has been described as distinctive; less steady, arising from changes in posture; slowness of movement (bradykinesia); and a shortened stride. This can be characterised by a combination of the following:

- A tendency to lean unnaturally forwards in a stooped position when walking and in some cases to lean sideways when standing.
- The head dropped forwards, with shoulders drooped, knees bent.

- Steps taken on the balls of the foot.
- Shuffling.
- Reduced stride length.
- Reduced arm swing.

Rogers *et al.* [6] estimate that the disease affects around 250/100,000 in the United Kingdom with diagnosis focussed on people presenting with motor symptoms including bradykinesia, rigidity, rest tremor and postural instability. Non-motor symptoms may also be prominent, including depression, cognitive impairment and autonomic disturbances. Parkinson's disease more commonly occurs in people over the age of 60 with prevalence higher in men than women [7]. As life expectancy increases, there is a general expectation that the prevalence and incidence of the disease will increase over time.

7.2 Application of gait analysis measurement technologies

Muro-de-la-Herran *et al.* [8] proposed an overview of common parameters used in the study of human gait, tabulated in Table 7.1.

Gait analysis is now an established research activity and routine clinical tool for many life science, medical and healthcare applications ranging from assessing the efficacy of orthoses, prostheses and surgical procedures; quantifying the impact of the progression of neuropathies; monitoring pathologic gait degradation or rehabilitation treatment improvements during clinic presentations. Some severe gait disorders can be observed without resorting to quantitative measures but this can result in subtle changes going un-noticed. The qualitative method typically exhibits significant inter- and intra-observer variabilities that may adversely affect disease severity assessments and subsequent treatment planning.

The quality and validation of quantitative measures are dependent on the suitability of the approaches applied to adequately assess human gait among TDA and in the presence of pathologies. For example, only acquiring cadence, gait speed and distance travelled may provide an over-simplified measure of a complex mechanism governed by the neuromuscular system.

Objective measurement of human gait can be achieved using a variety of different devices to acquire and derive the gait parameters described in Table 7.1. The methods are grouped into four distinct categories used in synchrony or alone to measure a subject or patient's kinematics, kinetics, energy consumption and muscle electrophysiological activity. The devices are designed to be applied directly to the skin, worn or carried by the user; on or in floor sensors to measure gait parameters and GRFs, or based on passive or active image processing technologies.

Bespoke laboratories and specialist centres using optical motion capture systems continue to dominate clinical gait analysis. Systems employed commonly use multiple, high sampling rate, fixed cameras acquiring body motion through reconstructing in three dimensions, the spatial and temporal location of

Table 7.1. Overview of gait parameters and applications

Gait parameter	Application			Description
	Clinical	Sports	Recognition	
Stride velocity	X	X	X	
Step length	X	X	X	Linear distance between two successive placements of the same foot
Stride length	X	X	X	Linear distance between the placements of both feet
Cadence	X	X	X	Number of steps per time unit
Step width	X	X	X	Linear distance between two equivalent points of both feet
Step angle	X	X	X	Direction of the foot during the step
Step time	X			
Swing time	X			Time from the moment the foot lifts from the floor until it touches it again, for each foot
Traversed distance	X	X		
Gait autonomy	X			The maximum time a person can walk, taking into account the number and duration of the stops
Stop duration	X			
Existence of tremors	X			
Fall	X			
Accumulated altitude	X	X		
Route	X	X		
Gait phases	X	X	X	
Body segment orientation	X	X		
GRFs	X	X		
Joint angles	X	X		Ankle, knee, hip
Muscle force	X	X		
Momentum	X	X		
Body posture	X	X	X	Inclination, symmetry
Long-term monitoring of gait	X	X		

detectable markers applied to anatomical landmarks. Such systems deliver highly accurate kinematic analyses, further enhanced if the facility also includes ground reaction force (GRF) plates to acquire synchronised kinetic information to derive joint moments and powers. The disadvantage of such systems is that they are relatively expensive and are restricted to laboratory environments so may not fully reflect gait variability in all real-world settings.

7.3 Wearable devices

The emergence of ubiquitous electronic miniaturisation and localised transmission protocols such as Bluetooth® in recent decades has resulted in the ready availability of small and wearable wireless electromyography (EMG) modules in many clinical laboratories. The development and launching of other wearable devices including inertial measurement units (IMUs) and insole pressure monitors have similarly offered the opportunity to add useful data to a patient or subject's gait analysis. The technologies are low cost, portable and versatile that provide a rich seam of data within both indoor and outdoor environments that offers the prospect of seamless analysis between the gait analysis clinic and routine life environments.

7.3.1 EMG

The addition of the routine incorporation of EMG into hybrid laboratory systems to describe the electrical activity within skeletal muscles [9] was simultaneously obtained during a gait analysis to be correlated with the kinematics and kinetics data. The resulting data are particularly useful for assessing gait disorders that include symptoms such as severe muscle weakness or abnormal activity.

Wearable sensors placed on pre-determined portions of a subject or patient's body such as the feet, knees, hips or sacrum offer the measurement of different characteristics of human gait. Such sensors may acquire forces, accelerations, rotations, inclinations, muscle electrical activity, etc. or be in the form of goniometers or extensometers to measure linear displacements.

7.3.2 Goniometers

The goniometer is a wearable sensor or clinical tool used to measure the angle of human joints most commonly at the ankle, knee, hip, metatarsals and any other joints where the technology can be applied (Figure 7.6). In its simplest form, single-axis or point or rotation mechanical goniometers are routinely used in physiotherapy, occupational therapy and sports medicine clinics to document changes in limb range of motion at patient presentations. The device is a particularly useful tool to provide objective measurements to assess the progress or otherwise of a rehabilitation programme. Should a patient exhibit a decreased range of motion, the therapist with a knowledge of the anatomy of the body can assess the joint to perform an intervention, using the device to quantify progress. As an example, when measuring the knee joint range of motion, the device axis would be co-located with the lateral epicondyle of the femur with the fixed arm lined up with the greater trochanter. The movable arm is aligned to the lateral malleolus of the fibula with the degree of measurement easily documented. Such simple devices are commonly used for medico-legal purposes during occupational injury and disability evaluations but inter- and intra-measurement errors can be of the order of 5°–10°.

In 1971, Larry Lamoureux used active goniometers applied to a metal exoskeleton to measure tri-dimensional angles at the hip and one-dimensional angles at

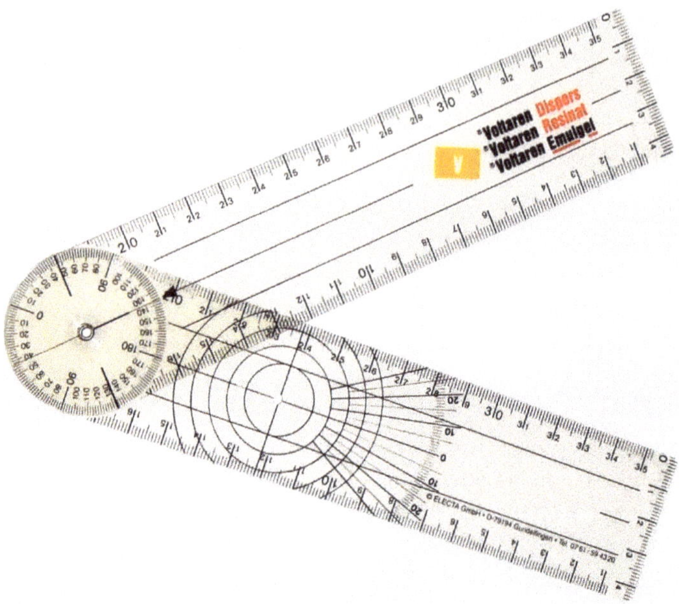

Figure 7.6 Goniometer (Physiopedia)

the knee and ankle [10]. By 1980, a tri-axial goniometer based on a gyroscopic principle and capable of generating tri-dimensional angle outputs [11] was developed, but cross talk problems introducing physical measurement artefact dogged the early designs. Inherent mechanical morphological inflexibility introduced challenges when attempting to correlate the surface measures with actual skeletal joint rotations that limited early application.

Significant technological developments have overcome many of the early disadvantages, resulting in researchers and health professionals continuing to use goniometers for measuring joint angles in many applications. Contemporary active offerings include those based on strain gauges, inductive sensing, potentiometers or digital angle encoders. The principle behind strain-gauge technologies is based on measuring changes in electrical resistance, proportional to the degree of sensor flexion that is directly correlated to angle. Commercial devices have achieved sufficient miniaturisation for fitment within shoes to measure ankle to foot angles; generate two-axis outputs; and measure torsional rotations [12].

7.3.3 Insole pressure and force sensors

Commercial pressure sensors are routinely used in the clinical measurement of foot plantar pressure distribution, under-foot ground reaction force, gait phase and step detection (Figures 7.7–7.9). Decisions related to pressure range, linearity and sensitivity define if resistive, piezoelectric, capacitive or piezo-resistive technologies are optimally suitable for a given measurement. Piezoelectric sensors offer

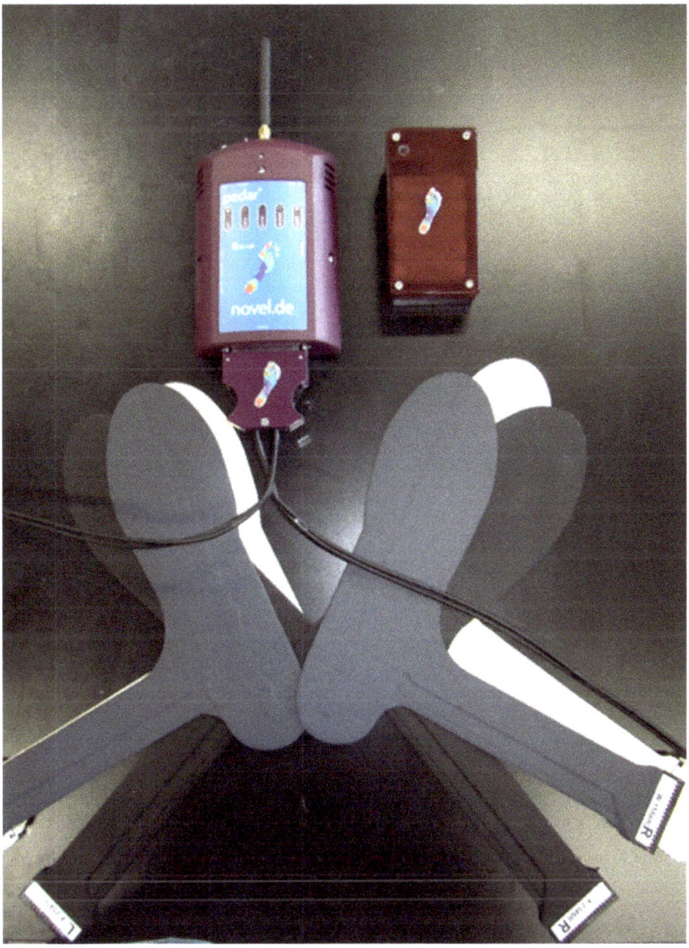

Figure 7.7 Insole pressure devices

excellent linearity and reactivity but because of their relatively large size are not suitable for application to some surfaces. Integrating piezo-resistive sensors and arrays [13] into instrumented insoles has proven to be useful to measure and analyse GRF distribution on patient's feet, to assess balance problems and to provide a warning of falls.

7.3.4 IMUs

Inertial navigation is a well-understood technique to calculate using dead reckoning the position, orientation, speed of motion and direction of an object without the need for any external reference such as the satellite-based ground positioning system. Such equipment are usually large complex systems found in aircraft, space

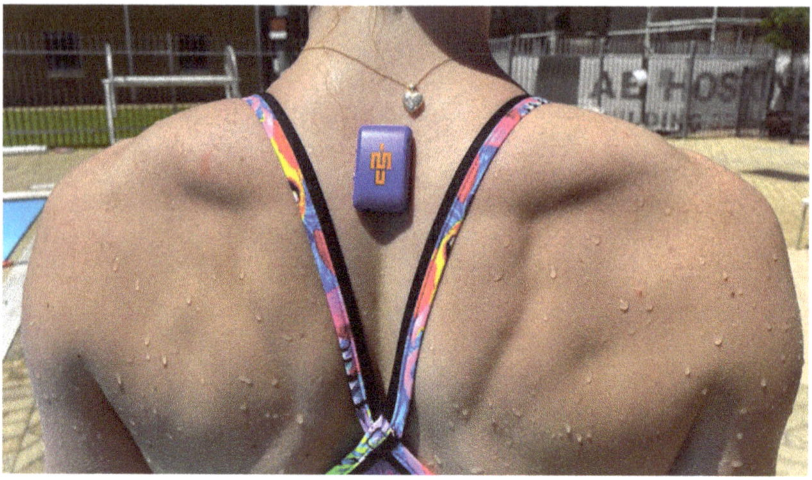

Figure 7.8 IMU human applications (IMeasureU Ltd)

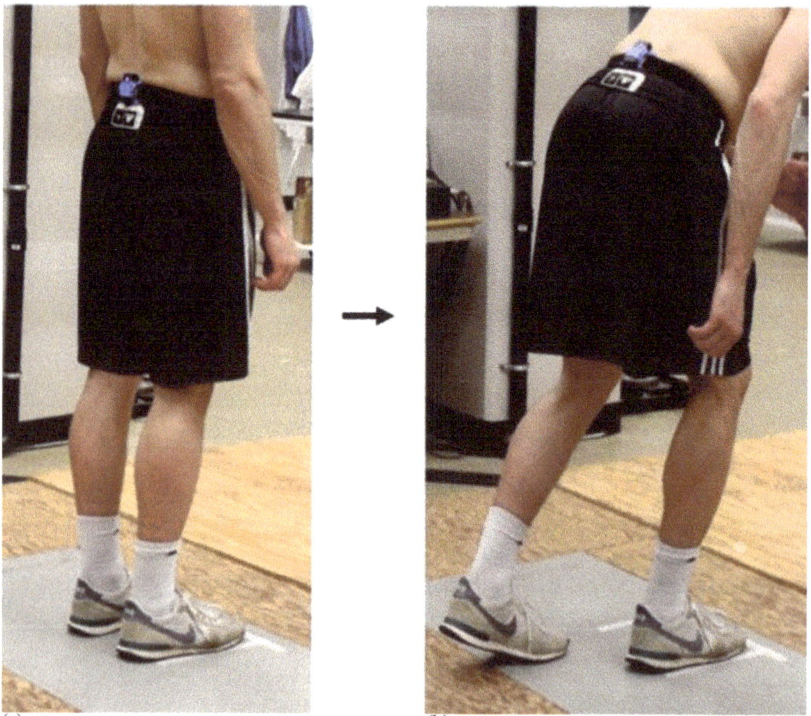

(a) (b)

Figure 7.9 Sacral IMU [14]

vehicles and rockets. An inertial measurement unit (IMU) is a mechanism that reports non-gravitational forces, angular rate and magnetic fields around a body to derive orientation (roll, yaw and pitch) and heading.

Opportunities to apply inertial measurement sensing too many life sciences applications have arisen in recent years due to the mass commercialisation and miniaturisation of the technology, now widely used in tablets, smartphones, games controllers and automobiles. IMU when used as single units or in combination are now well established as the most common wearable sensors for kinematic gait analysis. The technology is physically implemented by incorporating nine-axis micro-electro-mechanical system (MEMS) elements, packaged into a device the size of a small fingernail, to measure angular velocity, acceleration, orientation and gravitational forces.

MEMS devices and structures are manufactured using the technique of micro-fabrication and can vary from relatively simple structures having no moving elements to extremely complex electro-mechanical systems with multiple moving parts under the control of integrated micro-electronics. The common criterion to classify a MEMS device is that it converts a measured mechanical change to an electrical signal.

Inertial measurement devices use a combination of a three-axis accelerometer, three-axis gyroscope and a three-axis magnetometer or compass. The unit functions by detecting linear accelerations and rotational rates with the compass providing a heading reference. The output is usually expressed as a quaternion of the form $f(w, i\text{x}, j\text{y}, k\text{z})$ where w is correlated with rotation angle, x, y and z are real numbers and i, j, k can be interpreted as unit-vectors pointing along the three spatial axes, conceptually as the rotation axis.

Rate gyroscopes measure angular velocity ω and, when temporally integrated, provide the change in angle or orientation with respect to an initially known value:

$$GS_{\dot{q}_t} = \frac{1}{2} GS_{q_t} \otimes \Omega_t$$

where GS_{q_t} represents the quaternion describing the rotation from the sensor; (S) to global frame (G) at time t. $\Omega_t = \left(0, \omega_x, \omega_y, \omega_z\right)^T$ is the quaternion representation of the angular velocity ω_t.

The three-axis accelerometers measure the vector of acceleration (a) and gravitational acceleration (g) in sensor coordinates. The sensor output can be expressed in the global reference system if the orientation of the sensor GS_{q_t} is known:

$$G_{a_t} - G_{g_t} = GS_{q_t} \otimes \left(S_{a_t} - S_g\right) \otimes GS_{q_t}{}^*$$

After removing the gravity component, the acceleration can be integrated once to derive velocity v_t and twice to estimate position p_t, in the global frame:

$$G_{\ddot{p}_t} = G_{a_t}$$

Each axis accelerometer is implemented as a mass on two springs. If there is stasis, an electrode prong connected to the mass stays between two outer electrode pins to form a capacitor. Under applied and varying accelerations in the axis, the prong moves within the pins in the direction so changing the capacitance in direct correlation with the movement of the mass.

Gyroscopes use the property that all bodies revolving around an axis develop rotational inertia. MEMS devices rely on the 'Coriolis Effect' to measure angular velocity. The device employs a resonating mass within driving combs. A rotation of the gyroscope generates a force acting on the mass resulting in an alternate direction. Electrodes to derive the angular velocity in the affected axes measure the motion of the new direction.

The magnetometers or compasses utilise the 'Hall Effect' to generate a voltage difference across an electrical conductor, transverse to an electric current in the conductor, and to an applied magnetic field perpendicular to the current.

IMU used in research and clinical environments conventionally contains a battery with the inertial measurement device connected to and managed by a microcomputer usually described as a system on a chip or 'SoC'. The nine-axis or quaternion outputs are transmitted either by wire or by a short-range radio protocol such as Bluetooth for further processing and analysis.

The potential applications of a single IMU in everyday clinical practice are only limited by the need to establish the efficacy and validation of the approach coupled with the imagination of the medical practitioner, health professional or researcher. The following section describes an example of the use of an IMU for a specific condition to demonstrate the potential application of the technology in a routine clinical environment. The use of inertial sensors and accelerometers has become common practice in ambulatory motion analysis for a number of decades [15,16]. Interest in applying single IMUs to the lower trunk as an alternative method to assess gait outside the conventional use of expensive and less flexible optical [17] or multi-IMU [18] based gait analysis laboratories continues to be reflected in a range of academic publications [19–25] and a number of commercial implementations [26–29].

7.3.5 Example of a clinical application using a single IMU – diabetic peripheral neuropathy

Over the last decade, researchers at Oxford Brookes University in the United Kingdom have been investigating how spatio-temporal gait patterns could be derived from using the output of a single IMU. They applied a unit to the base of the spine monitoring changes in the location of the centre of mass (CoM) in ambulatory humans. The researchers tested the method on a large control group (>2,000) of TDA and hundreds of patients diagnosed with a number of different progressive neurological diseases. Through multi-centre trials and subsequent publication in peer-reviewed academic journals, they have demonstrated that there is a correlation between the patterns of CoM motion and disease severity, which is not prevalent in the control group.

Esser *et al.* were awarded international patents [30] to describe their invention of an apparatus for measuring the vertical movement of human, local frame of reference, CoM. They applied a single inertial sensor to the sacrum to derive the rotational matrix and first acceleration so as to estimate vertical displacement or a derivative of vertical displacement of CoM in a global reference frame. The inventor's claim was based on an independent estimation that vertical displacement of CoM represented greater than 65% of the mechanical work done in gait. Esser *et al.* concluded that pattern asymmetry of CoM displacement in normal and pathologic gait may assist in the diagnosis of an abnormality and monitoring of the impacts of disease progression.

Diabetic peripheral neuropathy (DPN) affects over 50% of persons with diabetes mellitus [31]. Timely intervention, preventative management, education and advice can reduce the rate of complications including ulceration, Charcot's feet or amputation. DPN results from nerve damage via poorly managed and chronically high levels of blood sugar. In patients who have type 1, which usually affects those 25 years and younger, there is insulin deficiency. For those diagnosed with type 2, their insulin production inadequately meets the body's daily need to metabolise sugar and starches following ingestion of some foods. Nerve damage may occur due to small nerve fibre compression from swelling Schwann cells that constitute the myelin sheath around neuronal axons or by neural ischaemia being a lack of adequate blood supply to maintain large nerve fibre vitality [32].

Alam *et al.* [31] also proposed a direct correlation with age and the duration of the disease (>25 years) potentially having a major impact on patient sensorimotor systems. The progression of the disease affects the sensory, motor and autonomic elements of the nervous system that can lead to alterations in gait due to a loss of protective sensation coupled with intrinsic dysfunction of the muscles of the foot. Similarly, the symptoms can manifest as aching, burning, sharp or pressure pains. A study by Brown *et al.* [33] found that DPN patients showed a greater deviation in their CoM during staircase and level walking, suggesting these patients may have difficulty maintaining dynamic stability and locomotive balance. Artefacts for those patients with a loss of sensation and muscular dysfunction were characterised by an increased number of steps and evidence of bradykinesia manifested by fewer acceleration patterns in their gait cycle.

Lalli *et al.* [34] reported a greater variation in step length and velocity among patients who also reported painful symptoms, observing that they tended to be more mobile than the insensitive group but also identified a fear of and an increased number of falls requiring hospitalisation.

Public Health England [35] has reported (2015) an estimated prevalence of 3.8 million people in the United Kingdom of both types of diabetes with 90% of cases being type 2. They further estimated that 940,000 individuals might be unaware that they have contracted the disease.

Hazari *et al.* [36] undertook a systematic and meta-analysis review of the literature to determine if there were any common themes in the research outcomes of common kinematic and kinetic measures to indicate the presence or otherwise of DPN compared with TDA subjects and patients diagnosed with type 2 diabetes without neuropathy. The most significant finding was that gait velocity (ms^{-1})

between aged-matched participants (128 DPN, 131 TDA) was significantly lower in the neuropathic case (−0.09; 95% CI − 0.13 to 0.05; $p < 0.0001$).

Esser *et al.* [37] compared the spatio-temporal parameters of 17 subjects (14 male) with diagnosed DPN (aged 63±9.2 years), using a monofilament applied to the sole of the foot together with blood and microvascular evidence, against 42 (30 male) TDA controls (aged 61±4 years). No significant difference in age, gender ratio, height or BMI between groups was observed ($p > 0.05$). Considering an area under a Receiver Operator Curve, ROC, (area under the curve) AUC), gait velocity or walking speed gave the highest discrimination between affected and control groups (0.975 AUC ROC $p > 0.001$) which aligned well with the findings of Hazari *et al.* [36].

DPN severity scoring methodologies are widely used and well understood within routine clinical practice:

- Neuropathy Symptom Score (NSS): calculated by burning, numbness or tingling, fatigue, cramping, aching or nocturnal exacerbation.
- Neuropathy Disability Score (NDS): calculated by ankle reflex, pinprick sensation, temperature sensation, vibration sense tested using a tuning fork and monofilament.
- Diabetic Neuropathy Examination Score (DNE): The DNE score consists of eight items, two testing muscle strength, one a tendon reflex, and five sensations. The maximum score is 16. A score of >3 points is considered abnormal. Only the right leg and foot are tested.
- Diabetic Neuropathy Symptom Score (DNS): described by Meijer *et al.* [38]. The authors compared score characteristics of the generally accepted NSS with the DNS score, and tested construct validity, predictive value and reproducibility with the DNE score, Semmes–Weinstein monofilaments and Vibration Perception Threshold (clinical standards) in 73 patients with diabetes (24 Type 1, 49 Type 2; 43 male/30 female; mean age 57 years (19–90); mean diabetes duration 15 years (1–43)). They observed a good correlation between NSS and DNS and the score was high (Spearman $r = 0.88$). Patient scores were more differentiated on the DNS score. The relation of the NSS and DNS scores, respectively, with clinical standards was good (Spearman $r = 0.21$–0.60). Reproducibility of the DNS score was high (Cohen weighted kappa 0.78–0.95).

Using NDS scoring as an example, patient assessments are divided into four categories: no symptoms (NDS ≤ 2); mild neuropathy (NDS 3–5); moderate neuropathy (NDS 6–8); and severe neuropathy (NDS 9,10).

The discriminatory performance results of two studies [35,36] gave confidence that a possible correlation might exist between the reported gait assessment findings and routine clinical neuropathy disability severity scoring. One possible solution to test this hypothesis might be to apply contemporary machine learning algorithms such as a multi-category cluster classification methodology using a support vector machine (SVM) to generate unique signatures with a similar granularity. This thought is not original as Kazemi *et al.* [39] have applied an SVM [40] to standard NDS data acquired from a questionnaire given to independently diagnosed diabetic patients (aged 53.26 ± 14.46 years; mean disease duration 9.26 ± 14.46 years). In their study, 175 patients did not exhibit neuropathy; 265 were clinically classified with mild neuropathy; 127 patients with moderate neuropathy; and 33 patients with severe neuropathy. The authors selected 13 features to classify by machine learning. They achieved a 76% accuracy in class prediction when using a classifier built from a balanced dataset and radial basis kernel function with a one-against-one strategy.

If a reliable, verifiable and validated correlation can be found between DPN severity classes predicted from IMU CoM gait assessment data and routine and well-understood clinical diagnostic measures, then the idea has the potential to become a very useful and potentially widely applied tool. The wider acceptance of the IMU in clinical and general motion analysis is rapidly gaining footholds in multiple applications [40–46].

7.3.6 Example of a single or dual limb IMU – sports science

Movement analysis is an area of sports science that has risen in popularity over the last decade (Figure 7.10). Coaches have built their careers on subjectively analysing movement to optimise a skill or technique; but formal, objective sport movement analysis has been confined to the motion capture laboratory until the

Figure 7.10 Sports science inertial measurement (IMeasureU Ltd)

emergence and commercialisation of miniature and radio capable IMU such as the IMeasureU (IMeasureU Ltd, Auckland, New Zealand) Blue Trident product offerings [47].

Sports biomechanics is one of the fundamental disciplines in the larger field of sports science. The area incorporates the detailed analysis of the fundamental movements in sport with the aim of reducing injury risk and improving athletic performance [48]. Researchers looking at the physical aspects of performance use biomechanics analysis to try to understand the relationship between kinematics, kinetics and athletic achievement. Biomechanical analysis can have a huge impact on performance by creating models for optimal movement and load analysis. This is used to determine the safest or most efficient way to move or perform a skill, and investigate the impact a different environment can have on how the body moves or reacts.

Sport biomechanics often focusses on lower limb biomechanics due to the role our legs play in locomotion. One area of research occupying scientists over recent years is in the understanding of fatigue and the impact it has on athletes' bodies. Researchers use biomechanical assessment tools such as force plates or inertial sensors to gain an understanding of the forces acting on the body together with the interaction of kinetics and kinematics measures. Muscle mechanics come into play when athletes are being analysed for dynamic movements. How their muscles react to certain pressures can reveal clues about their injury risk and the duration of their recovery. Inertial sensors measure the granular metrics of displacement, velocity and speed of movement in a straight line or a rotary direction. Findings from these tests inform proper load-bearing techniques to help preserve overall athlete health.

7.3.7 Capture of human ambulatory motion using IMU groups

The use of groups of IMU without any dependence on external light sources or cameras has become an established alternate ambulatory motion analysis method especially in the measurement of body joint angles exhibiting multiple degrees of freedom. One commercial example is the Xsens MVN system (Xsens Technologies B.V.) consisting of 17 IMU (38 $\times$ 53 $\times$ 21 mm 30 g) installed into a bespoke suit for ready placement of the sensor units and interconnecting cables [18]. Data are transmitted by a wireless connection to a support computer for processing and visualisation (Figure 7.11). Similarly, the same technology can be affixed to anatomical locations identified by palpation. Roetenberg *et al.* [18] identified that biomechanical modelling of the human body cannot be restricted to a simple kinematic chain of pre-defined joints as they allow laxity in both position and orientation and is independent of the direction of motion. Complex joints such as the shoulder and non-rigid body segments such as the back must be described with up to six degrees of freedom that requires an orientation measurement capability. Body segment position and orientation using IMU can be estimated by integrating gyroscope data and double integrating the accelerometer data over time. Even with the contribution of magnetometer data, the technology is susceptible to body segment and joint location drift with respect to a global coordinate system. A second limitation is that it is not possible when using only IMU data to measure the

Figure 7.11 Xsens MVN system (Xsens Technologies B.V.)

clearance of feet during running and jumping motion captures. The solution adopted is to integrate devices such as acoustic or magnetic trackers [45–47] to align the IMU data to the world coordinate system. The resultant system is fully ambulatory without any imposed environmental restrictions and locations when used to track and capture dynamic motion in three dimensions. The equipment corrects for drift and other errors, notably of segments connected by joints, through applying pre-defined constraints within a biomechanical model of the human body to the raw sensor outputs. The published [18] maximum update rate of all kinematics is 120 Hz.

7.4 Floor-mounted kinematic and kinetic capture technologies

Floor-mounted equipment can be grouped into kinematic and kinetic device categories with the former consisting of instrument walkways to capture basic gait parameters such as cadence, stance and swing time or 'Pedobarography' being the study of pressure fields acting between the plantar surface of the foot and a supporting surface. Such an equipment cannot derive horizontal or shear components unlike the latter category that are capable of acquiring the pattern of forces applied when walking, known as the GRF, being the exact opposite of the initial forces under Newton's Third Law. GRF values vary during the time the foot is in contact with the floor. Maximum pressure occurs when the heel touches the floor and when the toe pushes off for the next step, forming a pattern similar to a butterfly wing. The values may reach up to 120–150% of the patient's body mass. Commercial force platforms generate three-force and the three-moment components to derive a temporal three-axis force vector to assist a health professional in making a diagnosis from the evolution of a patient's foot pressure on the floor in real-time.

7.4.1 Instrumented walkways

Instrumented walkways offer a simple method to capture individual steps and averaged spatial and temporal gait parameters. Independent research by Webster *et al.* [48] was undertaken to compare the performance of a commercial system GAITRite (CIR Systems Inc., Franklin, NJ, USA; Figure 7.12) with an optical motion capture system VICON 512 (Vicon Motion Systems Ltd, Oxford, UK) [49]. Ten post-operative knee replacement patients aged 54–83 years (mean 66.5) walked at comfortable and fast speed across the device at the same time as the

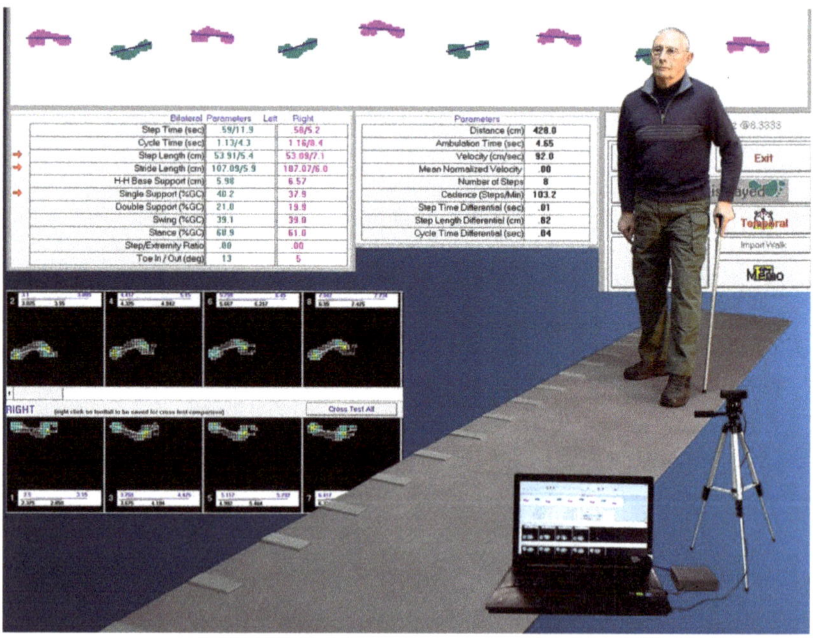

Figure 7.12 GAITRite instrumented walkway (CIR Systems Inc.)

optical system recorded the motion of their shoes. Walking speed, cadence, step length and step time variables, averaged across one walk for both systems, showed an excellent level of agreement with intra-class correlation coefficients (ICC) of between 0.92 and 0.99 and repeatability expressed as coefficients (RC) between 1.0% and 5.9% of mean values. Step length and step time variables recorded for each footstep also showed good agreement between the systems at both speeds (ICCs between 0.91 and 0.99; RC between 2.6% and 7.8%). Recorded frequency distributions demonstrated that individual step values between systems were within 1.5 cm and 0.02 s on the majority (80–94%) of occasions.

An alternate offering is the OptoGait (Microgate Corporation, Bolzano, Italy) which is a system of optical floor-mounted transmitting and receiving bars placed on either side of a walkway. Each transmission bar contains 96 infrared light emitting diodes (LED) communicating with the same number of detectors on the opposite receiving bar. Once positioned on the floor or a treadmill, the system detects the interruptions of the communication between the bars – caused by the patient's movement to calculate the duration and position. The system operates at a frequency of 1,000 Hz and with an accuracy of 1 cm, detecting the relevant spatial and temporal gait parameters, for walking, running or other measurement scenarios.

7.4.2 Pressure measurement systems (pedobarograph)

Used most often for biomechanical analysis of gait and posture, pedobarography is employed in a wide range of applications including sports and gait biomechanics. The spatial and temporal resolutions of the images generated by commercial systems range from approximately 3 mm to 10 mm and 25 Hz to 500 Hz, respectively. Such resolutions yield a contact area of approximately 500 sensors (based on a typical adult human foot with a surface area of approximately 100 cm^2) (Figure 7.13). Equipment is routinely used to monitor and quantitatively assess the progressive changes of foot deformity over time [50,51]. Commercial offerings include the emed® pedography system (Novel GmbH), the Foot Scan system (Tekscan Inc., Boston, MA, USA) and footscan® (RS Scan International, Beringen, Belgium).

7.4.3 GRF plates

The measurement of the kinetic characteristics of Ground Reaction Force during human walking is an important indicator of pathological gait, easily obtained during routine clinical gait analyses after complex surgery or in the treatment planning for hemiplegic, cerebral palsy, muscular dystrophy and stroke patients (Figures 7.14 and 7.15).

A force plate relies on the use of load cells to determine forces. The load cells may contain piezoelectric elements, strain gauges or beam load cells usually located in the corners of a machined flat metal plate of known mechanical characteristics. As downward force is applied to the surface of the plate, the sensors generate measurable voltage or electrical charge changes that are proportional to the applied force. Placing the sensors in different orientations enables the direction

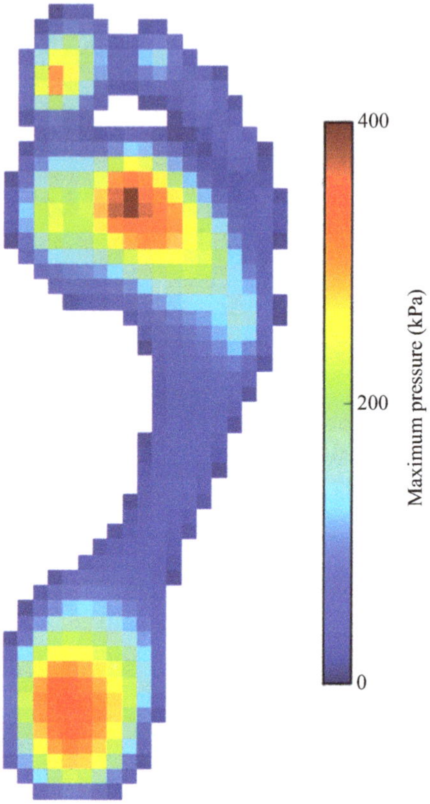

Figure 7.13 Pedobarograph

and magnitude of forces in three dimensions to be obtained. Other information that can be derived includes the 'Centre of Pressure' (CoP), the centre of force and the moment around each of the axes of the force vector. Commercially, manufacturers include Advanced Mechanical Technology, Inc. (Watertown, MA, USA), Kistler Group (Winterthur, Switzerland) and Bertec® (Columbus, OH, USA).

When standing, humans have a natural postural sway backwards and forwards (anterior and posterior) and from side to side (medial and lateral). As a subject sways, there are horizontal forces acting, as well as the vertical force. The CoP is the point at which the force is acting beneath the feet. During postural sway, the CoP moves forwards, backwards and side to side between the locations of the subject's feet.

Webster and Darter [52] describe how the GRF vector is initially located posterior to the ankle at initial contact, creating a plantarflexion moment. During the loading response, the plantarflexion moment continues at the ankle, and ankle dorsiflexors must resist this torque to prevent foot drop. During mid-stance, the GRF vector moves anteriorly, so a strong dorsiflexion moment at the ankle is

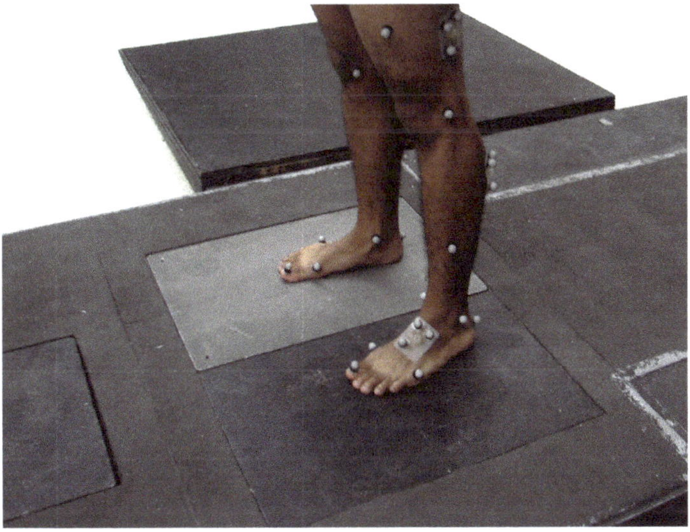

Figure 7.14 P GRF plates (Vicon Motion Systems Ltd)

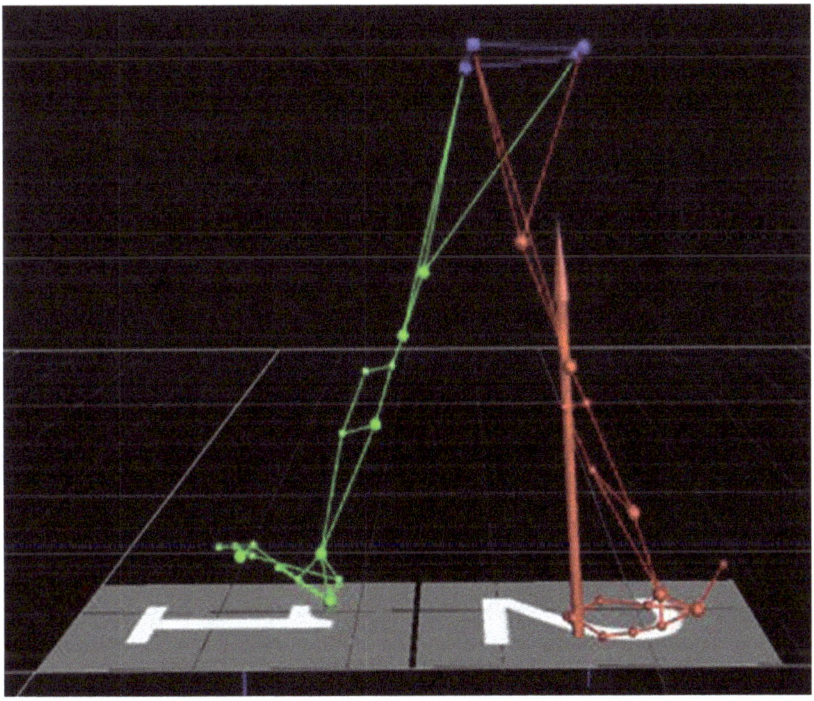

Figure 7.15 GRF (Vicon Motion Systems Ltd)

produced in a terminal stance. This dorsiflexion moment is opposed by the ankle plantar flexors to limit the forward progression of the tibia. In pre-swing, a dorsiflexion moment remains at the ankle. Thus, ankle motion is created by concentric contraction of the ankle plantar flexors, and this helps to propel the stance limb forward.

GRF of the foot in contact with a force plate [52] generates anteroposterior (F_X), vertical (F_Y) and medio-lateral (F_Z) components of the GRF at a single point in the single-support phase of the applied leg. When walking forward, the medio-lateral component due to little side-to-side movement of the body is smaller relative to the anteroposterior and vertical components of the GRF. The movement of the centre of gravity during the gait cycle is determined by the resultant force acting upon it. During a period of single-support, the resultant force acting on the centre of gravity is determined by body weight and the GRF exerted on the grounded foot. During a period of double-support, body weight and the GRFs exerted by both feet determine the resultant force acting on the centre of gravity. A force plate will present force–time curves to describe the GRF vectors exerted on the applied leg during a gait cycle.

7.5 Image processing motion capture

Optical motion capture uses bespoke cameras to reconstruct the body posture of a subject, patient or performer. The most common implementations employ multiple synchronised cameras in known locations and orientations to capture markers applied to anatomical landmarks on the body through palpation or when using close fitting body suits. Commercial offerings can use either passive retro-reflective or active light source marker solutions (Figure 7.16).

7.5.1 Passive optical motion capture systems

Passive optical motion capture systems use markers coated with a retro-reflective material to reflect light back to a strobed source surrounding a camera lens (Figures 7.17 and 7.18). In clinical applications, markers are applied directly to the skin by hypoallergenic tape through palpation to identify the location of bony landmarks. Motion capture cameras are shuttered to open only during a light strobe flash and contain filters optically matched to the light source. Only relatively bright markers will impinge upon the camera sensor, ignoring skin, fabric and other objects within the field of view. The centre of a marker is estimated as a position within the two-dimensional (2D) image of a camera frame by analysing the intensity of the light impinging upon each pixel and applying a circle of best-fit calculation (or if less than 3 pixels, a centroid). Before each capture session, a calibration object with markers attached at known positions is used to establish the scaled global coordinate system of the measurement volume and to determine the physical position and orientation of each camera. The optical distortion of the cameras (particularly from the lens at the corners) is also calculated and a correction is applied to improve overall system accuracy.

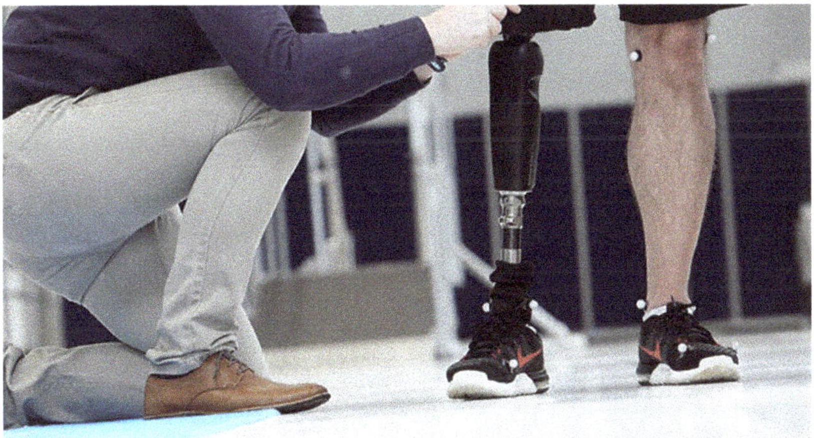

Figure 7.16 *Passive retroreflective markers (Vicon Motion Systems Ltd)*

Figure 7.17 *Cameras and strobes (Vicon Motion Systems Ltd)*

Providing a minimum of two calibrated cameras observe a marker so as to calculate a ray from the centre of each 2D image and normal to each sensor image plane, then the third dimension can be reconstructed using well-understood photogrammetric algorithms. The presence of a third or more cameras further improves the confidence in the reliability of the spatial and temporal marker location reconstructions within the defined measurement volume.

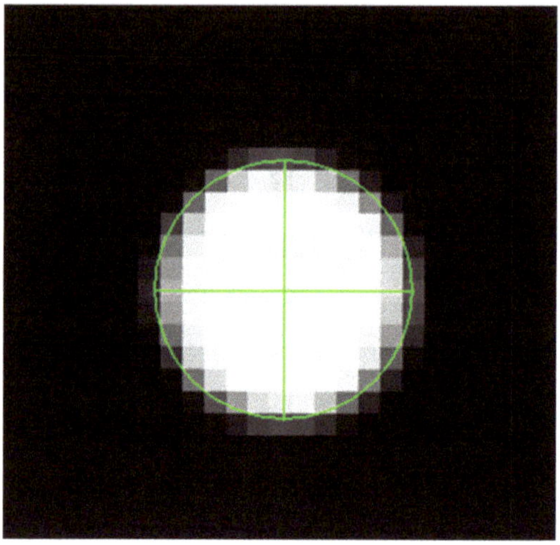

Figure 7.18 Marker centres (Vicon Motion Systems Ltd)

7.5.2 Retro-reflectivity

Passive optical motion capture systems remain wholly dependent on retro-reflectivity to identify the unambiguous location of anatomical landmarks in video camera images. The physical phenomenon is exhibited by a device or material that reflects radiation, usually light, directly back to its source with minimal scattering. Unlike a planar mirror placed exactly perpendicular to a radiation wave front with a zero angle of incidence, a retro-reflector will exhibit a wide angle of incidence or acceptance angle. The acceptance angle of a single retro-reflector or material supporting an array of retroreflectors is the incident angle at which a reflected ray is present. A common type of retroreflector consists of multiple miniature spherical, high refractive index optical elements arranged uniformly upon a flexible and reflective surface. One commercial example is 3M™ Scotchlite™ Sheeting # 7610 [53] that has been successful when used to coat passive marker shapes. The material has published retro-reflectivity characteristics 900 times greater than a white paper and still at 700 times at $\pm$ 45° so exhibiting a wide acceptance angle.

Clinical anatomical markers are conventionally hard spheres fitted to a flat base that can be affixed to a patient's skin using hypoallergenic double-sided tape. In animation applications, the markers are usually soft and hemispherical and affixed to tight-fitting suits using Velcro® materials. Either the markers, usually 14 mm in diameter, are coated in the retro-reflective sheeting or high refractive index elements are sprayed onto the outer surface under controlled conditions to ensure an even coating. Importantly, the wide acceptance angle means that the diameter of the reflected circle is consistently the diameter of the marker sphere, a factor that is very helpful in achieving acceptable three-dimensional system reconstruction accuracy.

When illuminated, the markers appear as bright circular objects from any 2D perspective such as that viewed by a camera. Performance of the retro-reflective elements is defined by the equations:

$$\sin\alpha = n\sin\beta$$

$$\gamma = 4\sin^{-1}\left(\frac{r'}{r}\frac{1}{n}\right) - 2\sin^{-1}\left(\frac{r'}{r}\right)$$

where n is the refractive index of the material.

7.5.3 Confirmation of retro-reflective marker performance

An experiment was developed by the author to measure the relative intensity of reflected rays from 3M #7610 microsphere array sheeting marketed as Scotchlite™ at different angles to an incident ray. In the experiment, a 630 nm Class II laser was used as the incident ray source. The ray was passed through a semi-silvered mirror delivering 50% intensity to the surface of the material under test. The material was rotated between $+50°$ and $-50°$ at $10°$ increments. The reflected ray was passed through a 1 mm diameter attenuation aperture and 630 nm optical notch filter before illuminating a photocell. The output of the photocell was in volts, therefore, providing a relative performance value for the material at each angle.

The material exhibited a relatively flat response over the range $\pm$ 30° that aligns well with the published specifications [53]. The author concluded that the slight increase in performance at the greater the angle from normal is likely to be an experimental artefact. The incident laser beam is of a finite diameter when striking the material suggesting that more microspheres are illuminated as the angle of the test sample increases so contributing to the summed intensity of the reflected rays.

In a second experiment, a Vicon VANTAGE V16 16 MPixel camera (Vicon Motion Systems Ltd, Oxford, UK) was placed normal to 14 mm spheres (Vicon) and Pearl markers (B&L Engineering, Santa Ana, CA, USA) coated with an array of high refractive index microspheres at a 5 m distance, lens set at infinity, aperture adjusted to generate a grayscale video image. Performance was described in terms of the histogram of image intensities (0–255, where 0 = black and 255 = white) obtained within a cropped scene. An annulus light source surrounding the camera lens set at f-stop (F8) illuminated the scene to provide a grayscale return image for comparison.

The average number of pixels/intensity for the VICON marker was 188 ($\pm$ 409 1σ) and for the B&L Pearl 180 ($\pm$ 418 1σ). The former offering demonstrated the presence of higher intensity returns above grayscale level 185 but a smaller standard deviation implying the material has a potentially narrower acceptance angle compared to the B&L offering. Mean intensities were similar across the intensity range (note Black = 0 was excluded from the calculation). Both markers are suitable for use with passive motion capture systems for most clinical applications but the offering using the #7610 had better reflectivity performance for those experiments requiring detections at an extra distance for a given maker size.

7.5.4 Motion capture cameras

In order to capture human motion during particular fast transitions such as heel strike and small body segments such as feet in all environments demands specific camera performance requirements. For maximum usefulness, cameras must operate at greater than 100 frames per second, with pixel resolutions greater than at least one MPixel with a high complementary metal oxide semiconductor (CMOS) sensor quantum efficiency peaking within or near the light spectrum of the stroboscopic light source.

7.5.5 Illuminating markers and removing background from images

Most commercial offerings use arrays of LED in an annulus shape designed to generate as much light as close to the centre of a camera lens for optimum marker reflections. Motion capture cameras are designed to support whole image frame shuttering such that an instant in time can be acquired with minimal chance of image smear particularly when attempting to capture fast marker motions. The light output from LED arrays can be multiplied past their maximum permissible continuous ratings if strobed. If the light flash can be timed to coincide with the camera shutter opening, the resultant reflected marker images do appear circular.

The signal-to-noise ratio of the camera image can be further improved if a narrow-bandpass optical filter closely matching the emission spectrum of the strobe light source is placed between the lens and the imager.

7.5.6 Detecting marker images in two-dimensions

The combination of an optimum sensor quantum efficiency, strobing, shuttering and optical filtering results in the markers appearing as bright circles on a black background when viewed in a video image (Figures 7.19–7.22). The pixels whose location on the sensor is known in two dimensions generate a voltage directly proportional to the intensity of the light impinging upon them. All motion capture systems use this characteristic of CMOS and charge-coupled device (CCD) sensors to translate the analogue image into a digital format where each pixel can be described of the form: *f* (event time, horizontal location, vertical location, edge).

Event time can be recorded at the instant the camera is stimulated to open the shutter and illuminate the strobe. The next event will be at the inverse of the selected frame rate during a capture session (e.g. at a frame rate of 120 Hz the next temporal measurement will be ~ 8.3 ms later).

Early motion capture system designs used simple electronic 'comparator' circuits to identify the presence of a marker image impinging onto the imager. The approach, called binary slicing, used a manual threshold voltage comparison where any pixel value above a pre-determined level generated a digital '1' and any below a '0'. A second feature of the early designs was identifying the transition from '0' to '1', described as a 'leading edge' and conversely from '1' to '0' being a falling edge. At this point, the function f (frame number, horizontal pixel location, vertical pixel location, Leading Edge/Falling Edge flag) was stored by the system digitally.

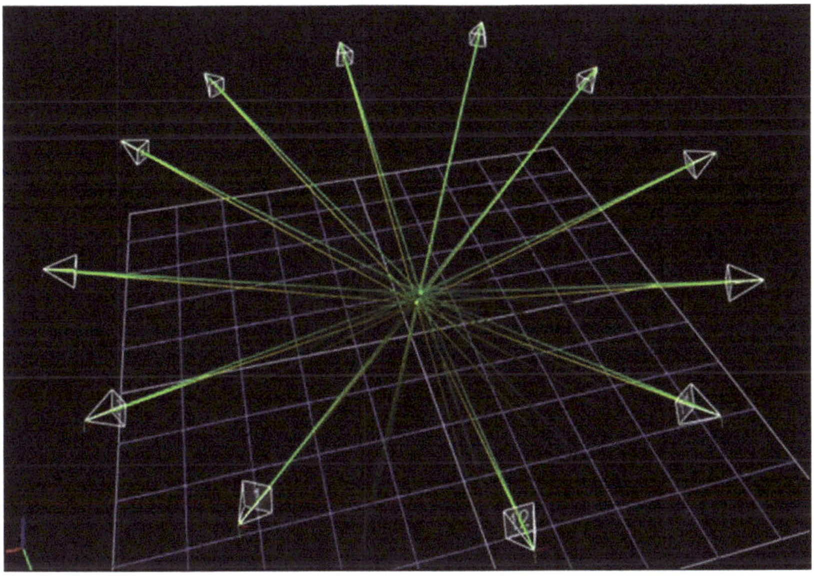

Figure 7.19 Detecting markers in three dimensions using photogrammetry (Vicon Motion Systems Ltd)

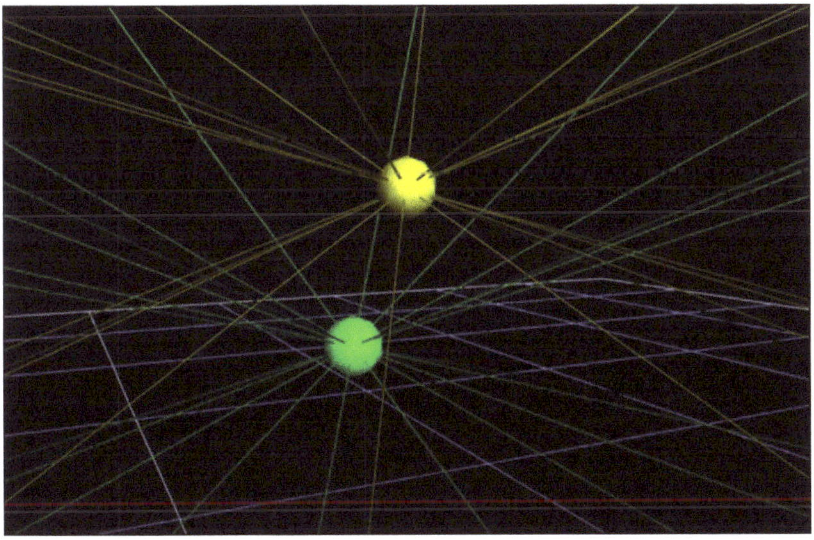

Figure 7.20 Identifying marker locations in a volume (Vicon Motion Systems Ltd)

In such systems, post-processing within a host computer was applied to each digitised 2D array of values to calculate the centre of a 'circle of best fit' for each marker represented by groups of leading and falling edge points. There are a

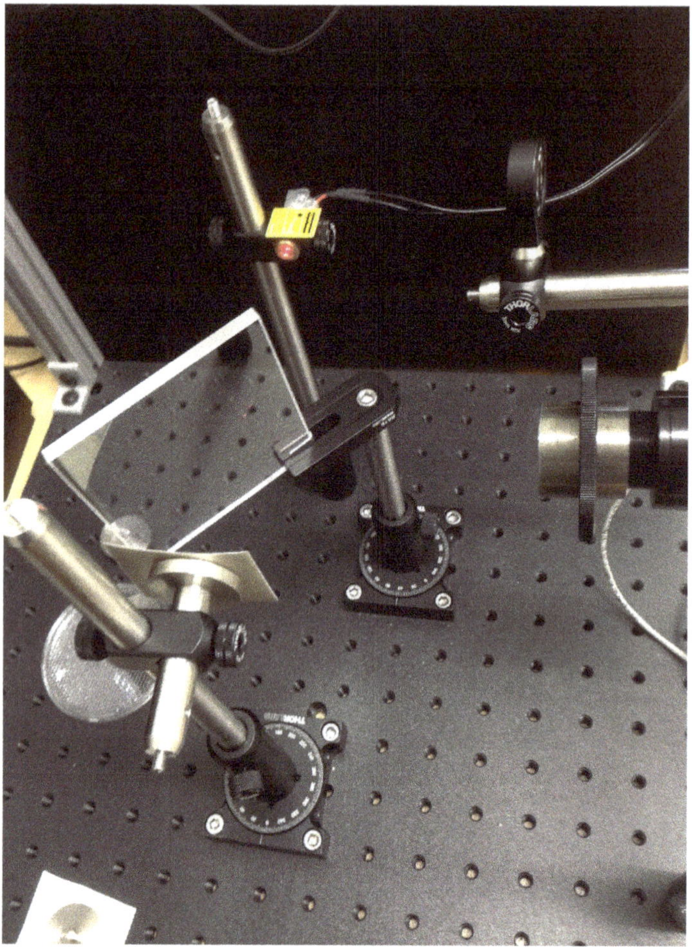

Figure 7.21 Marker material performance measurement (Vicon Motion Systems Ltd)

number of suitable algorithms with the most common being the least squares method:

$$\text{Sum of square of errors}(x_c, y_c) = \sum_{i=1}^{n} \left(r - \sqrt{(x_i - x_c)^2 + (y - y_c)^2} \right)^2$$

$$r = \frac{1}{n} \sum_{i=1}^{n} \sqrt{(x_i - x_c)^2 + (y - y_c)^2}$$

where r = *radius*, x_c, y_c = *centre of circle estimate*.

In those cases where the marker is either small or far away such as when described by less than three pixels, the centre was estimated by using a simple centroid calculation.

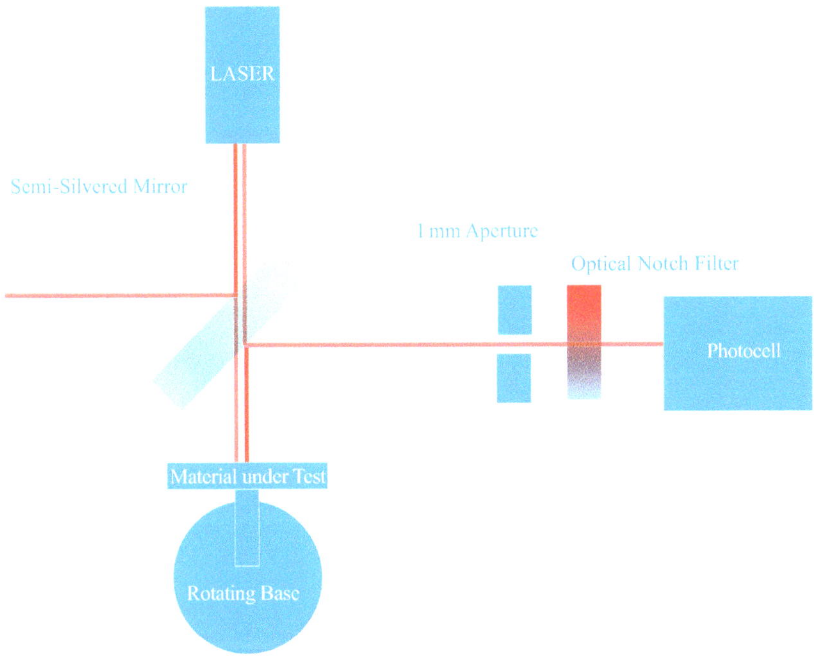

Figure 7.22 Material marker performance apparatus (Vicon Motion Systems Ltd)

A disadvantage of the binary slice method is that when markers are physically close together such as on the foot, it is possible that they will be merged in the detection. A solution employed in later systems is to consider the acceptance angle characteristics of the retro-reflective material and the inherent optical and electronic filtering upon the theoretical 'impulse function' of the marker image that in actuality is a Gaussian function of the form:

$$G(u, v) = \frac{1}{2\pi\sigma^2} e^{-\frac{u^2+v^2}{2\sigma^2}}$$

where σ defines the shape of the object and is analagous to standard deviation in a planar bell curve.

Improved close marker discrimination can be achieved by including Gaussian greyscale values in the calculation of marker centres and radii. Since multiple pixel samples are used to calculate centres and radius, the resolution of these measures will achieve sub-pixel resolution performance (Figures 7.23 and 7.24).

7.5.7 Compensating for lens non-linearity

All camera lenses exhibit distortion that is a physical phenomenon which can be modelled as a radial distortion, the most prominent component, and a second, with a lesser effect, an offset distortion that has both radial and tangential components. All motion capture systems must correct for these distortions either as a separate

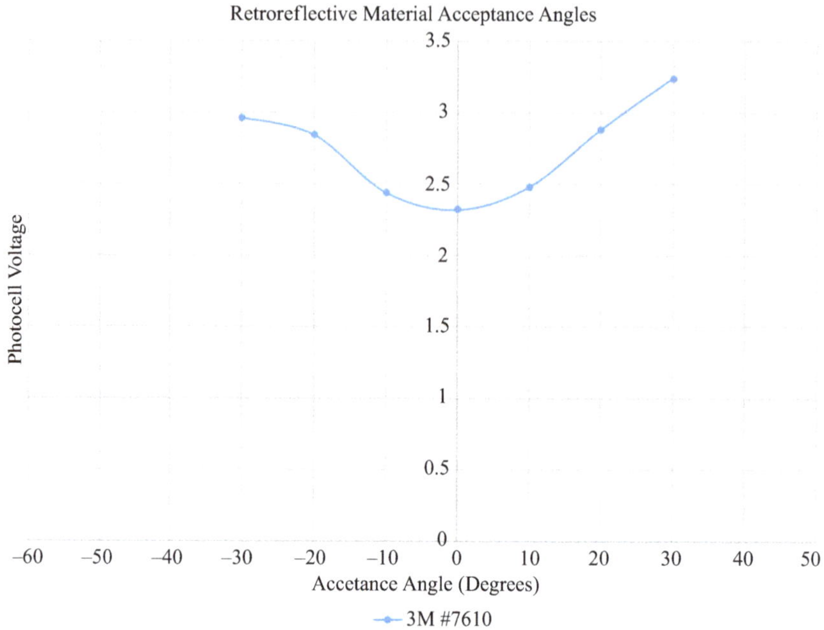

Figure 7.23 Retroreflective material acceptance angles (Vicon Motion Systems Ltd)

Figure 7.24 Sensors (Vicon Motion Systems Ltd)

step or during a calibration procedure to ensure there is no degradation in performance correlated to the location of any marker reflections upon a camera image sensor. The shapes of the lens optical components cause radial distortion with tangential distortion being caused by centralising errors and non-orthogonality of the components with respect to the optical axis.

7.5.8 Three-dimensional calibration

All motion capture systems have to undertake a calibration phase to resolve the marker motions into the third dimension within a pre-defined measurement volume. Early approaches employed markers in known positions to derive real-world volume coordinates and camera view scaling using the 'Direct Linear Transformation' method. A disadvantage was that the method to measure actual locations by independent means was very time-consuming thereby limiting the usefulness of motion capture for everyday clinical gait analysis and character animation.

Contemporary methods employ a two-stage dynamic calibration approach where an object of known dimensions described as a 'wand' is placed in the centre of a measurement volume or on the edge of a force plate in a specific orientation to set the world coordinates (e.g. x, y, z). In the second step, the relative location and orientation of cameras and lens non-linearity are identified and minimised by waving the object in the volume until the system reports an acceptable calibration has been achieved.

Following calibration, the photogrammetric reconstruction of the marker centres during static and dynamic motion as long as two or more cameras view the motion will meet a manufacturer's specifications.

7.5.9 Marker labelling and trajectory management

Markers within a measurement volume can only be identified if the user or an automated model has some insight into their relative location with respect to others placed on a patient, subject or actor. The usefulness of motion capture systems is also dependent on how quickly markers and their trajectories can be identified and de-conflicted during a capture trial. One method is to use Kalman filtering [54] to review a number of previous and labelled marker locations including their velocity and direction to estimate a box where a marker is likely to be located in a subsequent frame. The assumption can be made that for most biomechanical measurement scenarios, co-incident markers are each less likely to change direction suddenly within a single temporal and spatial slot, which assists in de-conflicting cases where two trajectories are crossing. Once such decisions have been taken, each applicable marker can be labelled and added to the trajectory solutions (Figure 7.25).

7.5.10 Performance of motion capture systems

'Accuracy' is defined as how close measurements are to a known value; while 'precision' refers to how close measurements are to each other. In other words, accuracy describes the difference between the measurement and actual value, while precision describes variations observed. Precision can be further described in terms of:

- *Repeatability*: The variation observed when the same system measures the same parameter repeatedly.
- *Reproducibility*: The variation observed when different systems measure the same parameter repeatedly.

7.5.11 Independent protocol for quantifying the accuracy of motion analysis systems

In 1999, the North American-based Gait and Clinical Motion Analysis Society (GCMAS) Standards Council proposed a protocol for quantifying the accuracy of a

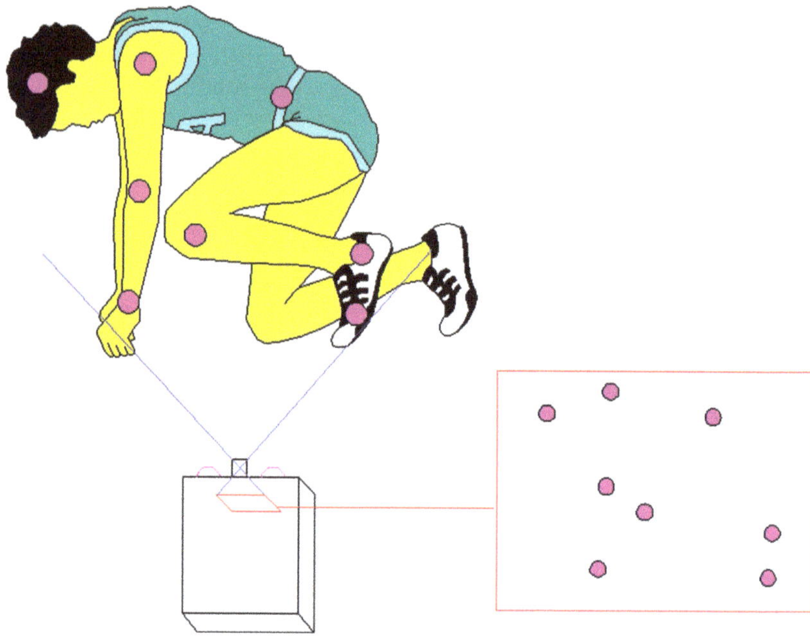

Figure 7.25 View of anatomical landmarks (Vicon Motion Systems Ltd)

motion analysis system (i.e. how accurate the system is at locating markers in three-dimensional space). The resulting Standard Assessment of Motion System Accuracy protocol (SAMSA) was intended to test video-based systems that employ passive markers [55]. The device consists of a beam fitted with markers and rotated at 60 RPM by a motor. The protocol is designed to test the ability of a motion system to (1) track moving markers; (2) resolve markers using a subset of the cameras; and (3) resolve markers that pass close to one another during a trial. The protocol has been tested in the laboratories of seven standard council members and the results have been used to formulate an accuracy statement in terms of acceptable error thresholds. This testing and the error thresholds were presented at the 2007 GCMAS meeting in Springfield [56] and found to be between 0.21 mm and 3.50 mm mean; 0.08 mm and 3.86 mm SD, independent of equipment supplier. Merriaux *et al.* [57] used a very similar apparatus to study the positioning performance of an 8 camera Vicon 'T40 Series' system (Vicon Motion Systems Ltd, Oxford, UK). Their protocol included evaluations of static and dynamic performances. Mean error as well as positioning variabilities were studied with calibrated ground truth equipment that was not based on other motion capture modalities. They introduced a new protocol that enabled direct estimation of the absolute positioning accuracy for dynamic experiments contrary to state-of-the-art experiments that relied on inter-marker distances. The test system performed as expected on static experiments with a mean absolute error of 0.15 mm and a variability lower

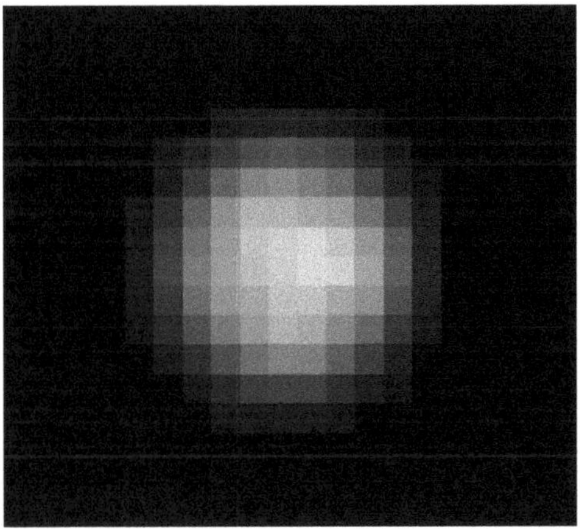

Figure 7.26 2D marker image on a sensor (Vicon Motion Systems Ltd)

than 0.025 mm. The dynamic experiments were undertaken at marker speeds likely to be found in real applications. The authors reported the following for eight dynamic experiments: cumulative distribution function errors of 0.39–0.47 mm at 1σ or 68.27% to 0.8–1.2 mm at 3σ or 99.73% at < 1 ms^{-1} to >3 ms^{-1} marker speeds (Figure 7.26).

The ASTM International, formerly known as the American Society for Testing and Materials has recently published a standardised test method (ASTM Standard E304-16 'Test Method for Evaluating the Performance of Optical Tracking Systems that Measure Six Degrees of Freedom (6DOF) Pose'). The standard defines metrics and procedures for measuring, analysing and reporting the relative pose error of optical tracking systems that compute the pose (i.e. position and orientation) of a rigid object while the object is moving. Following the standard protocol, internal indicative experimental results using two different sensor resolution, 26 camera Vicon Vantage based systems (V16 and V5 models) over a two day period in a 9 m $\times$ 5 m $\times$ 1.5 m height measurement volume (Vicon Motion Systems Ltd, Oxford, UK) was undertaken. Twenty sets of data were captured over four calibrations, over 2 days, to establish dynamic distance errors values using a NIST E3064 measurement object comprising two clusters of passive markers forming a rigid body at each end. This object was moved around the volume in a defined way in compliance with the standard, and the two clusters were tracked independently by the system. The distance between the two tracked clusters was analysed and compared to the published reference value. The outcomes using this method in the representative test volume had a mean measurement within -0.017 mm (ideal volume) of the reference value, with a mean root mean squared error of 0.2 mm, respectively. These findings were consistent with those independently reported by Bostelman *et al.* [58] so giving confidence that the standard was generically useful in establishing dynamic three-dimensional point reconstruction performance of motion capture systems.

7.5.12 Biomechanical modelling

In the clinical gait analysis environment, trajectory solutions are applied to a conventional gait model [59] or in the entertainment sector, to a character animation model for further processing. One practical limitation of marker-based motion capture systems for use in clinical gait analysis is that the technology is attempting to derive skeletal joint locations from palpated surface anatomical landmarks which as an estimation does introduce errors so elegantly described by Dr David Sutherland as attempting '...to measure the movements of a broomstick within a gunny sack?' [60].

A second limitation is that during dynamic motion, the location of some markers at the joints can move described as 'the soft tissue artefact'. The artefact arises from movement or deformation of the subcutaneous tissues associated with muscular contractions, skin movement and inertial effects. The extent of the artefact for any movement depends upon the physical characteristics of individuals, marker locations and the nature of the movement task being performed. Charlton *et al.* [61] estimated that when developing human lower limb biomechanical models, marker location covariance was estimated to be 10 mm^2 in all co-ordinate directions (Figures 7.27 and 7.28).

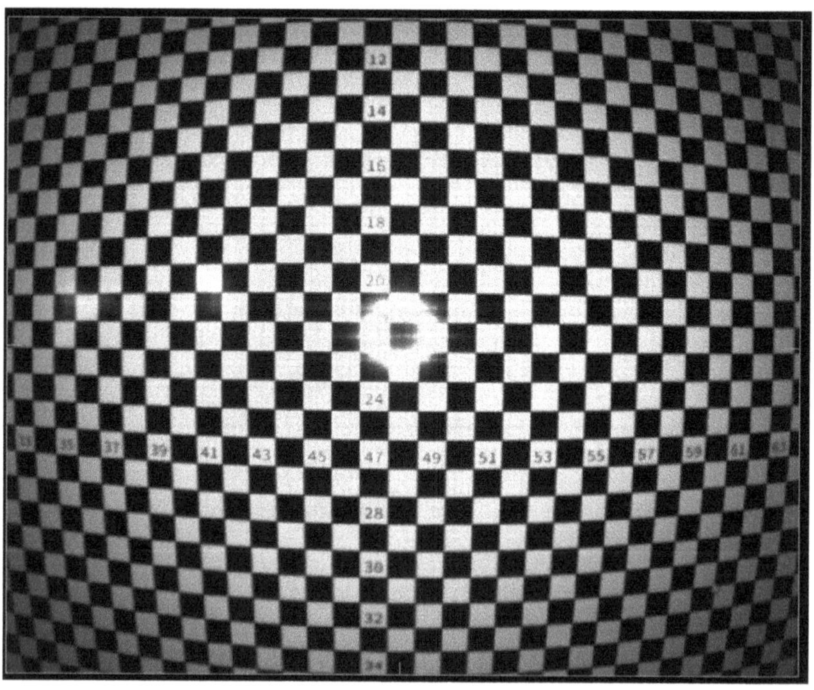

Figure 7.27 Compensating for lens non-linearity (Vicon Motion Systems Ltd)

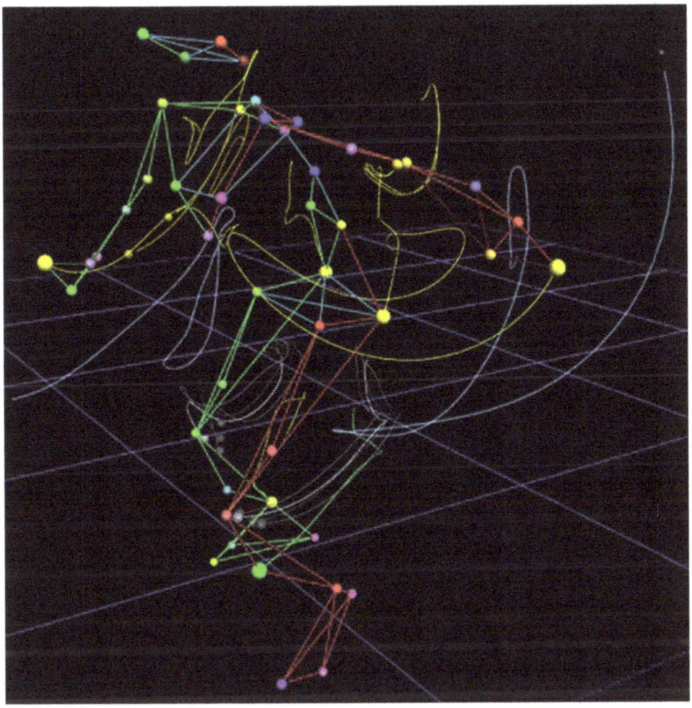

Figure 7.28 Trajectories (Vicon Motion Systems Ltd)

Peters *et al.* [62] undertook a systematic review to evaluate the quantification of the artefact in lower limb human motion analysis. The research had a specific focus on assessing the quality of previous studies and comparing quantitative results. The published search strategy identified 20 public domain articles or abstracts that fulfilled the selection criteria. The quality of the articles was evaluated using a customised critical appraisal tool with data extraction tools employed to identify key aspects reported in the articles. Most studies had small sample sizes of mostly young, slim participants. Eleven of the reviewed articles used physically invasive techniques to assess the error. The artefact was found to reach magnitudes of greater than 30 mm on the thigh segment, and up to 15 mm on the tibia. The range of soft tissue artefact reached greater than 25 mm in some cases when comparing the results of reviewed studies.

Many approaches have been proposed to reduce the effects of this error in the measurements, but most share a common technique of 'fitting' a model of the skin-mounted markers to the data based on a simplified model of the artefact. In addition, towards the end of the 1990s, researchers proposed moving away from predictive equations for locating joint centres, as used in the Helen Hayes Model [55], to focus on estimating joint centres and axes by functionally exercising the joints and then fitting the data to an idealised joint model [63]. Many research and gait laboratories have since proposed techniques of a broadly similar nature. The combination of this 'optimised model calibration' with some form of soft tissue artefact compensation is the basis of work to be described in a later chapter on future developments.

Marker-based systems can be prone to lacking sensitivity in quantifying axial rotation in some joints such as the thigh [64] which has been overcome by placing markers on short rods to use the resulting moment arm to amplify the rotation or through the use of additional markers or clusters (Figures 7.29–7.33).

Passive marker-based motion capture system manufacturers include BTS (BTS S.p.A. Milan, Italy), Qualysis (Qualisys AB, Göteborg, Sweden), Motion Analysis (Motion Analysis Corp., Rohnert Park, CA, USA), OptiTrak (NaturalPoint Inc., Corvallis, OR, USA), Vicon (Vicon Motion Systems Ltd, Oxford, UK).

Figure 7.29 SAMSA device [55,56]

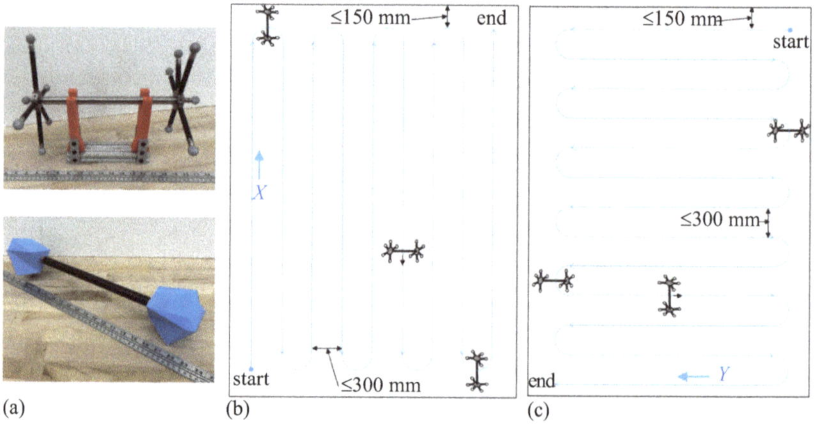

Figure 7.30 ASTM E304-16

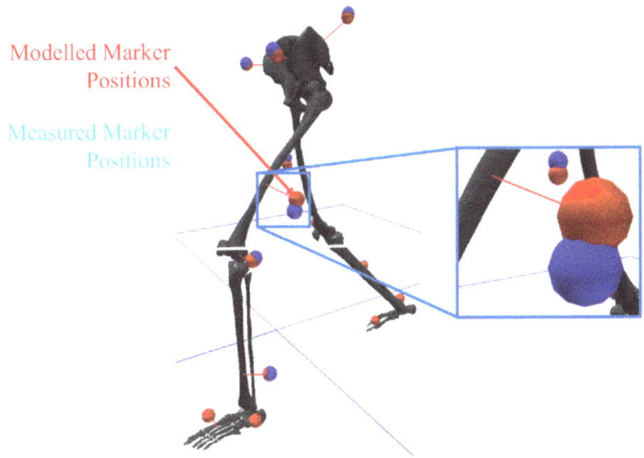

Figure 7.31 Soft tissue artefact (Vicon Motion Systems Ltd)

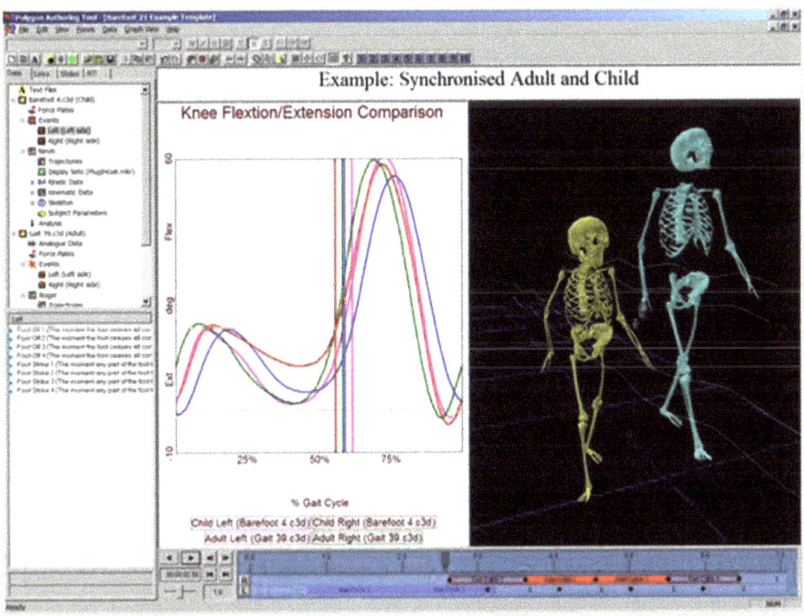

Figure 7.32 Gait analysis reporting – gait cycle visualisation (Vicon Motions Systems Ltd)

7.5.13 Active marker motion capture systems

The advent of efficient high-intensity LED emitters with small, low cost, high power density, lithium-ion batteries and electronic device miniaturisation has resulted in the re-emergence of active marker-based motion capture systems, both

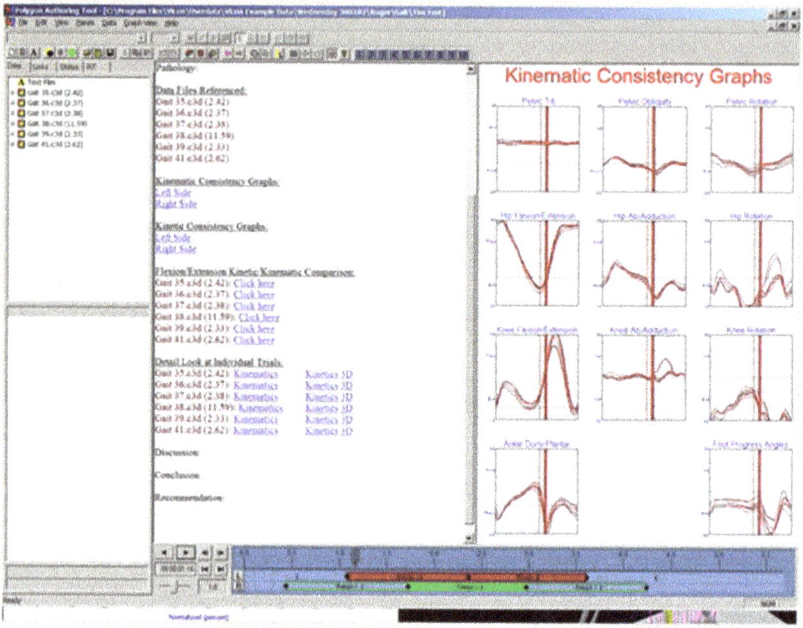

Figure 7.33 Gait analysis reporting – kinematic consistency graphs (Vicon Motions Systems Ltd)

wired and wireless based. The approach is particularly useful when measurement objects are highly reflective or for those environments with high ambient lighting of a level that might interfere with a passive system. Examples include the Qualysis active short- and long-range markers (Qualisys AB, Göteborg, Sweden), Simi Aktisys (Simi Reality Motion Systems GmbH, Unterschleissheim Germany) and Origins (Vicon Motion Systems Ltd, Oxford, UK).

7.6 Energy expenditure and oxygen consumption

The objective of many treatment decisions is to improve, through intervention, the biomechanical efficiency of a patient's gait to ensure that their energy expenditure remains below their available resource and capabilities [65]. Steele *et al.* [66] reported that following their research they concluded that children diagnosed with cerebral palsy (CP) expend more energy to walk compared to typically developing peers (TDP). They evaluated oxygen consumption and gait kinematics for 573 children with bilateral CP. The researchers found that the average net non-dimensional oxygen consumption during gait of the children with CP was 2.9 times that of speed-matched TDP.

Estimation of energy expenditure through calculations of mechanical work and measurements of oxygen consumption and metabolic cost, pre- and post-treatment offer additional insight into whether the patient's walking has been made easier or not with intervention [67]. One common approach to assess expenditure was reported by Ryan *et al.* [68]. The researchers captured resting and activity oxygen consumption, carbon dioxide concentration and turbine flow using a Cosmed K5 (Cosmed Srl, Rome, Italy) portable metabolic system. The protocol defined a 10-min rest period in a semi-recumbent position followed by 6 min of ground walking at a self-selected speed measured using gait analysis.

7.7 Clinical gait analysis, review and reporting

The goal of clinical gait analysis is to assist in treatment decision-making for a patient with complex and not easily understood walking problems. Quantitative analysis is a process whereby gait characteristics are measured, abnormalities are identified, causes are postulated and treatments are proposed [69]. It is important to appreciate that while quantitative gait analysis does use technology to measure gait patterns, it does not replace the human observer (i.e. the observational gait analysis approach that a clinician might employ in the clinic or office setting). Rather, the analysis serves as an adjunct to aid in understanding – more precisely and at times more accurately to the visual impressions of a patient's gait impairment. Through the thoughtful use of technology, quantitative gait analysis provides an opportunity to appreciate the details of complex movement patterns that may involve a number of lower (and upper) extremity joints and segments simultaneously and include movements that occur in several planes of motion concurrently, as well as to understand and correlate the associated kinetics, muscle activity and energy expenditure.

It is also important to understand that while treatment decision-making is assisted by the analysis, they must always be made in the context of the clinician's experience in the overall management of the particular disorder presented by the patient. Although patients seen most often for gait analysis present with complex and sometimes subtle gait deviations that may suggest surgical intervention if necessary, other recommendations for management may include physiotherapy exercises, bracing (orthoses), spasmolytic medications (such as Botulinum toxin, baclofen) or condition monitoring.

A typical gait analysis session can take from 1 h to 2 h, depending on the level of evaluation performed and on the cooperation, behaviour and gait complexity (e.g. involvement) of the patient. Usually, a clinical scientist, physiotherapist or kinesiologist works directly with the patient and a more technically oriented person such as an engineer or technician manages the operation of the measurement system during the test. Dependent on laboratory capabilities, the process can yield the following:

- video recordings,
- clinical measures,
- stride and temporal gait data, such as step length, cadence and walking speed,
- three-dimensional joint and segment motion plots (kinematics),
- three-dimensional joint torque or moment and power (kinetics),

- electromyographic tracings, and
- measurement of metabolic energy expenditure.

These parameters can then be evaluated to identify abnormalities by comparing outcomes against a database of normal typically developed subjects and a local knowledge of normal gait biomechanics as a baseline. Deviations from normal are interpreted in the context of their relative impact on gait function. The multiple sources of data do provide useful redundancy, allowing corroborating information to be identified and conflicting observations to be understood. At major clinical laboratories, gait data are interpreted by a team that consists of the orthopaedic surgeon to whom the patient was referred and the clinical scientist, physiotherapist and/or kinesiologist who collected data. At times, the engineer or technician, who assisted in the data collection, or the biomechanist who developed the mathematical models used to process the data, is involved if there are questions of data quality or if some previously unseen walking mechanism is observed. Most laboratories will use multi-centre internationally validated conventional gait models for routine clinical analyses [70]. In general, it is important that the team have at least a rudimentary understanding of the gait model used to produce the results, in addition to a well-developed understanding of normal gait patterns. This knowledge base, underpinned by experience gained from the examination of many pre- and post-treatment cases, is essential to produce a proper interpretation and treatment decision. The use of validated models helps in the exchange of experience through multi-centre collaboration by using a common language and methodology.

7.8 Systematic review of the efficacy of clinical gait analysis

Wren *et al.* [71] published a systematic review in 2011 that evaluated and summarised the current evidence base related to the clinical efficacy of gait analysis. They conducted a literature review to identify references related to human gait analysis published between January 2000 and September 2009 plus relevant older references. Four reviewers using a hierarchical model of efficacy adapted for gait analysis assessed the references independently, and final scores were agreed upon by at least three of the four reviewers. One thousand five hundred and twenty-eight references were identified relating to human instrumented gait analysis. Of these, 116 original articles addressed technical accuracy efficacy, 89 addressed diagnostic accuracy efficacy, 11 addressed diagnostic thinking and treatment efficacy, 7 addressed patient outcomes efficacy and 1 addressed societal efficacy, with some of the articles addressing multiple levels of efficacy. This body of literature provides strong evidence for the technical, diagnostic accuracy, diagnostic thinking and treatment efficacy of gait analysis. The existing evidence also indicates efficacy at the higher levels of patient outcomes and societal cost-effectiveness, but this evidence is sparse and does not include any randomised controlled trials. The researcher's conclusion was that the current evidence supports the clinical efficacy of gait analysis, particularly at the lower levels, but recommended a need for additional research to strengthen the evidence base at the higher levels. An updated systematic review was published in 2020 by Wren *et al.* [72] covering the period 2009–2019 on the clinical efficacy of three-dimensional instrumented gait analysis (3DGA). The researchers identified English language articles published from September 2009 to October 2019 reporting primary research that used typical motion analysis laboratory methods to study human walking. Five gait laboratory experts classified articles according to the highest type of efficacy they addressed: type 1 (technical), 2 (diagnostic accuracy), 2b (outcome prediction), 3–4 (diagnostic thinking and treatment), 5 (patient outcome), 6 (societal). Articles classified into type 3–4 and higher were rated for quality using LEGEND. Of 2712 articles related to the efficacy of 3DGA, over 99% addressed technical ($n = 313$), diagnostic ($n = 1{,}466$) or outcome prediction ($n = 927$) efficacy. Six type 3–4 studies showed that 3DGA changes treatment plans, increases clinicians' confidence in their treatment decisions and increases agreement among clinicians. Two type 5 articles based on a randomised controlled trial demonstrated that patient outcomes improved only when 3DGA data were available and its recommendations were followed. A population-based type 5 study found that the incidence of severe crouch gait dropped from 25% to 4% following practice changes including the addition of 3DGA. The strength of evidence was mainly 3b (lesser quality prospective cohort studies) but also included stronger studies (three level 2 controlled clinical trials). Literature related to the clinical efficacy of 3DGA has grown substantially over the last decade. Thousands of articles contribute to the continued improvement of data collection and interpretation, as well as an understanding of gait pathology and treatment. A smaller number of studies clearly demonstrate the efficacy of 3DGA in changing and reinforcing treatment decisions, increasing clinicians' confidence in treatment planning and increasing agreement among clinicians, as well as the potential to improve patient outcomes. Although the focus of the research lay within the United States of America gait analysis community, the clinical data research outcomes are directly transferrable to the European population as the intended purpose and common equipment from all manufacturers are employed.

References

[1] Canemaker J. (1987), *Winsor McCay – His Life and Art*. New York, NY: Abbeville Press.

[2] Igarashi T., Moscovich T. & Hughes J.F. (2005), Spatial Keyframing for Performance-driven Animation. In *ACM SIGGRAPH/Eurographics Symposium on Computer Animation*.

[3] Alexander G. E. (2004), Biology of Parkinson's disease: pathogenesis and pathophysiology of a multisystem neurodegenerative disorder. *Dialogues Clin. Neurosci.*, 6(3): 259–280.

[4] Hausdorff J.M. (2009), Gait dynamics in Parkinson's disease: common and distinct behaviour among stride length, gait variability, and fractal-like scaling. *Chaos,* 19(2): 026113.

[5] https://www.epda.eu.com/about-parkinsons/symptoms/motor-symptoms/gait/. Accessed 25 February 2020.
[6] Rogers G., Davies D., Pink J. & Cooper P. (2017), Parkinson's disease: summary of updated NICE guidance. *BMJ*, 358: j1951.
[7] Kalia L.V. & Lang A.E. (2015), Parkinson's disease. *Lancet*, 386(9996): 896–912. Epub Apr 19.
[8] Muro-de-la-Herran A., Garcia-Zapirain B. & Mendez-Zorrilla A. (2014), Gait analysis methods: an overview of wearable and non-wearable systems, highlighting clinical applications. *Sensors (Basel)*, 14(2): 3362–3394. Published 2014 Feb19. doi:10.3390/s140203362.
[9] De Stefano A., Burridge J.H., Yule V.T. & Allen R. (2004), Effect of gait cycle selection on EMG analysis during walking in adults and children with gait pathology. *Gait Posture*, 20(1): 92–101.
[10] Lamoureux L. (1971), Kinematic measurements in the study of human walking. *Bull. Prosthet. Res.*, 10(15): 3–8.
[11] Chao E.Y.S. (1980), Justification of triaxial goniometer for the measurement of joint rotation. *J. Biomech.*, 13: 989–1006.
[12] http://www.biometricsltd.com/goniometer.htm. Accessed 26 February 2020.
[13] https://www.tekscan.com/products-solutions/embedded-force-sensors. Accessed 26 February 2020.
[14] Gurchiek R.D., McGinnis R.S., Needle A.R., McBride J.M., & van Werkhoven H. (2017), The use of a single inertial sensor to estimate 3-dimensional ground reaction force during accelerative running tasks. *J. Biomech.*, 61: 263–268.
[15] Morris J. (1973), Accelerometry – a technique for the measurement of human body movements, *J. Biomech.*, 6: 729–736.
[16] Bonato P. (2003), Wearable sensors/systems and their impact on biomedical engineering. *IEEE Eng. Med. Biol. Mag.*, 22(3): 18–20.
[17] Merriaux P., Dupuis Y., Boutteau R., Vasseur P. & Savatier X. (2017), A study of Vicon System Positioning performance. *Sensors (Basel)*, 17(7): 1591.
[18] Roetenberg D., Luinge H. & Slycke P. (2013), Xsens MVN: Full 6DOF Human Motion Tracking Using Miniature Inertial Sensors, XSENS TECHNOLOGIES – Version April 3.
[19] Lu H., Huang H, Lee P. & Lu T. (2017), Control of the body's centre of mass motion during over-ground and treadmill walking in the pelvic reference frame: implications for wearable sensing, *J. Biomech Sci. Eng*, 12(2): 17-00074.
[20] Tunca C., Pehlivan N., Ak N., Arnrich B., Salur G. & Ersoy C. (2017), Inertial sensor-based robust gait analysis in non-hospital settings for neurological disorders. *Sensors (Basel)*, 17(4): 825.
[21] Iosa M., Picerno P., Paolucci S. & Morone G. (2016), Wearable inertial sensors for human movement analysis. *Expert Rev. Med. Dev.*, 13(7): 641–659.

[22] Storm F.A., Buckley C.J. & Mazzà C. (2016), Gait event detection in laboratory and real life settings: accuracy of ankle and waist sensor based methods. *Gait Posture*, 50: 42–46.

[23] Trojaniello D., Ravaschio A., Hausdorff J.M. & Cereatti A. (2015), Comparative assessment of different methods for the estimation of gait temporal parameters using a single inertial sensor: application to elderly, post-stroke, Parkinson's disease and Huntington's disease subjects. *Gait Posture*, 42(3): 310–316.

[24] Bugané F., Benedetti M.G., Casadio G., *et al.* (2012), Estimation of spatial-temporal gait parameters in level walking based on a single accelerometer: validation on normal subjects by standard gait analysis. *Comput. Methods Prog. Biomed.*, 108(1): 129–137.

[25] Trojaniello D., Cereatti A. & Croce U.G. (2014), Accuracy, sensitivity and robustness of five different methods for the estimation of gait temporal parameters using a single inertial sensor mounted on the lower trunk. *Gait Posture*, 40(4): 487–492.

[26] http://www.btsbioengineering.com/products/g-walk/. Accessed 27 February 2020.

[27] http://www.mcroberts.nl/products/movemonitor/. Accessed 27 February 2020.

[28] https://www.actigraphcorp.com/support/activity-monitors/. Accessed 27 February 2020.

[29] http://www.biosensics.com/products/pamsys/. Accessed 5 November 2020.

[30] Esser, P., Howells K., Dawes H., *et al.* (2014), Patent Application WO 2010/073044 A1 dated 1st July 2010. US Patent US8868369B2, 21st October 2014.

[31] Alam U., Riley D.R., Jugdey R.S. & Azmi S. (2017), Diabetic neuropathy and gait: a review. *Diabetes Ther.,* 81: 253–1264.

[32] https://www.podiatrytoday.com/article/8781. Accessed 18 October 2018.

[33] Brown S.J., Handsaker J.C., Bowling F.L., Boulton A.J. & Reeves N.D. (2015), Diabetic peripheral neuropathy compromises balance during daily activities. *Care,* 38(6): 1116–1122.

[34] Lalli P., Chan A., Garven A., *et al.* (2013), Increased gait variability in diabetes mellitus patients with neuropathic pain. *J. Diabetes Complic.*, 27(3): 248–254.

[35] https://www.gov.uk/government/news/38-million-people-in-england-now-have-diabetes. Accessed 18 October 2018.

[36] Hazari A., Maiya A.G., Shivashankara K.N., *et al.* (2016), Kinetics and kinematics of diabetic foot in type 2 diabetes mellitus with and without peripheral neuropathy: a systematic review and meta-analysis. *SpringerPlus*, 5: 1819. https://doi.org/10.1186/s40064-016-3405-9.

[37] Esser P., Collett J., Maynard K., *et al.* (2018), Single sensor gait analysis to detect diabetic peripheral neuropathy: a proof of principle study. *Diabetes Metab. J.*, 42(1): 82–86. https://doi:10.4093/dmj.2018.42.1.82.

[38] Meijer J.W., Smit A.J., Sonderen E.V., Groothoff J.W., Eisma W.H. & Links T.P. (2002), Symptom scoring systems to diagnose distal polyneuropathy in diabetes: the Diabetic Neuropathy Symptom score. *Diabet. Med.,* 19: 962–965.
[39] Kazemi M., Moghimbeigi A., Kiani J., Mahjub H. & Faradmal J. (2016), Diabetic peripheral neuropathy class prediction by multicategory support vector machine model: a cross-sectional study. *Epidemiol. Health.*, 38: e2016011. https://doi:10.4178/epih/e2016011.
[40] Vapnik V.N. (1999), An overview of statistical learning theory. *IEEE Trans. Neural Netw.* 10: 988–999.
[41] Bachmann E. (2000), Inertial and magnetic tracking of limb segment orientation for inserting humans into synthetic environments, PhD Thesis, Naval Postgraduate School.
[42] Molet T., Boulic R. & Thalmann D. (1999), Human motion capture driven by orientation measurements, *Presence*, 8(2): 187–203.
[43] Zhu R. & Zhou Z. (2004), A real-time articulated human motion tracking using tri-axis inertial/magnetic sensors package. *IEEE Trans. Neural Syst. Rehab. Eng.*, 12(2): 295–302.
[44] Giansanti D., Macellari V., Maccioni G. & Macellari V. (2005), The development and test of a device for the reconstruction of 3-D position and orientation by means of a kinematic sensor assembly with rate gyroscopes and accelerometers. *IEEE Trans. Biomed. Eng.*, 52(7): 1271–1277.
[45] Vlasic D., Adelsberg R. & Vannucci G. (2007), Practical motion capture in everyday surroundings. In *Proceedings of Siggraph '07*, San Diego.
[46] Roetenberg D., Slycke P. & Veltink P. (2007), Ambulatory position and orientation tracking fusing magnetic and inertial sensing. *IEEE Trans. Biomed. Eng.,* 54: 883–890.
[47] https://imeasureu.com/imu-sensor/. Accessed 28 February 2020.
[48] Besier T.F., Pal S. & Draper C.E. (2015), The role of cartilage stress in patellofemoral pain. *Med. Sci. Sports Exercise,* 47(11): 2416–2422.
[49] Webster K.E. Wittwer J.E. & Feller J.A. (2005), Validity of the GAITRite walkway system for the measurement of averaged and individual step parameters of gait. *Gait Posture*, 22(4): 317–321. Epub 2004 Dec.
[50] Riad J., Coleman S., Henley J. & Miller F. (2007), Reliability of pediobarographs for paediatric foot deformity. *J. Children's Orthopaed.*, 1(5): 307–312.
[51] Alani A. & Taylor G. (2012), Describing the normal foot Pedobarograph, a prospective study of 250 volunteers. In *Orthopaedic Proceedings* 94-B: SUPP_XXXVII, pp. 286–286.
[52] Webster J.B. & Darter B.J. (2019), Principles of Normal and Pathologic Gait in Atlas of Orthoses and Assistive Devices (5th ed.). New York, NY: Elsevier.
[53] http://multimedia.3m.com/mws/media/169286O/3mtm-scotchlitetm-high-gain-reflective-sheeting-7610.pdf?fn=7610.pdf. Accessed January 2020.

[54] Kalman R.E. (1960), A new approach to linear filtering and prediction problems. *J. Basic Eng.,* 82: 35. doi:10.1115/1.3662552.

[55] Richards J.G. (1999), The measurement of human motion: a comparison of commercially available systems. *Human Move. Sci.*, 18: 589–602.

[56] Piazza S.J., Chou L.-S., Denniston N.L., *et al.* (2007), A proposed standard for assessing the marker-location accuracy of video-based motion analysis systems. In *Proceedings of the 12th Annual GCMAS*, Springfield, MA.

[57] Merriaux P., Dupuis Y., Boutteau R., Vasseur P. & Savatier X. (2017), A study of vicon system positioning performance. *Sensors (Basel),* 17(7): pii: E1591.

[58] Bostelman R., Hong T., Legowik S. & Shah M. (2017), Dynamic metrology and ASTME57.02 dynamic measurement standard. *J. CMSC.* 12(1): 314–315.

[59] Kadaba M.P., Ramakrishnan H.K. & Wootten M.E. (1990), Measurement of lower extremity kinematics during level walking. *J. Orthopaed. Res.*, 8: 383–392.

[60] Sutherland D.H. (2002), The evolution of clinical gait analysis. Part II – Kinematics. *Gait Posture*, 16: 159–179.

[61] Charlton I.W., Tate P., Smyth P. & Roren L. (2004), Repeatability of an optimised lower body model. *Gait Posture*, 20: 213–221. doi: 10.1016/j.gaitpost.2003.09.004.

[62] Peters A., Galna B., Sangeux M., Morris M. & Baker R. (2010), Quantification of soft tissue artifact in lower limb human motion analysis: a systematic review. *Gait Posture*, 31(1): 1–8.

[63] Leardini A., Cappozzo A., Catini F., *et al.* (1999), Validation of a functional method for the estimation of hip joint centre location. *J. Biomech.*, 32: 99–103.

[64] Lamoreux L.W. (1992), Errors in thigh axial rotation measurements using skin mounted markers. *J. Biomech.*, 25(7): 769.

[65] Waters R.L. & Mulroy S. (1999), The energy expenditure of normal and pathologic gait. *Gait Posture*, 9: 207–231.

[66] Steele K.M., Shuman B.R. & Schwartz M.H. (2017), Crouch severity is a poor predictor of elevated oxygen consumption in cerebral palsy. *J Biomech.,* 60: 170–174. doi:10.1016/j.jbiomech.2017.06.036.

[67] Sutherland D.H. (2005), The evolution of clinical gait analysis. Part III – Kinetics and energy. *Gait Posture*, 21: 447–461.

[68] Ryan J.M., Theis N., Kilbride C., *et al.* (2016), Strength Training for Adolescents with cerebral palsy (STAR): study protocol of a randomised controlled trial to determine the feasibility, acceptability and efficacy of resistance training for adolescents with cerebral palsy. *BMJ Open,* 6: e012839. doi:10.1136/bmjopen-2016-012839.

[69] Davis R.B., Õunpuu S., DeLuca P.A. & Romness M.J. (1999), Clinical gait analysis and its role in treatment decision-making. *MedGenMed 1(1), [formerly published in Medscape Orthopaed. Sports Med. eJ.*2(5), 1998]. Available at: http://www.medscape.com/viewarticle/440148.

[70] Leboeuf F., Baker R., Barré A., Reay J., Jones R. & Sangeux M. (2019), The conventional gait model, an open-source implementation that reproduces the past but prepares for the future. *Gait Posture*, 69: 126–129.

[71] Wren T.A., Gorton G.E. III, Õunpuu S. & Tucker C.A. (2011), Efficacy of clinical gait analysis: a systematic review. *Gait Posture*, 34(2): 149–153.

[72] Wren T.A.L., Tucker C.A., Rethlefsen S.A., Gorton G.E.II & Õunpuu S. (2020), Clinical efficacy of instrumented gait analysis: systematic review 2020 update. *Gait Posture*, 80: 274–279.

Chapter 8
Clinical gait analysis

Howard Hillstrom[1], Jennifer Jezequel[1], Bridget Assip[1], Silvia Zanini[1], Sherry Backus[1], Paulo Selber[2] and David Scher[2]

8.1 Introduction

8.1.1 Purpose of clinical gait analysis

The purpose of clinical gait analysis is to provide objective data from which a treatment plan designed to improve gait, in the setting of complex neuromusculoskeletal diseases and injuries, may be based. The impairments that result from these conditions often affect a patient's posture and locomotion as well as their ability to conduct other purposeful movements and activities of daily living. The essence of the clinical gait analysis process is to obtain an integrated understanding of a patient's atypical structure, challenged control, and biomechanically altered function. From this deeper understanding, a team skilled in the process will be able to develop a well-conceived plan to provide the appropriate combination of physical therapy, orthotic and prosthetic management, and surgical care to meet the patient's and physician's expectations.

Gait analysis refers to the assessment of human walking. In the broadest sense, the approach may be qualitative [1] or quantitative [2]. Furthermore, both lower and upper extremity movements (kinematics), or the forces that cause movement (kinetics), may be included.

8.1.2 Overview of the clinical gait analysis process

Clinical gait analysis starts with a referral from a health care professional – a question that needs to be answered. These questions can impact the critical quality of life issues, such as identifying the treatment plan to maintain an individual's ability to walk independently. Referrals are screened by a clinical member of the laboratory staff and are clarified when necessary. The information gleaned from gait analysis can also be used as an objective outcome measure to assess the

[1]Rehabilitation Department, Hospital for Special Surgery, New York, NY, USA
[2]Department of Orthopedic Surgery, Hospital for Special Surgery, New York, NY, USA

efficacy of treatment and help clinicians evaluate the relative effectiveness between treatments, both surgical and non-surgical.

The visit begins with a focused interview/intake to understand any pertinent information relevant to the gait pathology as well as to ascertain patient/caregiver expectations. A laboratory health care professional captures the patient's (and caregiver's) chief complaint(s), incorporates the physician's referral, and performs a clinical exam. Typically, such an exam will include assessments of joint range of motion, muscle strength, motor control, reflexes, limb alignment and limb equality, tenderness or pain upon palpation, and any imaging data if available. The reason for acquiring this structural data is that it will need to be integrated with the objective gait analysis data and the patient's complaints or concerns to best answer the question which initially prompted the analysis. Movement disorders and severity are documented. Functional motor scales and validated measures of global functional impairments and quality of life are often part of the assessments.

High-definition video cameras are utilized to obtain gait and physical examination video clips – in the sagittal and coronal planes. The health care professional prepares the patient for clinical gait analysis. This includes the placement of passive or active markers upon selected anatomical landmarks to quantify motion between the marker-based proximal and distal coordinate systems. Surface electrodes are attached over muscle bellies to record the electromyograms (EMG) of superficial muscles. Fine wire electrodes are inserted to detect deeper muscle EMG activity. The choice of electrode type for superficial muscles varies between laboratories, but smaller or deeper muscle (i.e., the posterior tibialis) will require a fine wire insertion into the muscle. Patients are placed in a variety of postures and upon several devices to provide additional data. Postural sway can be measured in bipedal and one-legged stance with the patient standing on either a force plate or plantar pressure measuring device. Temporal-distance footfall parameters (stance time, step length, walking velocity, etc.) may be measured directly from sensors integrated within a mat or computed from marker trajectories measured by an optical motion capture system. Kinematics (e.g., joint angular excursions in the sagittal, frontal, and transverse plane) may be measured in three-dimensions (3D) using motion capture technology. In conjunction with the ground reaction forces, motion capture data may also be used to compute 3D kinetics (joint moments, powers). Mechanical energy may be computed from 3D kinetics but if metabolic energy expenditure (MEE) is required, then a special device is used to measure inspired O_2 and expired CO_2 and compute MEE. Several gait variables are optimized when we walk (e.g., speed, step width, and vertical displacement of the center of mass) to minimize energy expenditure. The distributed forces beneath the foot during gait can be measured by a plantar pressure measuring device. On occasion, isometric or isokinetic strength is required at one or more joints and acquired with a muscle dynamometer.

Following data collection, data analysis is conducted, and a report is generated that summarizes the findings and interpretations, presents all graphical and tabular data, and provides treatment recommendations.

The time needed for the entire gait analysis visit, including clinical measures takes several hours depending on the patient involvement and level of cooperation.

Abbreviating and expediting the acquisition process is always desirable especially in the setting of decreased endurance and gait limitations. Standardized protocols for data collection are required for adequate quality assurance and consistency between examiners and across time.

8.1.3 The team

A variety of health professionals are commonly involved in the collection, interpretation, and reporting of 3D gait analysis data. The clinical gait analysis team typically consists of a combination of physicians, physical therapists, occupational therapists, kinesiologists, athletic trainers, and engineers. The collection of data relies on the combined expertise of therapists, engineers, and kinesiologists. Orthopedic surgeons have been an integral part of 3D gait analysis laboratories around the world since the inception of these technologies. Physiatrists, neurologists, neurosurgeons, podiatric physicians, occupational therapists, and other specialists related to motion sciences are now routinely part of the clinical and technical teams as well.

Individuals involved in 3D gait analysis should attend gait analysis courses, conferences, and educational opportunities and remain in touch with the current and related literature. There is still much to discover and learn, the learning curve is steep and unending.

8.1.4 Laboratory accreditation

When available, the use of an accredited laboratory is advisable. For example, in the US a multi-disciplinary organization of physical therapists, physicians, and engineers has been formed for this purpose. The Clinical Motion Laboratory Accreditation (CMLA) was formed by three professional societies (the Gait and Clinical Motion Analysis Society (GCMAS), the American Academy of Cerebral Palsy and Developmental Medicine (AACPDM), and the American Physical Therapy Association (APTA)). The candidate laboratory submits a comprehensive application that details its standard operating procedures. Upon successful review, the laboratory becomes accredited for 3 years. To maintain CMLA accreditation beyond that period, the laboratory must apply for renewal which is also a rigorous process. To date, 14 clinical gait laboratories have achieved CMLA accreditation in the US.

An example of a clinical gait analysis laboratory is the Leon Root, MD Motion Analysis Laboratory (LRMALab) of the Hospital for Special Surgery (HSS) in New York City, USA (Figure 8.1).

8.2 Components of clinical gait analysis

8.2.1 Postural stability

Gait and posture are two important tasks of daily living. Being able to support our body weight and maintain an upright posture is a prerequisite for performing gait.

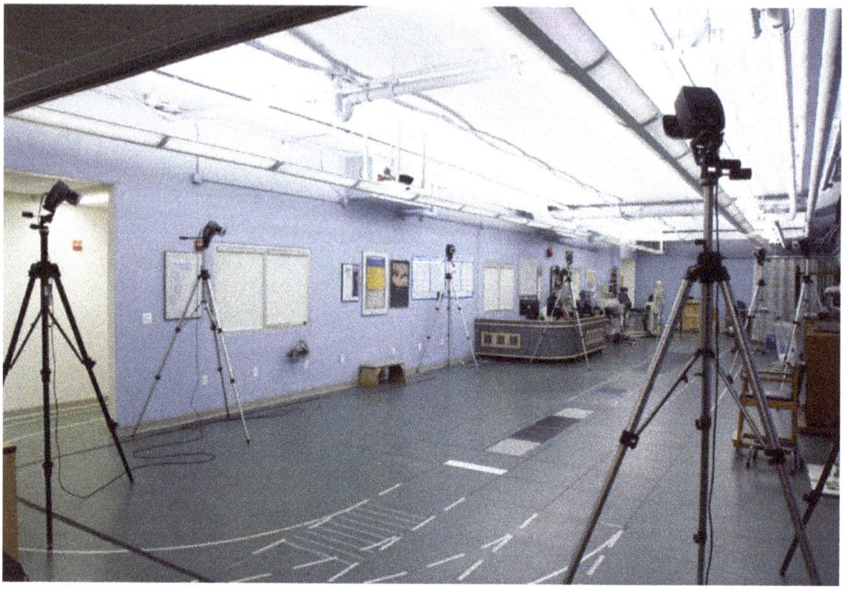

Figure 8.1 *LRMALab at HSS. The laboratory was designed to perform clinical motion analysis and related research. It is a 93' × 26' laboratory plus 4 single offices, 1 shared office, and 9 study carrels. The laboratory can adjustably reposition its 4 force plates (2 AMTI (Watertown, MA), 2 Bertec (Columbus, OH)), has a mechanically isolated data collection zone from the building to minimize vibration-induced errors, a 14 Kestrel/Raptor camera-based motion capture system from Motion Analysis Corp (Rohnert Park, CA), a 16 channel Motion-Labs (Baton Rouge, LA) MA400 dynamic EMG system, a Novel (Saint Paul, MN) emed-x plantar pressure measuring system, a Novel pedar-x in-shoe plantar pressure measuring system, a GAITRite (Franklin, NJ) temporal-distance gait pattern measurement system, and a JAKTool (Cranbury, NJ) arch height index device.*

Static equilibrium or Newton's third law (*for every action there is an equal and opposite reaction*) implies that our mass in the gravitational field elicits a downward directed force (body weight) which is met by an equal and opposite force (the ground reaction force vector). An adult consists of 206 bones, 360 joints, ~900 ligaments to stabilize those joints, and over 650 muscle-tendon systems to actuate those joints to perform purposeful movements, hence as expected, we sway during quiet standing. As our body sways, so does the ground reaction force vector. Several methods of assessing this postural sway involve measuring the movement of the origin of the ground reaction force vector (center of pressure), or at least the vertical component. Other methods measure the sway of the center of mass or gravity of the individual.

8.2.1.1 Stability

Stability is defined as not falling. There are four systems involved in maintaining stability: neuro-muscular, vestibular, auditory, and visual. The neuromuscular system efferently controls the muscle-tendons to develop the appropriate joint torques to maintain balance. Afferently this system contributes information (muscle length and velocity via spindles, force via Golgi tendon organs) to inform these real-time adjustments in sway.

8.2.1.2 Vestibular system

The vestibular system is composed of three semicircular canals filled with endolymph fluid that moves in the direction of the corresponding body's motion. Cilia-like structures stimulated by the moving endolymph send a train of neural action potentials down the vestibular nerve informing the brain that acceleration has occurred in one or more directions. If hyper-stimulated, this system may elicit motion sickness or nausea. In normal operating conditions, the vestibular apparatus is responsible for our sense of balance.

8.2.1.3 Visual and auditory systems

The visual system provides a map of 3D data as each eye focuses the incoming light rays upon the retina which sends trains of action potentials down the optic nerve. Each eye collects two-dimensional (2D) data which is transformed into a 3D image because we have two eyes [14]. Our visual system functions over a wide range of contrasts from darkness to bright light by dilating or constricting our pupils to control the amount of incoming light. The retina in the posterior eye has millions of tiny detectors known as rods and cones. Rods are responsible for vision at low light levels. Cones are active at higher light levels, capable of color vision, and responsible for high spatial acuity.

The auditory system is contained in the cochlea and contains the spiral organ of Corti – the receptor for hearing. It consists of tiny hair cells that transduce the vibration of sounds from its surrounding ducts into action potentials that travel to the brain via the auditory nerve. Human hearing is sensitive to sounds between 20 Hz and 20,000 Hz. As we age, the upper frequency diminishes.

The cervico-ocular and vestibulo-ocular reflexes are examples of how the visual system participates in balance. A postural stabilogram informs clinicians about a patient's ability to maintain balance, i.e., successfully integrate the neuromuscular, vestibular, auditory, and visual systems so that the patient does not fall. As shown in Figure 8.2a, a healthy individual is asked to stand upon a plantar pressure measuring device (or force plate) with their arms to both sides looking straight ahead for 60 s. The center of pressure (COP) sway is recorded in both the anterior–posterior and medial–lateral directions. The most common metrics include the sway area with a 95% confidence interval best fit ellipse to compute $SWAY_{area}$, COP_{AP}, COP_{ML}, and COP Path Length. The clinician could acquire a trial for each of the following conditions: eyes-open, eyes-closed, eyes-closed and ears-plugged, and the Rhomberg position. This would identify which system was most relied upon for balance and by providing the Rhomberg challenge (feet closest together while standing), test for the

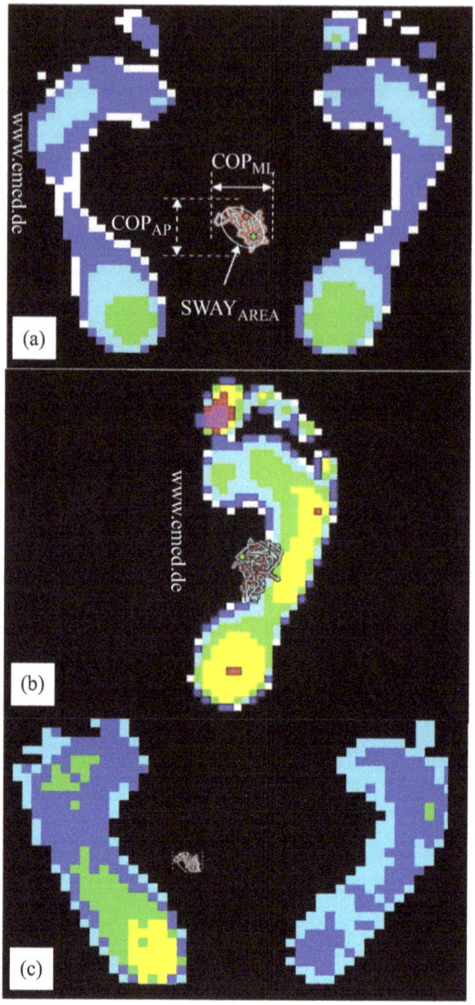

Figure 8.2 Postural stabilogram: (a) bipedal, (b) one-legged, and (c) bipedal with limb length discrepancy

impact of an upper motor neuron lesion. In more athletic or well-conditioned patients, you could conduct a one-legged postural stabilogram (Figure 8.2b). If the COP is not centrally located between the right and left feet in a bipedal postural stabilogram, this may denote a limb length discrepancy (Figure 8.2c).

8.2.2 Temporal-distance parameters

8.2.2.1 Central pattern generator (CPG)

CPGs are neurons interconnected by synapses that produce rhythmic outputs without a rhythmic input [3]. The CPGs responsible for locomotion are distributed throughout the lower thoracic and lumbar regions of the spinal cord [4]. CPGs can be modulated to adapt to the organism's environment. In addition to CPGs employing centrally generated rhythmicity and patterning, they can respond to sensory feedback by altering the patterns in behaviorally appropriate ways.

Thomas Brown's experiments in 1911 [6] demonstrated that the basic pattern of stepping could be produced by the spinal cord without the need for descending commands from the cortex. Gottschall and Nichols [5] examined the hindlimb movements of a decerebrate cat during walking. Comparing the decerebrate cats with normal cats demonstrated similar EMG patterns during level walking and EMG patterns during downhill walking with the head tilted up and uphill walking with the head tilted down. The investigators proved that neck proprioceptors and vestibular receptors contribute sensory feedback that alters the gait of the animal.

In 1998, Dimitrijevic *et al.* showed that the human lumbar pattern generating networks could be activated by large-diameter sensory afferents of the posterior roots. When tonic electrical stimulation was applied to these fibers in motor complete spinal cord injured individuals, rhythmic, locomotor-like movement of the lower limbs was elicited. Subsequent studies showed that these lumbar locomotor centers could form a large variety of rhythmic movements by combining and distributing stereotypical patterns to the lower limb muscles. CPGs promote our basic gait patterns (e.g., reciprocal arm swing) and their symmetry. These concepts are foundational for developing new treatments for neurologically based gait disorders.

8.2.2.2 Gait pattern

The human gait pattern develops from 9 to 15 months of age with the implicit goal of helping us move from point A to point B stably (without falling) and efficiently (minimizing energy demands). The COG translates in all three planes during walking. In the sagittal plane, the COG will move vertically in a sinusoidal pattern at a frequency established by the reciprocal of each stride length. In the transverse plane, the COG will move laterally in a sinusoidal pattern at a frequency established by the reciprocal of each step length which is half that of the sagittal plane. Projected upon the frontal plane, the COG adopts a Lissajous pattern (a horizontally oriented figure eight). The trajectory of the COG during gait must remain within the base of gait (distance between your feet) to ensure stability. An important contributing mechanism for maintaining stability during gait is the use of reciprocal arm swing. In the transverse plane, as the left leg advances the right arm advances. Similarly, as the right lower limb advances the left upper limb advances. The upper limbs establish an opposing moment about the long axis of the body to that of the lower limbs. These opposing moments serve to harmonically balance the system much like an internal combustion car engine. As a result, the COG in the frontal plane maintains a 1–2 cm excursion in the medial-lateral and superior-inferior directions, an example of dynamic stability.

Human gait is characterized by the temporal-distance footfall parameters (Figure 8.3). One stride, measured in meters, is from initial contact (or heel strike) to the next initial contact (or heel strike) on the same limb. A stride, also called gait cycle, consists of the stance phase (initial contact to toe-off) and swing phase

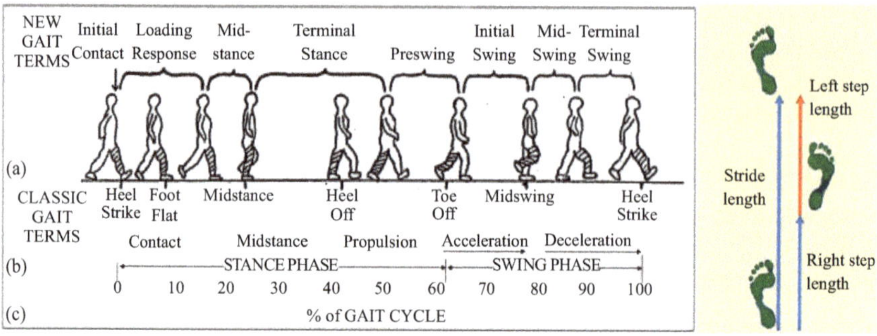

Figure 8.3 Gait pattern – temporal-distance footfall parameters (adapted from Wikipedia)

(toe-off to initial contact). A stride length is composed of a left step length and a right step length. The stance phase (expressed in seconds or % of stride) is divided into four primary events: initial contact, foot flat, heel-off, and toe-off.

The period between initial contact and foot-flat is called the contact phase; the period between foot-flat and heel-off is called mid-stance; the period between heel-off and toe-off is called propulsion. The swing phase (expressed in seconds or % of stride) is divided into three primary events: toe-off, mid-swing, and initial contact. The period between toe-off and mid-swing is called the acceleration phase while the period between mid-stance and initial contact is called the deceleration phase. The lower limbs move like a pair of pendulums.

8.2.2.3 Measurement approaches

Although simpler methods exist (e.g., measuring the distances between one's footprints from inking the plantar surface of the foot), these methods are time-consuming and ignore the temporal aspects of the gait pattern. Contemporary methods use two different approaches: (1) direct computer-based data capture of a matrix of switch closures within a mat or (2) through computations based upon marker trajectories using a multi-camera motion capture system. There may be some economy of computing temporal-distance footfall parameters from motion capture measurements because the 3D kinematics, 3D kinetics, and temporal-distance parameters may all be acquired from the same technology. On the other hand, if 3D kinematics and kinetics are not required it is likely to be far more time and cost-efficient to acquire the measurements directly using a mat-based switch matrix.

This technology typically uses a matrix of switches that are normally opened (Figure 8.4). When stepped on, the switches close. A multiplexor circuit scans all of the switches at a high sampling rate so that the location and time of each switch closure are known. Based on the timing of closures and the location of the switches, temporal-distance parameters (stance time, swing time, single support time, double support time, stride length, step length, width of the base of support, walking speed, etc.) are calculated. Values for these parameters in healthy controls have been published. Asymmetry of these variables has also been associated with structural

Figure 8.4 Temporal-distance footfall parameter measurements

(e.g., limb length discrepancy) and functional (e.g., antalgic gait in osteoarthritis) outcomes [27]. One may use the asymmetry index (%) as a metric for determining how much asymmetry is present in one's gait pattern [28]. For example,

$$\begin{aligned} ASI_{\text{stance time}}(\%) = |100 * (\text{stance time}_r - \text{stance time}_l)/2 \\ * (\text{stance time}_r + \text{stance time}_l)| \end{aligned} \tag{8.1}$$

Any gait variable that is measured bilaterally may be substituted into this equation to assess asymmetry. An alternative method of assessing variability is to use the Gait Variability Index (GVI) which is a conglomerate measure based upon nine spatial-temporal footfall parameters [7]. Gouelle *et al.* obtained GVI values of 100.3±8.6 for healthy and 70.4±7.9 for individuals with Friedreich's ataxia.

8.2.3 3D kinematics

8.2.3.1 Introduction

Kinematics, the quantitative study of motion, is measured as the relative rotation of the distal coordinate system about a fixed point (i.e., the origin of the joint) in the proximal coordinate system. In clinical gait analysis, each segment is considered a rigid body.

8.2.3.2 Marker set (definition of limb segment coordinate systems)

Clinical marker sets were introduced with an initial focus on the lower body (pelvis, hip, knees, and ankles). Marker sets used now also include the torso and the upper extremity. The goal of the early marker sets was to use a limited number of markers to minimize the burden of marker placement and tracking so that clinical gait analysis could be done as efficiently as possible. The Cleveland Clinic marker set (Figure 8.5) was one standard marker set that permitted a three-degree of freedom (DOF) analysis at the knee, hip, and pelvis and a 2DOF analysis (sagittal and transverse plane rotations) of the ankle [8]. The Helen Hayes or modified Helen Hayes marker sets are other standards that permitted a 3DOF rotational analysis (sagittal, frontal, and transverse planes) at the ankle, knee, hip, and pelvis [9,10]. All of these clinical marker sets used a single foot segment that was assumed to be a rigid body. The majority of clinical gait analyses currently use one of the standard marker sets unless multi-segment foot or complex upper extremity kinematics and kinetics are needed. Custom and more complex marker sets that can perform true 6DOF kinematic analysis (3 rotations and 3 translations) may be implemented but are not typically used for routine clinical analysis. Skin motion artifact (where the marker and skin move with respect to the bone) is generally considered to be too large for accurate translation measurements.

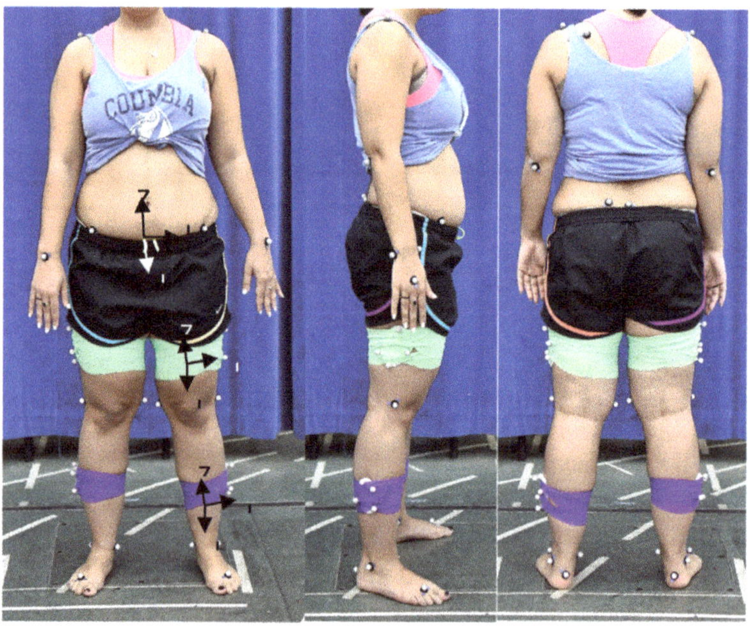

Figure 8.5 Cleveland Clinic marker set from Pelvis Down. Anterior view (left), lateral view (center), and posterior view (right). Markers include: right and left ASIS, sacrum, sacral offset (optional), bilateral thigh clusters, bilateral medial femoral epicondyles, bilateral lateral femoral epicondyles, bilateral shank clusters, bilateral medial malleolus, bilateral lateral malleolus, bilateral calcaneus, bilateral dorsal second metatarsal head.

8.2.3.3 Kinematic equations

Most solutions for 3D kinematics either use Euler angles [10] or a specific sequence of Euler angles known as the Grood and Suntay coordinate system (e.g., [11]). Euler angles are three finite rotations that occur in a specific sequence that collectively describe the 3DOF rotation from a reference frame to a subsequent coordinate system. To describe a position of a passive retroreflective marker, one must start with an orthogonal coordinate system $\{X_A, Y_A, Z_A\}$. Any point may be located with a 3 × 1 position vector. The leading superscript indicates the vector's coordinate system {A}. Components of ^{A}P have values which are the distances along the axes of {A} – like projections onto a unit vector (8.1). Reference frame {A} is represented as three mutually orthogonal unit vectors (Figure 8.6):

$$^{A}\boldsymbol{P} = \begin{bmatrix} p_x \\ p_y \\ p_z \end{bmatrix} \tag{8.1}$$

To represent the orientation of a body in space, we embed a coordinate system and describe this coordinate system {B} relative to the reference system {A} as shown in Figure 8.7. The unit vectors give the principal directions of the coordinate system {B} as $\mathbf{X}_B$, $\mathbf{Y}_B$, $\mathbf{Z}_B$. In terms of {A}: $^{A}\mathbf{X}_B$, $^{A}\mathbf{Y}_B$, $^{A}\mathbf{Z}_B$. If the 3 × 1 vectors are stacked together, they form a 3 × 3 rotation matrix $^{A}_{B}\boldsymbol{R}$.

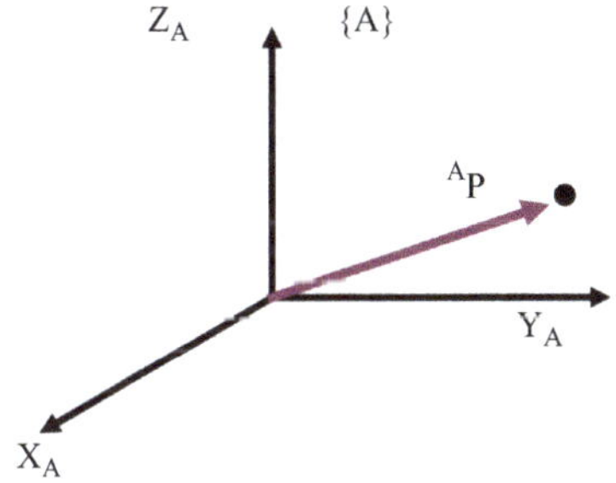

Figure 8.6 Description of a marker's position in reference frame {A}

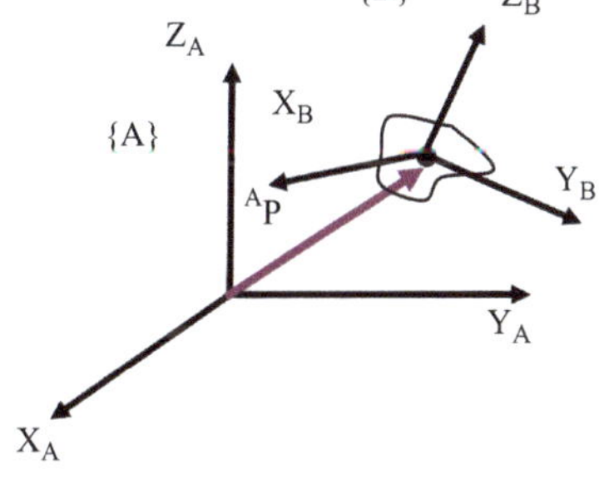

Figure 8.7 Description of an orientation

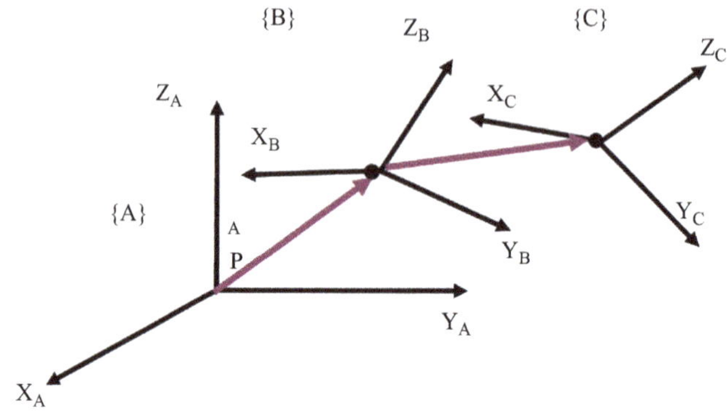

Figure 8.8 3D kinematics – rotation and translation

Three vectors, or a matrix, may specify any body's orientation (8.2). The position of a point may be represented by a single vector. In the rotation matrix, note that each element r_{ij} is a scalar:

$$
{}^{A}_{B}\boldsymbol{R} = {}^{A}X_B\,{}^{A}Y_B\,{}^{A}Z_B = \begin{bmatrix} r_{11} & r_{12} & r_{13} \\ r_{21} & r_{22} & r_{23} \\ r_{31} & r_{32} & r_{33} \end{bmatrix} = \begin{bmatrix} X_B\cdot X_A & Y_B\cdot X_A & Z_B\cdot X_A \\ X_B\cdot Y_A & Y_B\cdot Y_A & Z_B\cdot Y_A \\ X_B\cdot Z_A & Y_B\cdot Z_A & Z_B\cdot Z_A \end{bmatrix} \tag{8.2}
$$

Scalars are obtained since the components of any vector are the projections of that vector onto the unit directions of its reference frame. For a compact notation, one may append the position vector on the end of the rotation matrix to form the homogeneous transformation matrix (a 4 × 4 matrix). The homogeneous transformation matrix (8.3) can provide the rotation and translation between any coordinate systems (e.g., during a movement across time such as gait). Figure 8.8 illustrates how you progress from frame (coordinate system) A to B to C each having a rotation and translation to fully describe its motion:

$$
{}^{A}_{B}\mathbf{T} = \begin{bmatrix} X_B{\cdot}X_A & Y_B{\cdot}X_A & Z_B{\cdot}X_A & p_x \\ X_B{\cdot}Y_A & Y_B{\cdot}Y_A & Z_B{\cdot}Y_A & p_y \\ X_B{\cdot}Z_A & Y_B{\cdot}Z_A & Z_B{\cdot}Z_A & p_z \\ 0 & 0 & 0 & 1 \end{bmatrix} \tag{8.3}
$$

Leonhard Euler (1707–83) was a Swiss mathematician who made numerous contributions including the primary method for describing kinematics which is also widely used in navigation. Begin with the reference frame {A} as shown in Figure 8.9 which is often known as the global or laboratory reference frame. One approach is to first rotate {B} about Z_{B} by an angle α, then rotate about Y_{C} by an angle β, and then rotate about Z_{D} by an angle γ. Each rotation (α, β, γ) is about an axis of the moving system and the solution is sequence-dependent.

Typical for clinical gait analysis, the pelvis rotations are expressed in the reference coordinate system {**A**} of the laboratory. The *X*-axis is in the direction of progression,

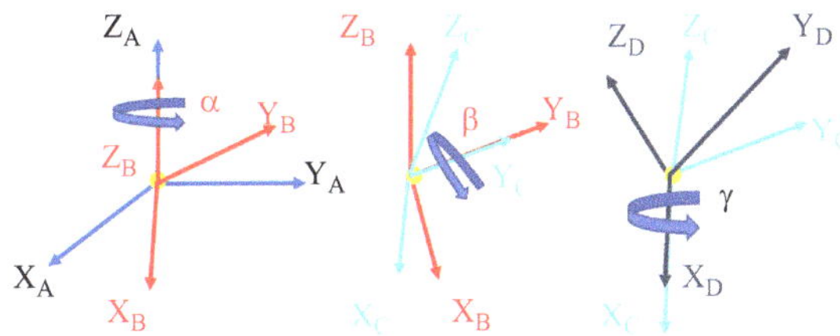

Figure 8.9 Euler angles (ZYX sequence) for the determination of joint rotations (α, β, γ)

the *Z*-axis is vertical, and the *Y*-axis is perpendicular to *X* and *Z* forming a right-handed coordinate system. For the thigh segment, the reference axes are the pelvic embedded axes. For the shank, the reference axes are the thigh embedded axes. For the foot, the reference is the shank embedded axes. When a segment rotates in the right-handed direction through an angle α about the *Y*-axis in a four-segment model, the pelvis tilts upward, the hip extends, the knee flexes, and the ankle plantar flexes. The new orientation of the embedded axes of the moving segment is denoted frame $\{\mathbf{B}\} = \mathbf{X}_B, \mathbf{Y}_B, \mathbf{Z}_B$. When the segment rotates by an angle β about the rotated axis X_B, the rotations become pelvic obliquity, hip ab/adduction, and knee varus/valgus. The new orientation of the moving body is denoted by $\{\mathbf{C}\} = \mathbf{X}_C, \mathbf{Y}_C, \mathbf{Z}_C$. When the segment further rotates an angle γ about axis Z_C, the pelvis is in rotation, hip in rotation, knee in rotation, and ankle in rotation. The final orientation of the embedded axes is $\mathbf{X}_D, \mathbf{Y}_D, \mathbf{Z}_D$. The Euler angle solution is expressed in (8.4a) and (8.4b):

$$
\begin{aligned}
{}^{A}_{D}\boldsymbol{R}(\alpha, \beta, \gamma) &= \boldsymbol{R}_Z(a)\boldsymbol{R}_Y(\beta)\boldsymbol{R}_X(\gamma) \\
&= \begin{bmatrix} Ca & -Sa & 0 \\ Sa & Ca & 0 \\ 0 & 0 & 1 \end{bmatrix} \begin{bmatrix} C\beta & 0 & S\beta \\ 0 & 1 & 0 \\ -S\beta & 0 & C\beta \end{bmatrix} \begin{bmatrix} 1 & 0 & 0 \\ 0 & C\gamma & -S\gamma \\ 0 & C\gamma & C\gamma \end{bmatrix}
\end{aligned} \tag{8.4a}
$$

where $Ca = \cos a$, $C\beta = \cos\beta$, $C\gamma = \cos\gamma$, $Sa = \sin a$, $S\beta = \sin\beta$, $S\gamma = \sin\gamma$.

$$
\begin{bmatrix} \boldsymbol{X_D} \\ \boldsymbol{Y_D} \\ \boldsymbol{Z_D} \end{bmatrix} = \begin{bmatrix} C1 * C3 + S1 * S2 * S3 & C2 * S3 & -S1 * C3 + C1 * S2 * S3 \\ -C1 * S3 + S1 * S2 * C3 & C2 * C3 & S1 * S3 + C1 * S2 * C3 \\ S1 * C2 & -S1 & C1 * C2 \end{bmatrix} \begin{bmatrix} \boldsymbol{X_A} \\ \boldsymbol{Y_A} \\ \boldsymbol{Z_A} \end{bmatrix} \tag{8.4b}
$$

where $C1 = \cos\alpha$, $C2 = \cos\beta$, $C3 = \cos\gamma$, $S1 = \sin\alpha$, $S2 = \sin\beta$, $S3 = \sin\gamma$ The rotational angles may be computed by (8.5a), (8.5b), and (8.5c):

$$\alpha = \arcsin\left[\frac{\boldsymbol{Z}_D \cdot \boldsymbol{X}_A}{cos\,(\beta)}\right] \tag{8.5a}$$

$$\beta = \arcsin\,(-\boldsymbol{Z}_D \cdot \boldsymbol{Y}_A) \tag{8.5b}$$

$$\gamma = \arcsin\left[\frac{\boldsymbol{X}_D \cdot \boldsymbol{Y}_A}{\cos\,(\beta)}\right] \tag{8.5c}$$

For the ankle, which has been limited to 2DOF in this solution, the following rotational angles may be computed by (8.6a) and (8.6b):

$$\alpha = \arcsin\left[(\boldsymbol{Z}_D \cdot \boldsymbol{X}_A)\right] \tag{8.6a}$$

$$\gamma = \arcsin\left(\boldsymbol{X}_D \cdot \boldsymbol{Y}_A\right) \tag{8.6b}$$

Note that Euler angles are sequence-dependent (in this gait analysis example, **YXZ**). The first rotation is flexion–extension because it is the largest, ab/adduction is the second rotation, and internal–external rotation is the third. There are 12 unique and meaningful ordered sequences of rotations (e.g., XYX, XYZ, XZX, etc.). Each will provide a correct but different solution for rotations. The Euler sequence for a specific movement is selected based on the range of motion for the task of interest and to avoid gimbal lock (where two axes become parallel and therefore a DOF is lost). The choice of the Euler sequence should be included in a description of the model. Typical normative kinematic data is depicted as grey bands (± 1SD of controls) along with the upcoming patient cases for comparison (e.g., Figure 8.21).

8.2.3.4 Gait deviation index (GDI) and the gait profile score (GPS)

The GDI is a multivariate measure of overall gait pathology. Based upon a comprehensive set of gait kinematics, a set of mutually independent joint rotation patterns, or gait features, are used to efficiently describe gait. Linear combinations of the first 15 gait features yielded a 98% correct reconstruction rate. The GDI is defined as a scaled distance between the 15 gait feature scores for a subject and the average of the same 15 gait feature scores for a control group of typically developing (TD) children. Concurrent and face validity for the GDI has been established [13]. GDI values of 100 are considered normal while values below 100 are associated with progressively more limited gait patterns. The GPS is the root mean square difference of the kinematic data for a given individual from the normative reference data across relevant kinematic variables throughout the entire gait cycle. The GPS serves as another summary measure of overall gait pathology associated with kinematic data [29].

8.2.4 3D kinetics

Kinetics is the study of the forces and moments that cause motion. Sir Isaac Newton's second law (law of momentum) provides the fundamental equation for dynamic analysis:

$$\boldsymbol{f} = m\boldsymbol{a} \tag{8.7a}$$

$$\boldsymbol{M} = I\boldsymbol{\alpha} \tag{8.7b}$$

where **f** is the force vector with components f_x, f_y, and f_z, m is the mass, and **a** is the acceleration vector with components a_x, a_y, and a_z, **M** is the moment vector with components M_x, M_y, M_z, I is the moment of inertia, and α is the angular acceleration vector with components α_x, α_y, and α_z.

In static equilibrium, $\mathbf{a} = \mathbf{0}$ and $\alpha = \mathbf{0}$, so the

$$\sum f = 0 \tag{8.8a}$$

$$\sum M = 0 \tag{8.8b}$$

8.2.4.1 Ground reaction force vector

In clinical gait analysis, the ground reaction force vector (Figure 8.10) results from the lower limb's contact with the ground from the initial contact to toe-off. The vertical force component, or load, appears like a "double-humped camel" where the first peak, $F_{\text{Load}}1$, corresponds to the end of weight acceptance, the valley, $F_{\text{Load}}2$, corresponds to mid-stance, and the second peak, $F_{\text{Load}}3$, corresponds to propulsion. $F_{\text{Load}}1$ and $F_{\text{Load}}3$ are typically equal and exceed body weight due to the acceleration of the body's mass into and away from the ground. The anterior-posterior force component, or AP, appears like an inverted sine wave. The first peak, $F_{\text{AP}}1$, corresponds to the posterior braking force while the second peak, $F_{\text{AP}}2$, corresponds to the anteriorly directed propulsion force. The medial–lateral force, or ML, is the smallest in magnitude. The first peak, $F_{\text{ML}}1$, corresponds to the medially directed force while the second peak, $F_{\text{ML}}2$, corresponds to the laterally directed force. Figure 8.10 is a typical ground reaction force vector for a healthy asymptomatic individual. Many changes can occur in the presence of pathology, e.g. an apropulsive gait may have a diminished second peak, $F_{\text{Load}}3$.

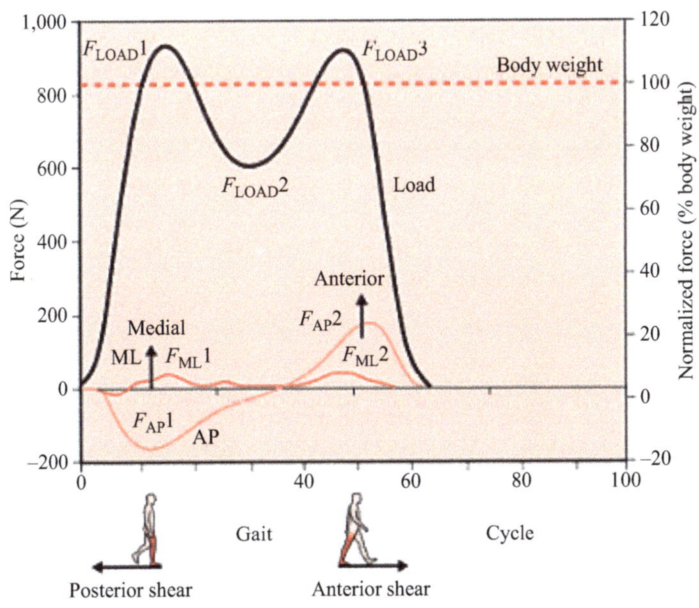

Figure 8.10 Ground reaction force vector components (load, AP, ML)

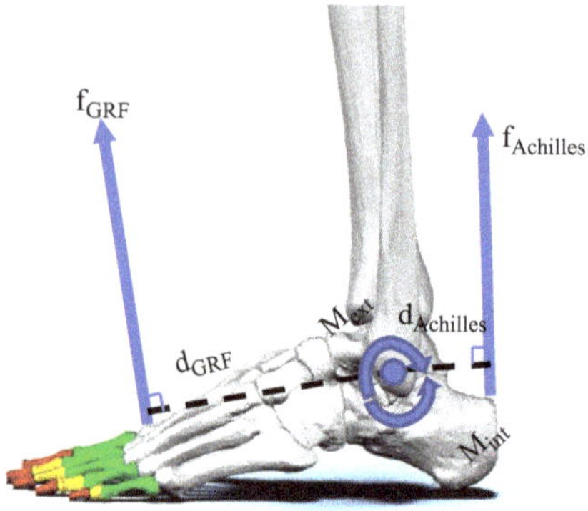

Figure 8.11 M_{ext} and M_{int} ankle moments

8.2.4.2 Definitions of joint moment and power

The solution for joint kinetics in the lower extremity is driven by the ground reaction force vector ($\mathbf{f}_{GRF}$). Since the structures within the lower limb are loaded sequentially in a kinetic chain, the ankle joint is typically the starting point. The internal moment (M_{int}) is composed of the force in the Achilles tendon gastrosoleus complex ($\mathbf{f}_{Achilles}$) crossed with the Achilles moment arm ($\mathbf{d}_{Achilles}$). The external moment (M_{ext}) from the $\mathbf{f}_{GRF}$ is crossed with the GRF moment arm ($\mathbf{d}_{GRF}$) (Figure 8.11). Assuming the mass and moment of inertia of the foot is negligible, then the internal and external moments are equal (8.9):

$$M_{ext} = M_{int} \tag{8.9a}$$

$$\boldsymbol{f} \times \boldsymbol{d}_{GRF} = \boldsymbol{f}_{Achilles} \times \boldsymbol{d}_{Achilles} \tag{8.9b}$$

The solution for the moments about the medial–lateral, anterior–posterior, and internal–external axes for each joint follows a similar construct except that mass and moment of inertia are not negligible for the knee and hip.

Power may occur in linear (force × velocity) or rotational movements (moment × angular velocity). In clinical gait analysis, we are concerned with the power generated or absorbed at a joint:

$$P = \boldsymbol{M} \cdot \omega = W/t \tag{8.10}$$

where **P** is the power, **M** is the moment, $\boldsymbol{\omega}$ is the angular velocity, **W** is the work, and **t** is the time. Power is a useful parameter to assess the role of the muscles. When power is negative, the body is absorbing energy with eccentric muscle contractions. When power is positive, the body is generating energy with concentric muscle contractions. Power is also equal to work or energy per unit of time.

8.2.4.3 Equations of motion

The equations of motion are derived from using superposition to analyze each segment (Figure 8.12). Appendix B of Vaughan *et al.* [23] details these equations and

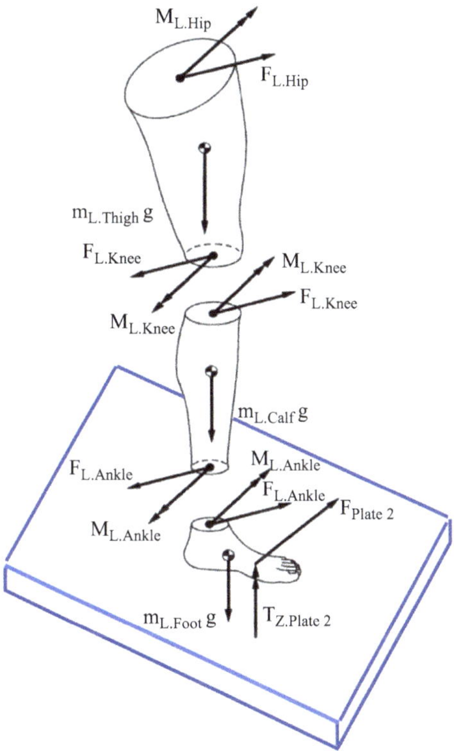

Figure 8.12 Superposition principle for inverse dynamics solution (modified from [23])

specifically how the limb segment masses and moments of inertia; the limb segment centers of gravity, displacements, velocities, and accelerations; ground reaction forces and moments (from force plates); joint centers and segment endplates; segment reference frames; and segment angular velocities and accelerations are integrated into a solution for the net resultant joint forces and moments. The importance of kinetics was best stated by David Winter – "the reason someone walks with an abnormal gait pattern is they can't produce the right torque, at the right joint, at the right time."

8.2.4.4 GDI – kinetics

Similar to the original GDI for kinematics, the GDI kinetics [28] is a composite index for kinetics [28]. The GDI kinetics identifies "features" of the raw gait kinetic data using singular value decomposition. Feature extraction provides a subset of features that explains a large percentage of the information in the raw gait kinetic data. An index is formed from a TD group of children as a linear combination of these features. The magnitude of the difference between the subject and the mean of the control is scaled and transformed to provide a simple, and statistically well-behaved, measure. Linear combinations of the first 20 gait features produced a 91% faithful reconstruction of the data. GDI kinetics has already been used to distinguish different presentations of cerebral palsy and has recently been used for individuals with knee osteoarthritis and other altered gait patterns. This tool, in addition to GDI kinematics, has shown promise for distinguishing complex multivariate data between individuals with healthy and altered gait patterns.

8.2.5 Dynamic EMG

8.2.5.1 Muscle physiology

Muscles are organized hierarchically large to small from the entire force-generating structure attached to the bony origin and insertion sites via tendons to the smallest functional units, the sarcomere [16]. A muscle contraction (eccentric or concentric) requires an electrical stimulus that begins with a neural action potential train propagating from the alpha motor neuron to the neuromuscular junction. Upon receiving a presynaptic action potential, acetylcholine (ACTH) is released from the presynaptic membrane to traverse the neuromuscular junction and elicit muscle fiber action potentials at the postsynaptic junction [15]. Those muscle fibers innervated by that single alpha motor neuron will have synchronized action potentials which, at the sarcomere level, will cause a shortening of the Z bands. The sliding filament theory describes how two proteins (actin and myosin) interact to reduce sarcomere length [17]. The overall electrical manifestation of the contracting muscle, as measured using skin-mounted electrodes or indwelling wire electrodes near the motor units, is known as the EMG [18]. The physiological process by which a motor unit produces the smallest unit of force (twitch) is known as the excitation–contraction coupling mechanism. To produce more force from a muscle, one may (1) recruit more motor units or (2) increase the firing rate of the neural action potential train.

8.2.5.2 Amplitude modulation (AM) model for EMG

Motor unit recruitment (multiple alpha motor neurons and their associated muscle fibers) elicit asynchronous electrical activity which appears like white noise ($W(t)$) [19]. As this noise is filtered by the electrode/tissue transfer function (H(S)), "colored noise" appears upon the epidermal surface. As muscle force is produced, a static memoryless functional yields a proportional signal to force (s). As shown in Figure 8.13, EMG may be modeled as the AM of s with $N(t)$:

$$\mathrm{EMG}(t) = \frac{1}{\sqrt{2\pi\sigma}} e^{\frac{-1}{2s(F)^2}}, \quad \mathrm{EMG}(s) = H(s)^2 \left|\sigma(F)^2\right|$$

8.2.5.3 Amplitude demodulation – linear envelope detection

Based upon the AM model for EMG, we can extract a signal proportional to force by doing amplitude demodulation. This process is known as linear envelope detection (Figure 8.14). Consider an instrumentation amplifier that acquires EMG from a pair of electrodes which in its raw form appears like random noise with both positive and negative values (bipolar). The first step in linear envelope detection is

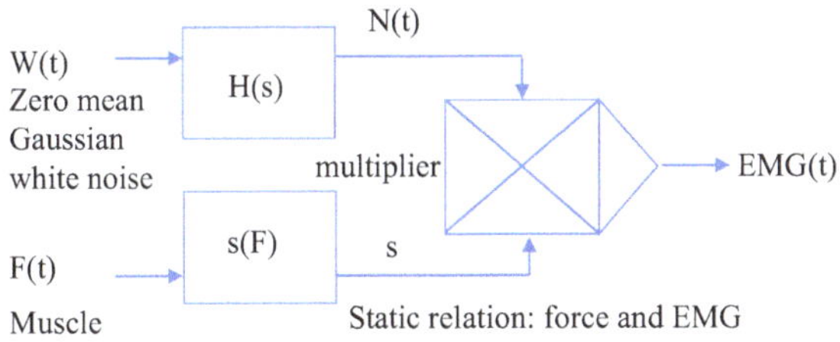

Figure 8.13 AM model for EMG

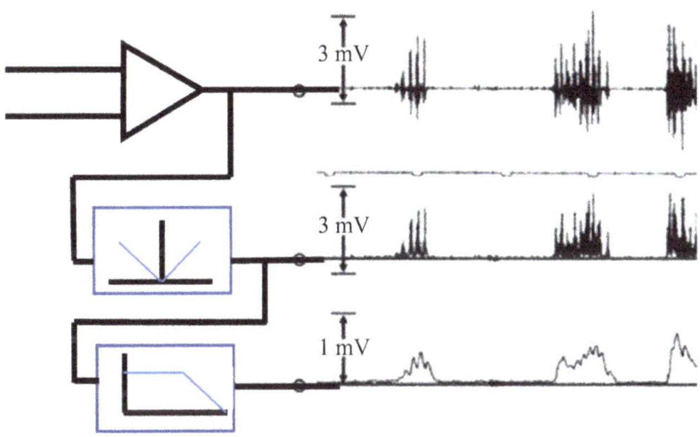

Figure 8.14 Amplitude demodulation process – linear envelope detection

to take the absolute value (full-wave rectification) of the raw EMG signal. As shown in Figure 8.14, the result is only the positive values of the EMG signal remain. Next, that signal is low pass filtered which strips the lower frequency EMG envelope from the higher frequency carrier signal. The result is a smoother curve that follows the shape, or envelope, of the EMG signal. This form of signal processing is exactly how an AM radio strips the higher frequency carrier signal from the music so it may be amplified and heard through a speaker.

Normative values of linear envelope-detected EMG signals across the gait cycle have been well characterized for the vast majority of lower extremity muscles [2]. Winter has provided the mean values of these signals in a table for every 2% of the gait cycle so anyone may compare the EMG data they have collected to these normative values. A general catalog of EMG studies has been performed across the body and described in Muscles Alive by Basmajian [18].

8.2.6 Plantar pressures

8.2.6.1 Foot type: another matter of structure and function

Plantar pressure measurement is an important tool for the assessment of biomechanically induced pathologies about the foot and ankle. Pressure, force per unit area, presents a concentrated, and potentially damaging, force to the plantar soft tissues. Plantar pressure measuring devices come in generally two form factors: a plate and an insole. These devices serve different but complimentary purposes. The insole plantar pressure measuring insole can determine the effectiveness of a foot orthosis for offloading an area of the foot (e.g., coincident with a diabetic neuropathic ulcer). The plate-based device is generally comprised of a matrix (e.g., 5 mm × 5 mm) of transducers that are used for objectively measuring barefoot function. As shown in Figure 8.15, the maximum pressure throughout stance phase plot, the highest peak pressure is beneath the first metatarsophalangeal (MTP) joint, there are also substantial forces beneath the hallux suggesting increased stiffness at the first MTP joint and taken together these medially deviated forces suggest the patient is over-pronating. To confirm or refute your suspicions about foot function, it is highly advisable to examine foot structure. One approach is to measure the arch height index (Figure 8.16) since low arches are associated with planus or flat feet,

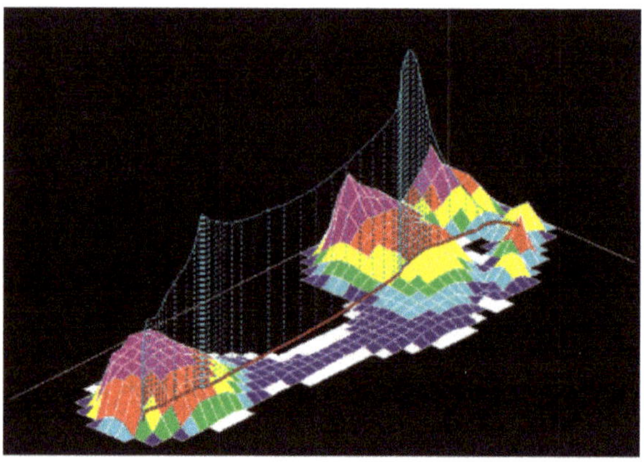

Figure 8.15 Maximum pressure throughout stance phase

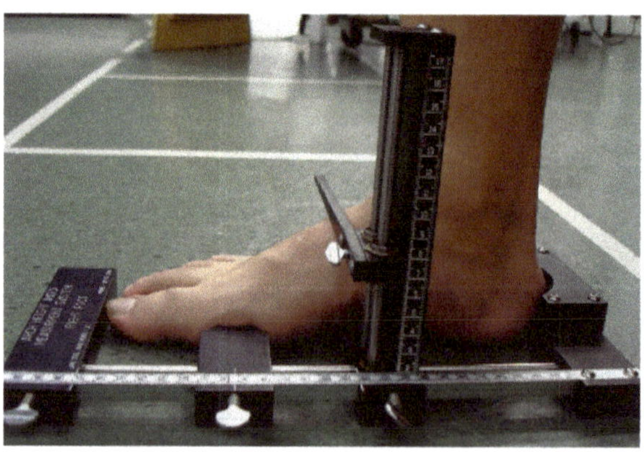

Figure 8.16 Arch height index

moderate arches are associated with rectus feet, and high arches are associated with cavus feet. Normative data for measures of foot structure and function has been published on a cohort of asymptomatic healthy adults [20,21]. It is clear that different foot structures elicit completely different biomechanical functions. This phenomenon is known as "Foot Type."

8.2.6.2 Masked plantar pressure parameters

Masking algorithms are important tools for computing loading parameters in different regions of the foot. Based upon the geometry of the foot undergoing analysis, the masking algorithm scales the template to fit the footprint and provide the desired number of stratified segments. For the mask shown in Figure 8.17, there are 12 distinct segments that one may compute specific parameters in (e.g., peak pressure, maximum force, pressure–time–integral, force–time–integral, and contact area). This is especially useful for pre–post-surgical analysis where the goal of the surgeon is to improve foot function through modification of foot structure.

8.2.6.3 Center of pressure excursion index

At each instant of the stance phase, there will be a plantar pressure pattern commensurate with the part of the foot that contacts the ground (e.g., the heel at heel strike, the hallux or second toe at toe-off). In the composite graph, the maximum pressure throughout the stance phase is depicted in Figure 8.18. The center of pressure curve demonstrates a reduced concavity from heel strike to toe-off when the foot is over-pronating (Figure 8.18 – right). Over-supinating feet exhibit an increased center of pressure concavity during the stance phase of gait (Figure 8.18 – left). Rectus feet have a center of pressure concavity between that of planus and cavus feet (Figure 8.18 – center). To quantify this phenomenon, the center of pressure excursion index (CPEI) was developed to measure the concavity of the center of pressure curve throughout the stance phase. The CPEI (%) calculation is given as:

$$\mathrm{CPEI}(\%) = (\mathrm{CPE}/\mathrm{FW}) * 100 \tag{8.11}$$

where CPE is the center of pressure excursion with respect to a construction line from the initial to final center of pressure points at the anterior third of the maximum pressure throughout the stance phase plot and FW is the foot width at that same region [20]. For a more detailed description of the CPEI algorithm and the exceptions for this calculation, see Diaz *et al.* [22].

Common clinical applications for this type of data include the following: identifying regions of excessive pressure in diabetic, neuropathic, and Charcot feet at risk for ulceration; estimating foot function to determine foot type; identifying abnormal sequences of loading throughout stance phase due to equinus; determining the pressures beneath lesser metatarsal heads in conjunction with painful digital deformities (e.g., hammertoes); measuring the distribution of plantar pressures due to pedal deformities (e.g., metadductus, club foot, etc.); and examining anatomical variations of the foot that may affect function (peroneal tubercle, retro-trochlear eminence, Morton's second toe, Haglund's deformity, etc.). If post-operative data

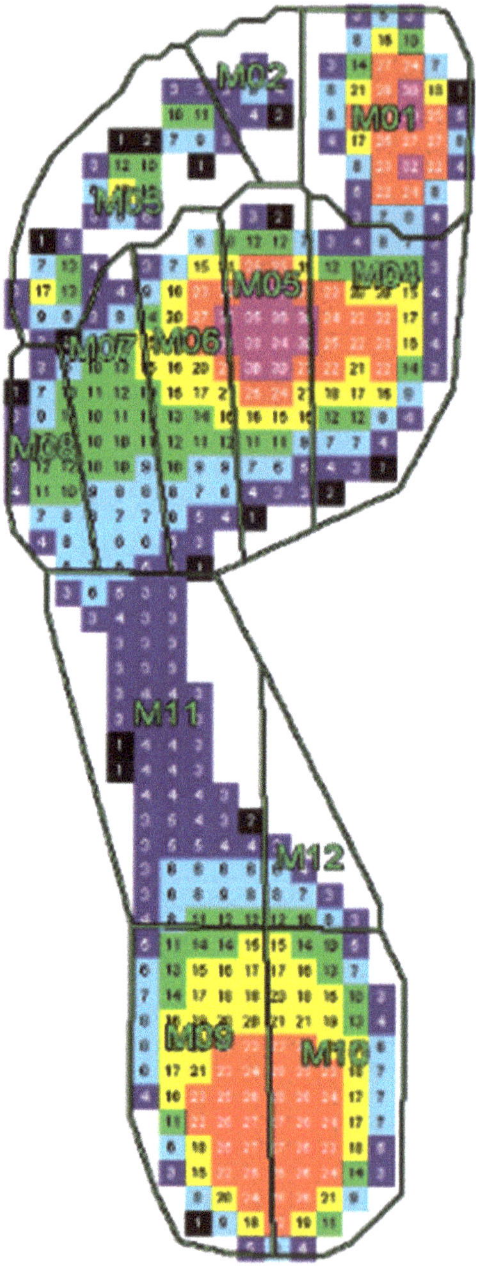

Figure 8.17 Masking algorithm overlayed with maximum pressure throughout stance phase plot

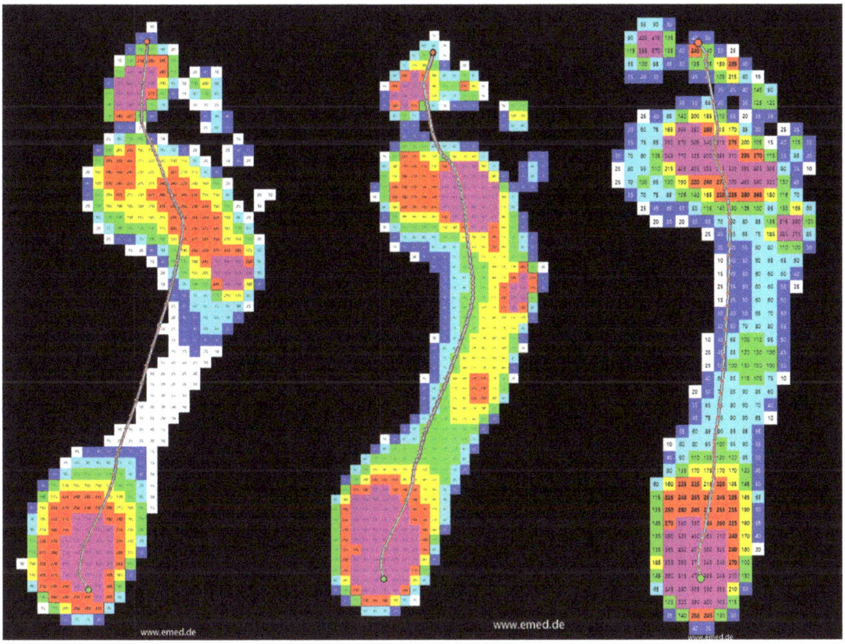

Figure 8.18 Cavus foot (left), rectus foot (center), and planus foot (right)

is also available, the changes that occur to the plantar pressure distribution due to surgical intervention are often substantial and informative (e.g., flatfoot reconstruction).

8.2.7 3D gait analysis interpretation

8.2.7.1 Clinical evidence consolidated by precision medicine

Pathological gait patterns found in patients with neurological disorders such as cerebral palsy can be precisely measured and documented through the use of 3D gait analysis technologies. The equipment used in 3D gait analysis laboratories has evolved significantly in the last four decades. The advent of personal computers, force platforms, surface EMG, high-definition video cameras, and 3D point localization technologies, deemed essential.

8.2.7.2 The reporting session

3D gait analysis data is comprehensive and multifaceted. Once the data is collected, collated, and formatted, as part of typical best practice, all of the components are presented for interpretation in a multidisciplinary reporting session. Typically, this includes laboratory personnel, the referring clinician, and other members of the health care team. Trainees (residents, fellows, graduate students, and post-doctoral researchers) are also welcome as this process affords many teaching moments. The discussion begins with a review of the pertinent history, the physical exam, and functional scores. The videos of standing and walking for all conditions are presented. This is followed by a review of the remainder of the kinematics and kinetics, pedobarography, dynamic EMGs, and metabolic testing. Any other specialized or custom analyses from the visit including the calculation of moment arms, multi-segment foot kinematics, and arch height index, are presented as well.

From this discussion, a treatment plan is formulated which may include a combination of surgical and nonsurgical treatment recommendations. Recommendations are made within the scope of practice and referrals for additional imaging and specialist are routinely made as needed. Orthopedic recommendations are provided by an orthopedic surgeon experienced in gait analysis. Neurosurgical recommendations or physiatry recommendations are suggested by the respective specialists. The complexity of 3D gait analysis data interpretation and its significance within the medical and surgical specialties makes a strong case for the inclusion of treatment strategy suggestions in the gait analysis report.

The 3D gait analysis laboratory team must have a comprehensive understanding of normal and pathological human gait and the various patterns commonly present in different disorders. The reporting team must understand the biomechanical models used to generate the data to enable recognition of data artifacts and the strengths and limitations of the models and measurement techniques used. Interpreting 3D gait analysis data is more complex and rewarding than just reading plots and graphs. 3D gait analysis interpretation sessions yield a unique opportunity for a concentrated effort and an in-depth understanding of atypical gait patterns, and their mechanisms, which are so disabling to patients.

"Over reading the data" is an expression sometimes used during a reporting session. Typical, mature, gait is consistent across the age span, and although a joint range of motion may differ at different gait velocities and among TD people, the patterns and sequences of motion do not. Therefore, we should carefully use the term "over reading the data" for the times the situation is rather one of data misinterpretation or lack of understanding than true pathology.

The comprehensive 3D gait analysis report includes all testing data, summarizes the collected evidence for pathological and compensatory mechanisms identified, and provides treatment recommendations. The ideal gait analysis report should be the result of precision medicine (3D gait analysis), clinical expertise, clinical evidence, and sensible reasoning. The report is sent to the referring physician.

8.2.7.3 Standardizing the reporting of kinematics, kinetics, and surface EMG

The ability to plot right and left gait cycles in one graph cannot be overemphasized. Overlapping right and left cycles on a single plot allow for the immediate comparison of both sides (lower limbs) during gait.

Ideally, kinematic, kinetic, and EMG data graphs format would be standardized to facilitate interpretation across laboratories. Every laboratory would use the same colors for depicting the right and left sides. Plotting the data on a universally agreed layout, e.g., sagittal, coronal, and transverse planes, using the same scales of magnitude on the *y*-axis, also facilitates immediate data interpretation across laboratories. A uniform output would give 3D gait analysis the needed sense of maturity regarding protocols. A consistent and standardized physical examination reporting layout should be pursued, as well as gait by video observation.

8.2.7.4 3D gait analysis follow-up exam

Postoperative follow-up gait analysis is done routinely at around 12 months. Longer-term follow-up gait analysis is routinely obtained in some countries. Unfortunately, patients' access to 3D gait analysis is still limited in most countries often due to the economic restriction of the health care systems. In an ideal clinical scenario, long-term follow-up with 3D gait analysis will allow for the identification of new gait deviations related to aging with cerebral palsy. 3D gait analysis will quantify the effectiveness or not of orthopedic and other interventions in the short-term, and more importantly, long-term. Ideally, repeating 3D gait analysis at 2, 5, and 10 years of post-intervention would enable us to understand the precise impact of treatments, which are often deemed appropriate in the short term and medium term, but of relatively unknown consequences, later in the lives of those patients.

8.2.8 Summary

The advent of computerized 3D gait analysis, more than four decades ago, has been essential to the understanding of pathological and compensatory mechanisms seen in atypical gait. This technology was initially used in cerebral palsy and neurological conditions such as spina bifida, but more recently, has expanded for use with other diagnoses. Research in many clinical disorders is now readily available. 3D gait analysis has enabled the healthcare team to quantify and qualify the changes occurring before and after interventions.

8.3 Clinical cases

The following clinical cases demonstrate the role of 3D gait analysis in children with cerebral palsy. Cerebral palsy is defined as a group of non-progressive disorders associated with the development of movement and posture resulting from an injury to the fetal or infant brain [25]. Individuals with cerebral palsy are classified based upon topographical distribution, such as hemiplegia or diplegia, or by functional motor ability. The Gross Motor Function Classification System (GMFCS) is a five-level ordinal scale used to classify the motor function of individuals with cerebral palsy. Children in Level I demonstrate the ability to walk without limitations, children in Level II walk without assistive devices though exhibit limitations in walking outdoors and in the community, children in Level III walk with the use of an assistive device, children in Level IV demonstrate limitations with self-mobility and are transported or use power mobility in the community, and children in Level V have severely limited self-mobility [26]. The Functional Mobility Scale (FMS) is an additional tool that is used to further describe functional mobility for children with cerebral palsy. The FMS rates the walking ability at three distinct distances (5 m, 50 m, and 500 m), which represents mobility within the home, at school, and in the community [30]. 2D video analysis was recorded using two video cameras in both the sagittal and frontal planes while the patient walked at a self-selected speed on a 10-m walkway in the motion analysis laboratory. The physical examination included assessment of the joint range of motion of the lower extremities, special tests (e.g., Duncan Ely test), and assessment of muscle strength using Manual Muscle Testing (MMT), when able [32]. In addition, abnormal tone, such as spasticity, was assessed using the Modified Ashworth Scale (MAS) [33]. The terms and description of joint range of motion and patterns are based upon the international expert consensus described by Nieuwenhuys *et al.* [24]. Phases of the gait cycle are based upon Dr Jacquelin Perry [31].

8.3.1 Case 1

8.3.1.1 Initial visit

History

The patient is a female, who was 6 years, 7 months old at the time of her first instrumented 3D gait analysis. The patient presented with cerebral palsy and spastic diplegia. Etiology is likely related to prematurity at 37 weeks. Cesarean section was performed due to breech presentation with left torticollis treated conservatively. The initial investigations included brain MRI, whole-genome sequencing, and electroencephalogram which were inconclusive.

Family's impressions and concerns at the initial visit

- frequent falling
- toe walking with overall in-toeing
- limited endurance

Mobility

GMFCS II, FMS 551

All findings described below (physical examination, gait by observation, kinematics, conclusions) are observed bilaterally unless otherwise stated.

Relevant physical examination data

- height: 117 cm; weight: 23.4 kg
- increased femoral torsion (45°) and internal hip rotation (70°)
- knee hyperextension (5°)
- positive Duncan-Ely test
- equinus contractures with positive Silverskold (2° with knee extended/14° with knee flexed)
- hip abductor weakness (3+/5 on MMT)
- selective motor control (Total score of 8/10 on the Selective Control Assessment of the Lower Extremity/SCALE)
- knee flexors/extensors spasticity (MAS of 1 for knee flexors and extensors)

2D video analysis – gait by observation (Figure 8.20)

- increased anterior pelvic tilt and a lordotic pattern throughout the gait cycle
- increased knee flexion at initial contact
- knee hyperextension in stance phase
- ankle equinus throughout the gait cycle

3D data

Kinematics (Figure 8.21)

Sagittal plane

- increased anterior pelvic tilt with increased range of motion
- hip extension deficit at terminal stance

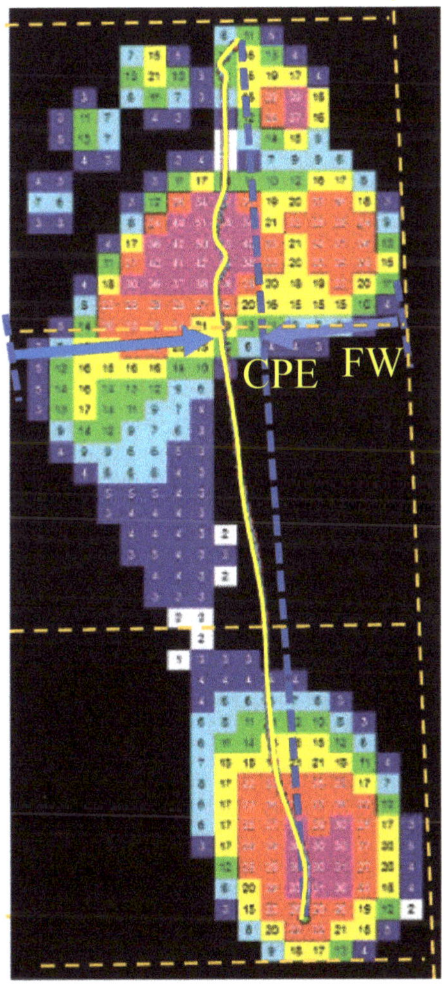

Figure 8.19 CPEI

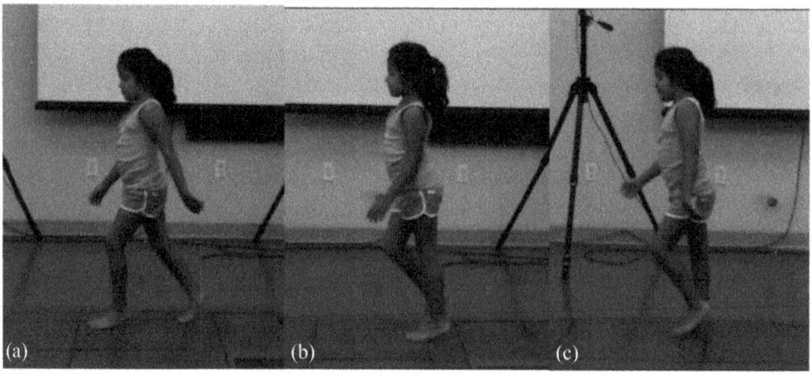

Figure 8.20 Initial visit 2D video analysis sagittal plane on the left leg: (a) initial contact, (b) stance phase, and (c) swing phase

- increased knee flexion at initial contact
- knee hyperextension in mid and terminal stance
- ankle plantar flexion at initial contact
- reversed first and second ankle rockers
- decreased and delayed peak knee flexion in swing phase
- decreased knee extension at terminal swing phase
- excessive ankle plantar flexion in swing phase and at initial contact

Coronal plane

- increased pelvic range of motion
- increased hip abduction at initial contact
- increased hip abduction in terminal swing phase

Transverse plane

- increased range of hip rotation: neutral to external at initial contact and loading response, then internal rotation in pre-swing phase
- in-toeing in stance phase, more significant on the right side

The overall GDI and GPS at each level are reported (Figure 8.22).

GDI
Right: 68.5; Left: 72.7

GPS

EMG – Surface EMG
EMG data were collected from the rectus femoris, vastus lateralis, medial hamstrings, tibialis anterior, and gastrocnemius muscles (Figure 8.23).

- rectus femoris – inadequate activity in pre-swing, early, and mid-swing phase
- vastus lateralis – adequate activity
- gastrocnemius – early and continuous activation in stance phase and terminal swing phase

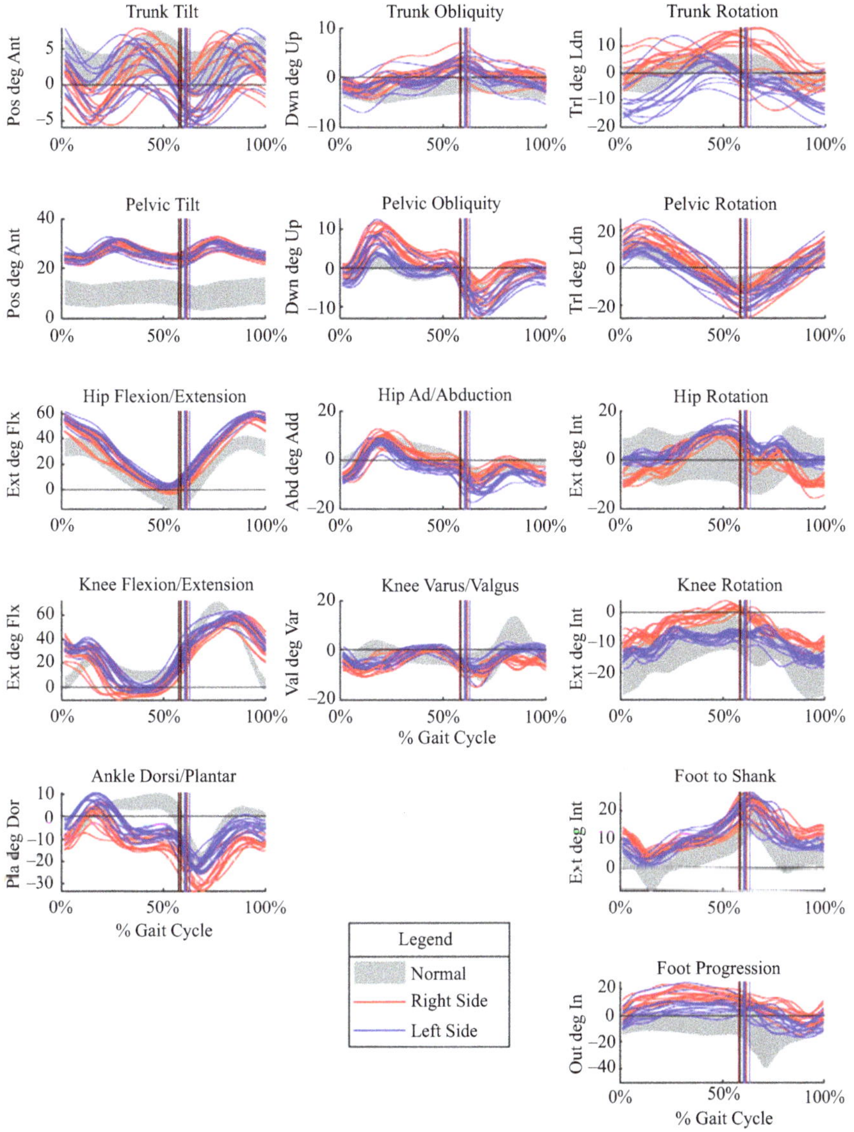

Figure 8.21 *Initial visit kinematics. Red curves: all trials, right side (HS to HS); blue curves: all trials, left side (HS to HS); grey bands: controls ± 1 SD; red/blue vertical lines: R/L toe-off.; black vertical lines: controls toe-off; all data is time-normalized to the gait cycle.*

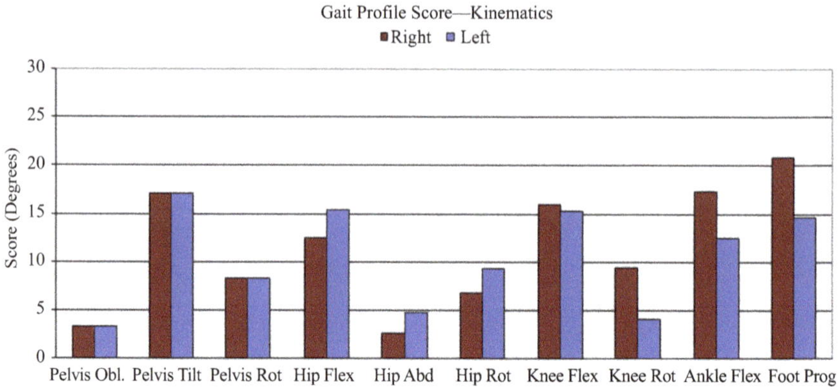

Figure 8.22 Initial visit gait profile score

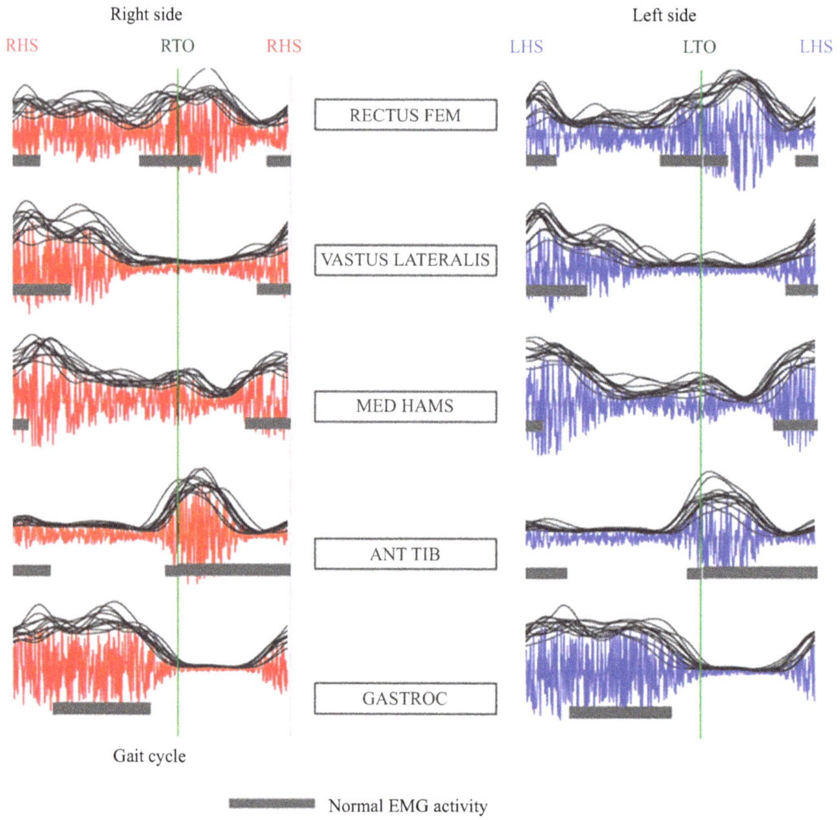

Figure 8.23 Initial visit surface EMG

Pedobarography

Decreased or absence of pressure in heel area at initial contact (Figure 8.24).

Conclusions

- exaggerated anterior pelvic tilt in the gait cycle
- sustained knee extension in mid to terminal stance phase likely related to excessive plantar flexion knee extensor coupling mechanism (gastrocnemius spasticity), excessive femoral torsion, and hip extensor weakness
- decreased knee extension at initial contact and terminal swing phase likely related to bilateral dynamic equinus, and mild hamstrings spasticity
- decreased and delayed peak knee flexion in swing phase likely related to excessive plantar flexion knee extensor coupling and mild rectus spasticity
- dynamic hip rotation in the gait cycle likely related to excessive femoral torsion and selective motor control at GMFCS II

Recommendations

- consider further quantification of femoral torsion
- increased femoral anteversion and equinus may contribute to excessive anterior pelvic tilt
- Strayer-type gastrocnemius recession, bilaterally
- solid ankle-foot orthosis in the first 6–12 months postoperatively

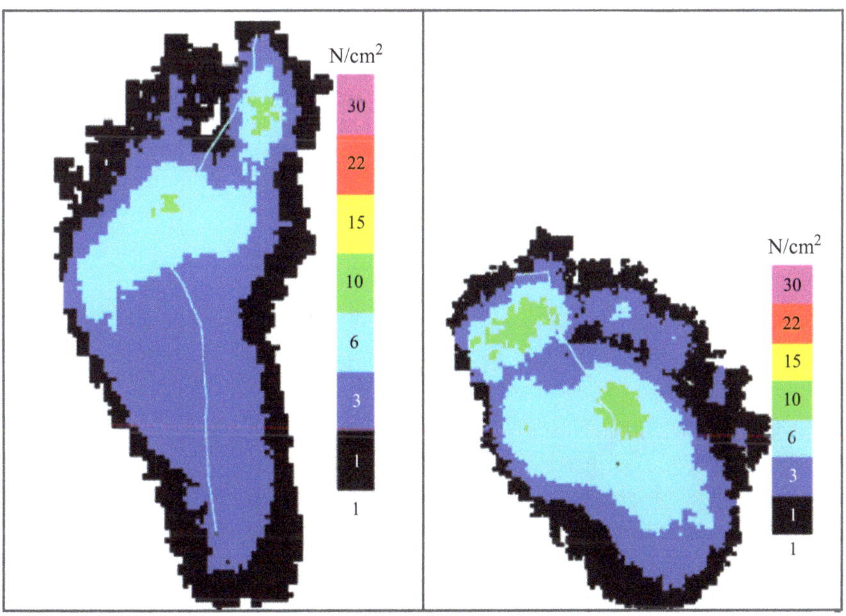

Figure 8.24 Initial visit pedobarography

Orthopedic interventions performed
Shortly after the initial visit, the patient underwent bilateral selective percutaneous myofascial lengthening (SPML) procedures to hip adductors, hamstrings, and ankle plantar flexors.

8.3.1.2 Follow-up visit

The patient returned to the motion analysis laboratory 2 years following the initial visit at the age of 8 years, 9 months.

Family's impressions and concerns at the second visit
- decreased in-toeing and improved heel strike
- decreased endurance when walking longer distances
- difficulty negotiating stairs
- increased tripping and falling, especially with increased fatigue

Relevant physical examination
- height: 131.4 cm; weight: 34.9 kg
- decreased hip extension in prone
- asymmetrical hip abduction range of motion with hips extended (right = 30° and left = 40°)
- Thomas test positive at the right hip (5°)
- quadriceps extensor lag (5–10°)
- positive Duncan-Ely
- ankle dorsiflexion range of motion with the knee extended/flexed (right = 10/20° and left = 15/22°)
- hip extensor weakness (MMT = 3/5)
- good selective motor control (SCALE scores = 10/10 bilaterally)
- knee extensor spasticity (MAS = 2/3)
- mild ankle plantar flexor spasticity (MAS score = 1)

2D video analysis – gait by observation (Figure 8.25)
- increased anterior pelvic tilt and lordotic gait
- decreased knee flexion in loading response
- knee hyperextension in stance phase
- decreased knee flexion in swing phase

3D data
Kinematics (Figure 8.26)

Sagittal plane

- increased anterior pelvic tilt with increased range of motion
- increased left knee flexion at initial contact
- bilateral knee hyperextension in mid and terminal stance phase
- excessive plantar flexion throughout the stance phase
- decreased and sustained peak knee flexion in swing phase

Transverse plane

- sustained and increased hip abduction in swing phase

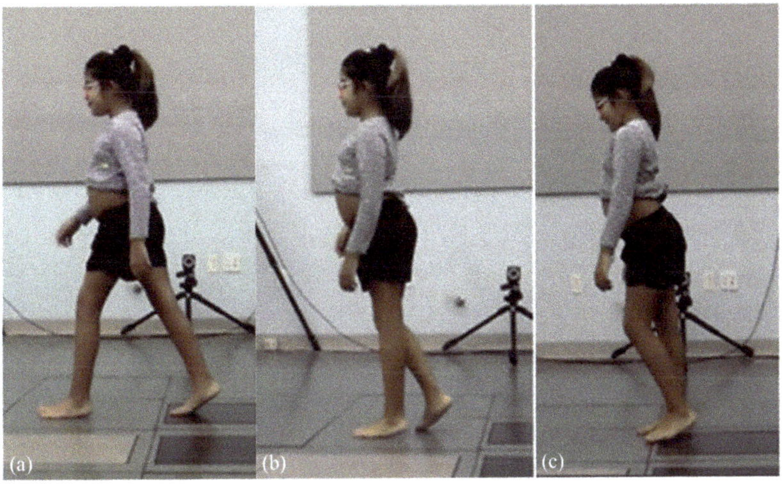

Figure 8.25 *Follow-up 2D video analysis sagittal plane of the left leg: (a) loading response, (b) mid-stance phase, and (c) swing phase. EMG for each muscle is time-normalized to the gait cycle and amplitude normalized to max across all trials: 0–100 (%).*

Coronal plane

- increased right side out-toeing in swing phase

The overall GDI and GPS at each level are reported (Figure 8.27).

GDI
Right: 79.3; Left: 83.0

GPS

EMG – Surface EMG
- unchanged, when compared to preoperative data.

Pedobarography
- improved plantar pressure distribution, when compared to preoperative data (Figure 8.28).

Conclusions
When compared to preoperative data, the follow-up visit shows the following:

- increased anterior overall trunk tilt
- increased anterior pelvic tilt likely related to excessive femoral torsion and hip extensor weakness
- knee hyperextension in stance phase likely due to ongoing excessive plantar flexor knee extensor coupling and hamstrings weakness
- stiff knee in swing phase likely related to imbalance between rectus spasticity and hamstrings weakness

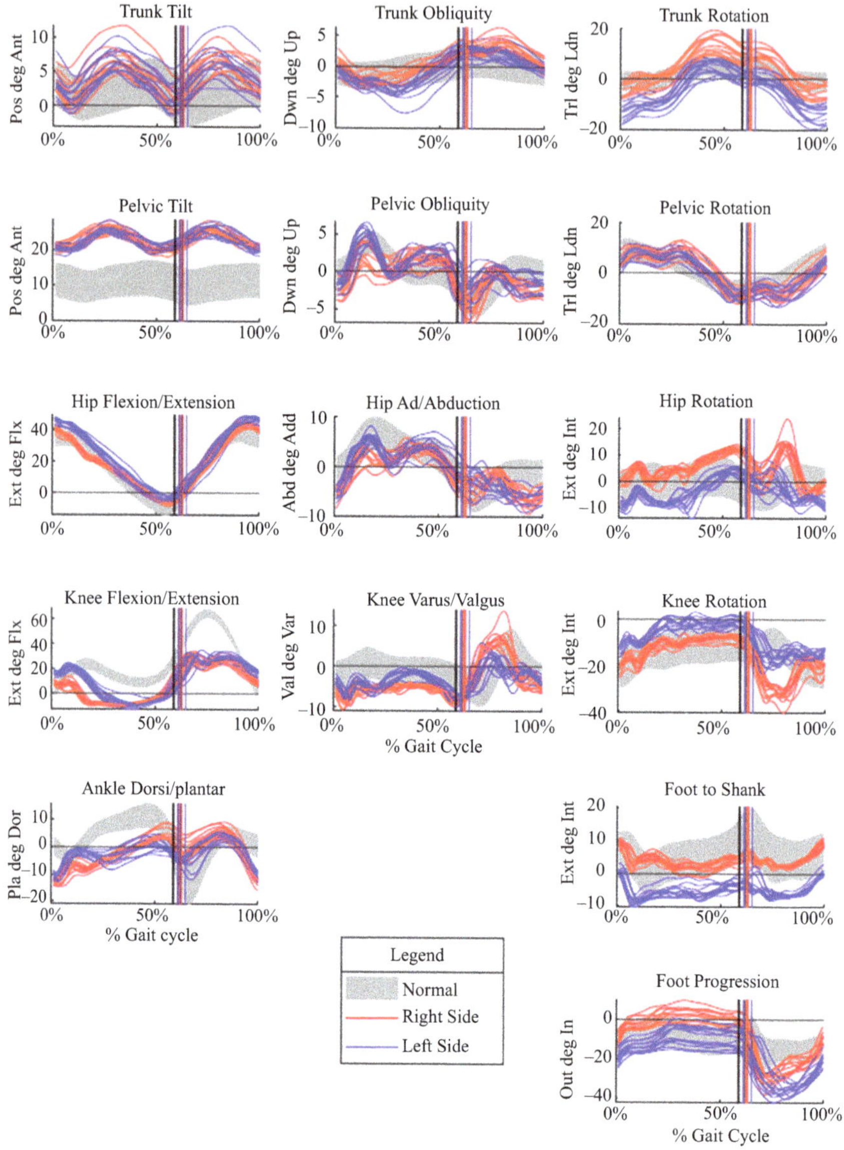

Figure 8.26 *Follow-up visit kinematics. Red curves: all trials, right side (HS to HS); blue curves: all trials, left side (HS to HS); grey bands: controls ± 1 SD; red/blue vertical lines: R/L toe-off.; black vertical lines: controls toe-off; all data is time-normalized to the gait cycle.*

- increased ankle plantar flexion in mid-stance phase likely due to a coping response to the contralateral decrease in peak knee flexion and ipsilateral anterior pelvic tilt

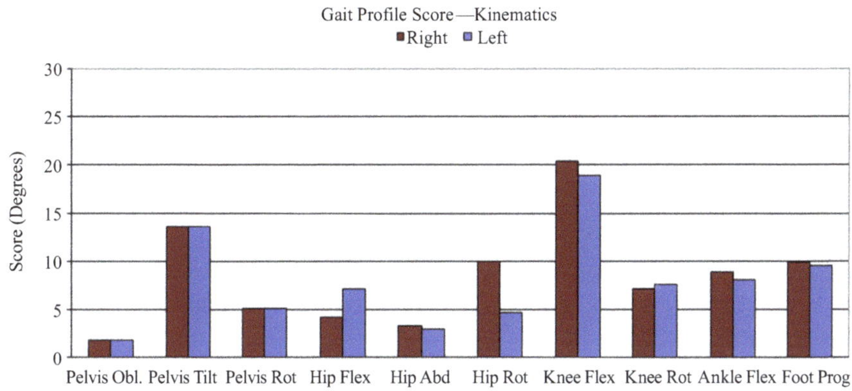

Figure 8.27 Follow-up visit gait profile score

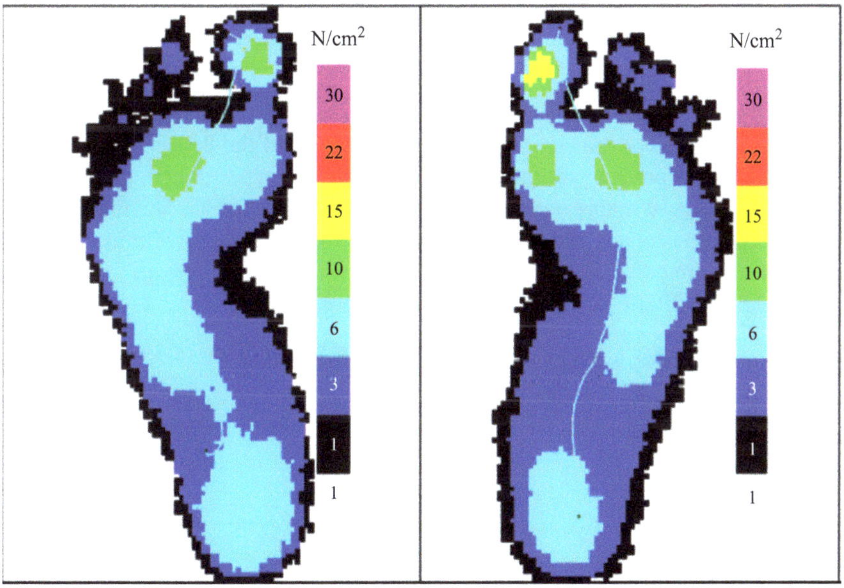

Figure 8.28 Follow-up visit pedobarography

Discussion

A comparison of the initial visit and the follow-up visit to the motion analysis laboratory offers several important findings to discuss. The most notable changes are the deterioration at the knees with hyperextension in stance phase and the diminished peak knee flexion. Following percutaneous muscles/tendon lengthening, the trunk seemed more anteriorly tilted in gait by observation and kinematics and the lordotic pattern was accentuated. Global weakness of hip extensors which was identified in the first gait exam was increased most likely due to the hamstrings percutaneous releases.

The principle of muscle power preservation in an already weakened patient with cerebral palsy should never be overlooked. To our knowledge, recovering patients with cerebral palsy from situations where muscles have been unnecessarily lengthened or over-lengthened are unfeasible with the orthopedic and/or rehabilitation techniques currently available. Long-term longitudinal studies and outcomes of orthopedic procedures that significantly decrease muscle power in patients with cerebral palsy are lacking. However, long-term reduction in strength and endurance often seen in several adults who have had extensive or repeated muscle lengthening procedures in childhood should not be attributed solely to the natural history of cerebral palsy. It should be conspicuous to all, that patients will be unable to walk efficiently in the long term, when several of their essential muscles (such as hip adductors, hamstrings, and triceps surae) are weakened beyond return.

8.3.2 Case 2

8.3.2.1 Initial visit

History

The patient is a male, 8 years, 4 months old at the time of the first instrumented 3D gait analysis. The patient presented with cerebral palsy, spastic hemiplegia on the left side. Etiology is likely related to prematurity at 34 weeks. Vaginal delivery with forceps assistance, breech presentation, twin pregnancy. Brain magnetic resonance imaging (MRI) demonstrating deep white matter lesions in the right hemisphere, consistent with periventricular leukomalacia.

Family's impressions and concerns at the second visit

- tripping/falling
- stiffness
- observable limp
- left ankle pain

Mobility

GMFCS II, FMS 666

Relevant physical examination data

- height: 128.3 cm; weight: 27.2 kg
- right knee hyperextension (+5°)
- decreased left knee extension range of motion (lacking 5° of extension)
- decreased left ankle dorsiflexion range of motion (−12/0° (R1/R2) with knee extension and +4° with knee flexion)
- left hip extensor weakness (MMT = 3+/5)
- impaired selective motor control left side (SCALE score 5/10 on the left compared to 9/10 on the right)
- mild left ankle plantar flexor spasticity (MAS score = 1)

Imaging (Figure 8.29)

- enlocated hips
- mild pelvic obliquity
- asymmetry proximal femoral physis tilt

2D video analysis: gait by observation (Figures 8.30 and 8.31)

- lateral trunk lean, to the right
- posterior pelvic tilt throughout the gait cycle
- increased left knee flexion throughout stance phase on the left
- left ankle equinus throughout the gait cycle

3D data

Temporal-spatial parameters

- decreased velocity – 56% of age-matched TD children
- decreased step length left side more significantly

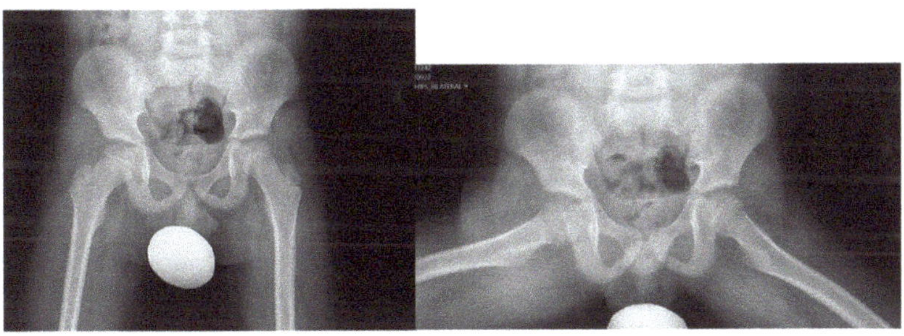

Figure 8.29 Anterior–posterior (AP) pelvis and frog-leg bilateral hip radiographs one-year before surgery

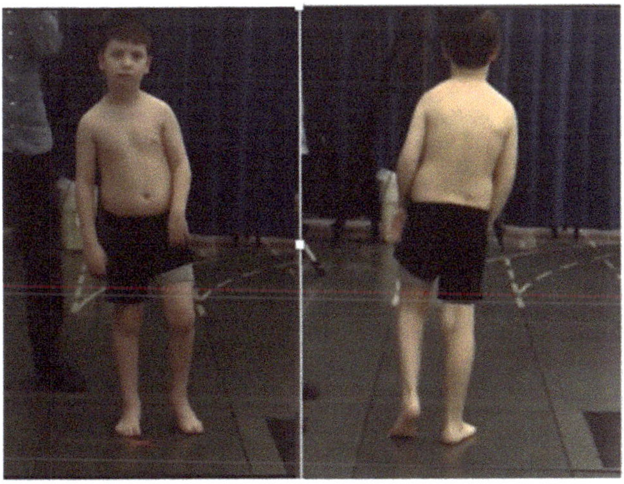

Figure 8.30 Initial visit 2D video analysis frontal plane standing

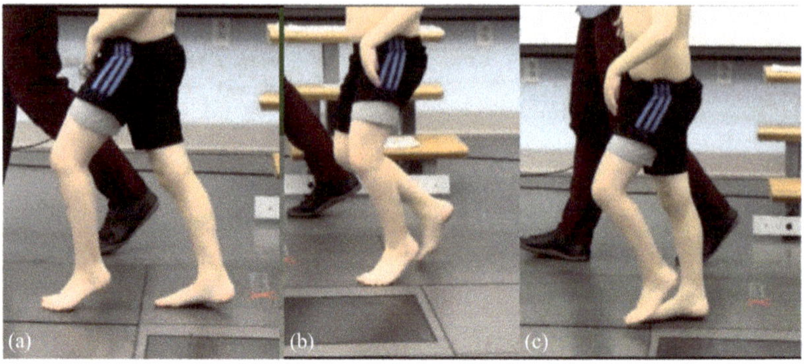

Figure 8.31 Initial visit 2D video analysis sagittal plane – left leg: (a) initial contact, (b) mid-stance, and (c) swing phase

Kinematics (Figure 8.32)
Sagittal plane

- posterior pelvic tilt throughout the gait cycle
- decreased range of motion hips to a greater degree on the left side
- left stiff knee gait with increased knee flexion in stance phase
- increased knee flexion in initial and terminal swing phase
- left ankle equinus throughout the gait cycle

Coronal plane

- pelvic and trunk obliquity (left side up) throughout the gait cycle
- left hip increased adduction throughout the gait cycle
- right hip abduction throughout the gait cycle

Transverse plane

- left pelvis externally rotated throughout the gait cycle
- asymmetric hip rotation – left neutral and right externally rotated
- bilateral out-toeing – right side in mid swing phase, left side throughout the gait cycle

The overall GDI and GPS at each level are reported (Figure 8.33).

GDI
Right: 84.3; Left: 75.6

GPS

Surface EMG
EMG data were collected from the rectus femoris, vastus lateralis, medial hamstrings, tibialis anterior, and gastrocnemius muscles (Figure 8.34).

- rectus femoris – continuous activity
- vastus lateralis – prolonged activity into terminal stance
- tibialis anterior – continued activity throughout the gait cycle
- gastrocnemius – premature activity in terminal swing and early stance phase

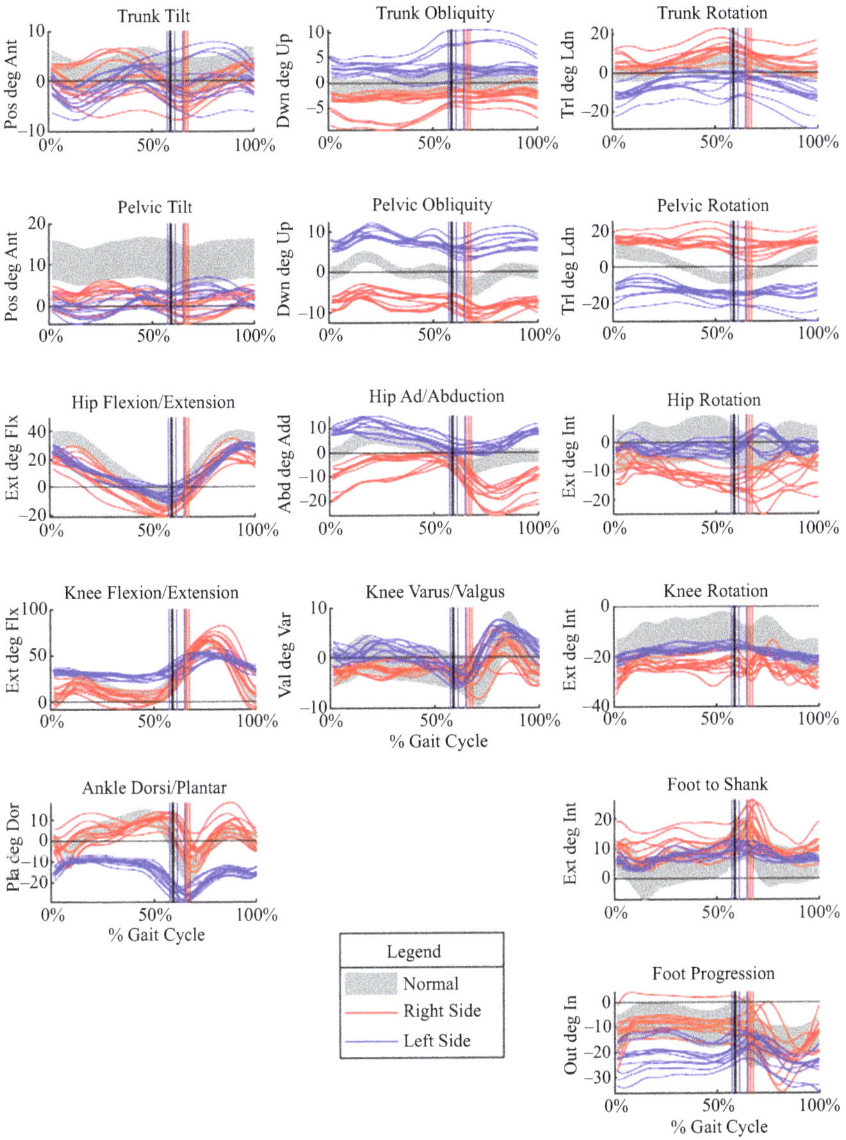

Figure 8.32 *Initial visit kinematics. Red curves: all trials, right side (HS to HS); blue curves: all trials, left side (HS to HS); grey bands: controls ± 1 SD; red/blue vertical lines: R/L toe-off.; black vertical lines: controls Toe-off; all data is time-normalized to the gait cycle*

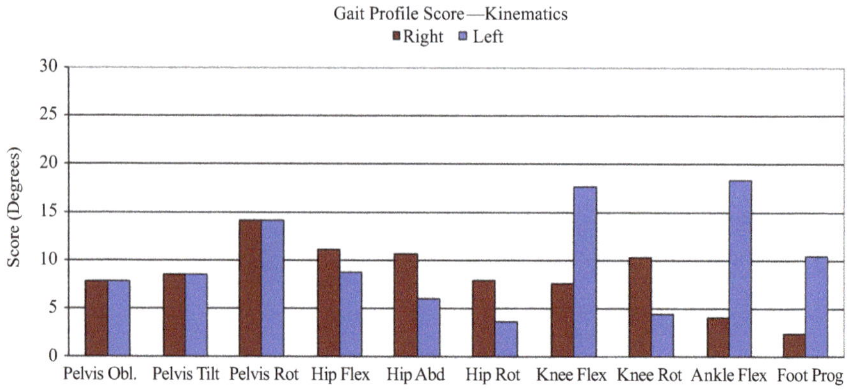

Figure 8.33 Initial visit gait profile score

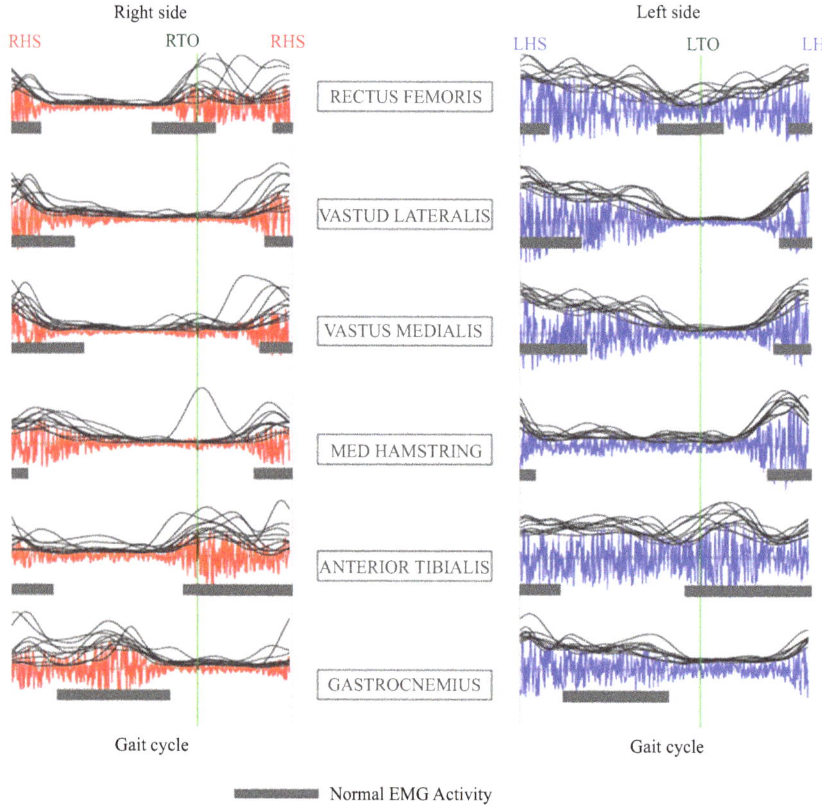

Figure 8.34 Initial visit surface EMG

Pedobarography (Figure 8.35)

- left foot – load bearing restricted to forefoot area
- right foot – increased load bearing on hind foot and lateral aspect of the forefoot

Conclusions

The data suggested this patient presented with left ankle fixed equinus due to gastrocnemius spasticity/contracture. Stiff and flexed knee gait due to rectus femoris spasticity, and medial hamstrings spasticity/contracture.

Recommendations

- left gastrocnemius-soleus recession
- solid ankle-foot orthosis in the first 6–12 months postoperatively
- left hamstring lengthening
- left rectus femoris transfer

Orthopedic interventions performed

- left rectus femoris transfer to semitendinosus
- fractional lengthening of the left semimembranosus
- tenotomy of left gracilis
- left gastrocnemius recession (Strayer technique)

8.3.2.2 Post-operative visit

The patient returned to the motion analysis laboratory for one-year post-surgical interventions, at the age of 9 years, 8 months.

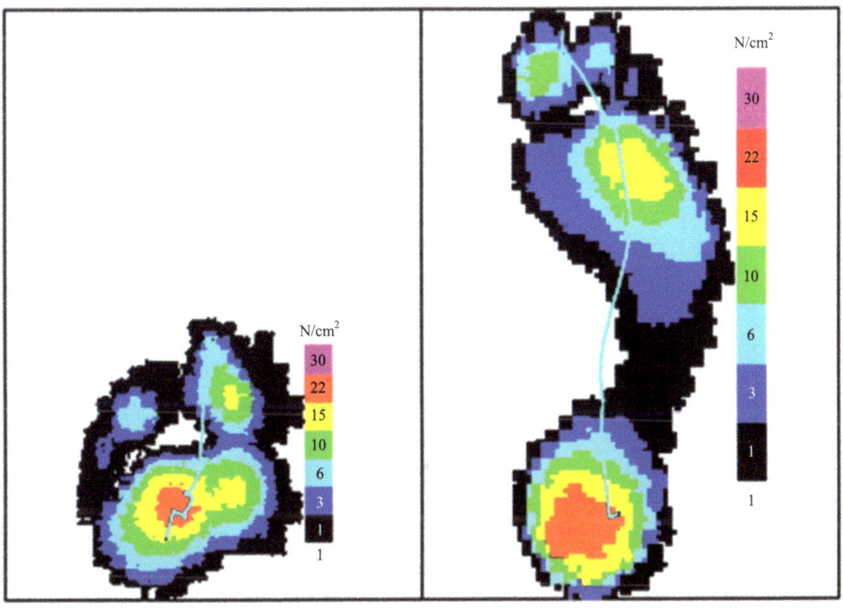

Figure 8.35 Initial visit pedobarography

Family's impressions and concerns at the second visit

- fewer episodes of tripping/falling
- less stiffness
- no longer limping significantly
- subsided left ankle pain

Relevant physical examination

- height: 142.2 cm; weight: 35.4 kg
- dorsiflexion range of motion left ankle (+8° with knee extension and +15° with knee flexion)
- left hip extensors weakness (MMT = 3+/5)
- left impaired selective motor control (SCALE score right = 9/10 and left = 5/10)
- left ankle plantar flexor spasticity (MAS score = 1)

2D video analysis – gait by observation (Figures 8.36 and 8.37)

- full left knee extension in stance phase
- left ankle dorsiflexion in stance phase and swing phase
- left external foot progression angle

3D data

Temporal–spatial parameters (Table 8.1)

When compared to the pre-operative gait analysis:

- increased velocity
- increased left step length

Kinematics (Figure 8.38)

Sagittal plane

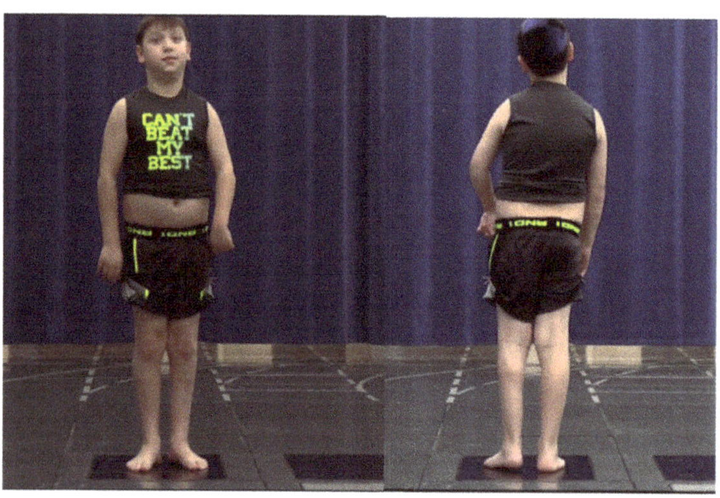

Figure 8.36 Post-operative visit 2D video analysis frontal plane standing

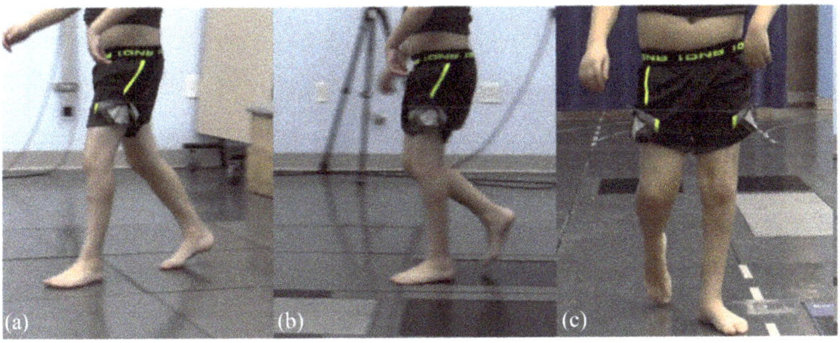

Figure 8.37 Post-operative visit 2D video analysis sagittal plane of the left leg: (a) initial contact, (b) mid stance frontal plane of the left leg, and (c) external foot progression angle

Table 8.1 Temporal-spatial parameters during examinations in initial and post-operative visits

Temporospatial parameters	Initial visit	Post-operative visit	% Normal initial visit	% Normal post-operative visit
Forward velocity (cm/s)	75.5	138.7	56	100
Stride length (cm)	63.8	104.0	52	83
Cadence (step/min)	141.5	160.0	106	120
Step length – R (cm)	37.6	51.8	62	83
Step length – L (cm)	26.1	51.8	43	83

- decreased left knee flexion at initial contact
- abnormal loading response of the left knee
- right knee hyperextension in terminal stance phase
- decreased left knee extension in terminal swing phase
- left ankle equinus at initial contact
- horizontal second rocker left side
- decreased dorsiflexion left ankle at initial contact

Coronal plane

- increased pelvic range of motion
- increased left hip abduction in mid-stance phase

Transverse plane

- increased pelvic range of motion
- increased right hip external rotation

The overall GDI and GPS at each level are reported (Figure 8.39).

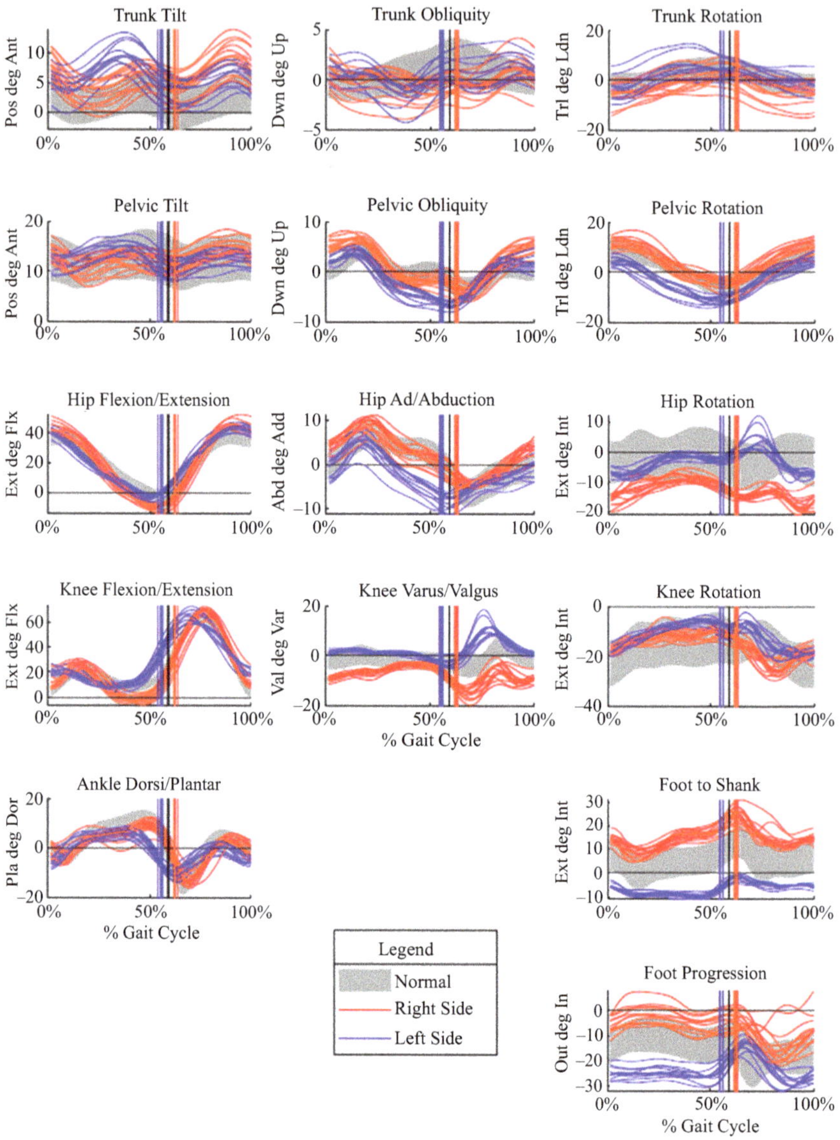

Figure 8.38 Post-operative visit kinematics. Red curves: all trials, right side (HS to HS); blue curves: all trials, left side (HS to HS); grey bands: controls ± 1 SD; red/blue vertical lines: R/L toe-off.; black vertical lines: controls toe-off; all data is time-normalized to the gait cycle.

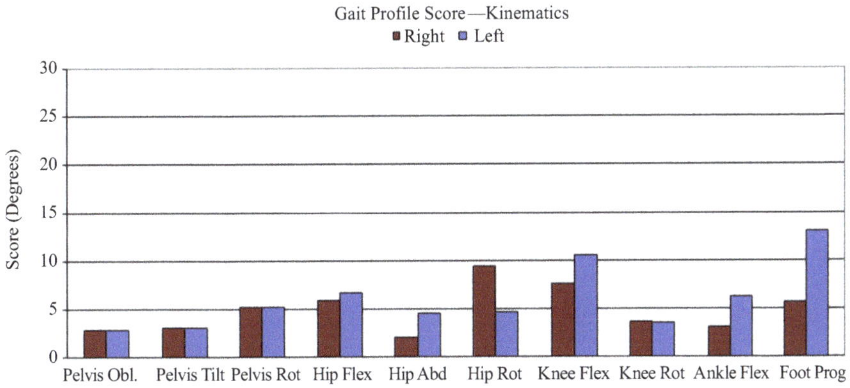

Figure 8.39 Post-operative visit gait profile score

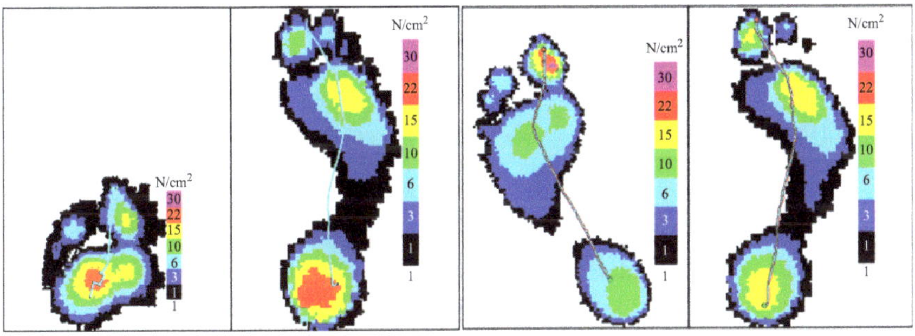

Figure 8.40 Post-operative visit gait profile score

GDI

Right: 101.9; Left: 95.6

GPS

Surface EMG

EMG was collected from rectus femoris, vastus lateralis, medial hamstrings, tibialis anterior, and gastrocnemius muscles.

- no change in patterns of activation between pre- and postoperative data.

Pedobarography

- increased area of load bearing at the left hind and midfoot (Figure 8.40).

Conclusions

When compared to the initial visit, the post-operative visit showed:

- pelvic tilt normalized
- left hip range of motion on the sagittal and coronal planes normalized

- left knee extension improved in stance phase
- abnormal loading response at the left knee
- improved time and range of left peak-knee-flexion in swing phase
- improved left ankle dorsiflexion
- improved pelvic obliquity (less than 5° asymmetry)
- left foot progression angle now in excessive external rotation

Discussion
3D gait analysis was useful in this case to determine a precise "orthopedic surgical dose" to correct the gait deformities on the left hemiplegic lower limb of this patient. The overall gait pattern is significantly improved 1 year after the interventions. The excessive left toe walking identified on the postoperative exam is related to underlying external tibial torsion which was "uncovered" when dorsiflexion was achieved via the gastrocnemius recession. Correction of the tibial torsion may be necessary to improve the lever arm dysfunctions commonly associated with external tibial torsion.

8.4 Summary

Computerized *three-dimensional gait analysis* was the term used to describe this emerging technology, 40 years ago. In recent years, with the dramatic advances in computer, optic, and video sciences, *three-dimensional motion analysis* seems a more appropriate description, for most moving objects or body segments even at high velocities are now studied e.g., trunk, head, neck, and upper limbs movements, as well, a variety of activities of daily living such as sports, lifting objects, ascending and descending stairs to mention a few.

The advent of 3D motion analysis and its historical and spread of use in the evaluation of patients with neuromuscular disorders like cerebral palsy has radically changed treatment options and strategies in those clinical scenarios. Precise quantification of cerebral palsy gait in pre- and post-orthopedic surgeries has contributed to the better understanding of the pathological gait patterns often seen in those patients and the development of more accurate treatment designs or surgical doses. Pre- and postoperative studies, using 3D motion analysis data, have set golden standards regarding achievable results for the sustainable improvement of patient's gait parameters, in the long term.

However, at least three long-term standing problems that have prevented broader use of 3D motion analysis in clinical practice, have not yet been resolved. First, equipment costs, even though drastically decreased in the last 20 years, equipment maintenance–upgrades–replacement, expert personnel, and the relatively large physical area required for the establishment of a useful laboratory, are certainly obstacles in many parts of the world. Second, there is a long and steep learning curve required by all professionals involved in these areas, from the engineers and technicians ensuring adequate data capture by the equipment, to the clinicians collecting and interpreting the patient's physical and anatomic data and those executing treatment interventions. Lastly, the bland interest of many segments of government and society, to promote and emphasize the state-of-the-art care for persons with disabilities such as cerebral palsy limits broader use.

Unfortunately, in the complex area of cerebral palsy where 3D gait analysis has a long-standing history and credibility, poor outcomes of clinical/surgical interventions are not uncommonly observed. It is our impression that, at this point in time, such phenomenon does not correlate with equipment or data collection errors, rather it could be related to data misinterpretation, inadequate treatment propositions, and sub-standard treatment executions. Furthermore, one cannot omit or overemphasize the multiple and tremendous complexities inherent to the patient populations commonly seen in most laboratories across the world.

3D motion analysis is the state-of-the-art and a consolidated science. It will continue to teach those interested. Dr James R. Gage often says: "Gait analysis talks to us." It is our duty to understand the language in depth and "listen to it."

References

[1] Perry, J. (1992). *Gait Analysis: Normal and Pathological Gait*. New Jersey: Slack International, ISBN:1-55642-192-3.

[2] Winter, D. (1991). *The Biomechanics and Motor Control of Human Gait: Normal, Elderly and Pathological*. Waterloo: Waterloo Biomechanics.

[3] Hooper S.L. (1999–2010). Central pattern generators. In *Encyclopedia of Life Sciences*. New York, NY: John Wiley & Sons.

[4] Kiehn O. and Butt S.J. (2003). Physiological, anatomical and genetic identification of CPG neurons in the developing mammalian spinal cord. *Prog. Neurobiol.* 70 (4): 347–361. doi:10.1016/S0301-0082(03)00091-1.

[5] Gottschall J.S. and Nichols T.R. (2007). Head pitch affects muscle activity in the decerebrate cat hindlimb during walking. *Exp. Brain Res.* 182 (1): 131–135. doi:10.1007/s00221-007-1084-z.

[6] Graham-Brown T. (1911). The intrinsic factors in the act of progression in the mammal. *Philos. Trans. R. Soc. Lond. B.* 84 (572): 308–319. doi:10.1098/rspb.1911.0077.

[7] Gouelle A., Megrot F., Presedo A., Husson I., Yelnik A. and Pennecot G. (2013). The Gait Variability Index: a new way to quantify fluctuation magnitude of spatiotemporal parameters during gait. *Gait Posture.* 38: 461–465.

[8] Ferrari A., Benedetti M.G., Pavan E., *et al.* (2008). Quantitative comparison of five current protocols in gait analysis. *Gait Posture.* 28 (2): 207–216. doi:10.1016/j.gaitpost.2007.11.009.

[9] Kaufman K.R., Miller L.S. and Sutherland D.H. (1996). Gait asymmetry in patients with limb-length inequality. *J Pediatr Orthop.* 16 (2): 144–150. doi:10.1097/00004694-199603000-00002.

[10] Kadaba M.P., Ramakrishnan H.K. and Wootten M.E. (1990). Measurement of lower extremity kinematics during level walking, *J. Orthop. Res.* 8 (3):383–392. doi: 10.1002/jor.1100080310.

[11] Grood E.S. and Suntay W.J. (1983). A joint coordinate system for the clinical description of three-dimensional motions: application to the knee. *J. Biomech. Eng.* 105(2):136–144. doi:10.1115/1.3138397.

[12] Wu G., Siegler S., Allard P., *et al.* (2002). ISB recommendation on definitions of joint coordinate system of various joints for the reporting of human joint motion – Part I: ankle, hip, and spine. International Society of Biomechanics. Standardization and Terminology Committee of the International Society of Biomechanics. *J. Biomech.* 35(4):543–548. doi:10.1016/s0021-9290(01)00222-6.

[13] Schwartz M.H. and Rozumalski A. (2008). ESMAC 2007 Best Paper Award, The gait deviation index: a new comprehensive index of gait pathology. *Gait Posture* 28: 351–357.

[14] Hubel D.H. and Torsten N.W. (2004). *Brain and Visual Perception: The Story of a 25-Year Collaboration.* Oxford: Oxford University Press. ISBN 0195176189

[15] Kuffler S.W. and Nichols J.G. (1976). From Neuron to Brain: A Cellular Approach to the Function of the Nervous System. Sunderland, MA: Sinauer Associates, Inc.

[16] Lieber R.L. (2010). *Skeletal Muscle Structure, Function, and Plasticity: The Physiological Basis of Rehabilitation*. Philadelphia, PA: Wolters Kluwer Health/Lippincott Williams & Wilkins.

[17] Gordon A.M., Huxley A.F. and Julian F.J. (1966). The variation in isometric tension with sarcomere length in vertebrate muscle fibres. *J Physiol.* 184 (1): 170–192. doi: 10.1113/jphysiol.1966.sp007909.

[18] DeLuca C. (1985). 'The Mathematics of EMG', Muscles Alive, Basmajian.

[19] Hogan N. and Mann R.W. (1980). Myoelectric signal processing: optimal estimation applied to electromyography – Part I: derivation of the optimal myoprocessor. *IEEE Trans. Biomed. Eng.* 27(7): 382–395. doi: 10.1109/tbme.1980.326652.

[20] Hillstrom H.J., Song J., Kraszewski A.P., *et al.* (2013). Foot type biomechanics. Part 1: structure and function of the asymptomatic foot. *Gait Posture.* 37(3): 445–451. doi:10.1016/j.gaitpost.2012.09.007.

[21] Mootanah R., Song J., Lenhoff M.W., *et al.* (2013). Foot Type Biomechanics Part 2: are structure and anthropometrics related to function? *Gait Posture*; 37(3): 452–456. doi:10.1016/j.gaitpost.2012.09.008.

[22] Diaz M.A., Gibbons M.W., Song J., *et al.* (2018). Concurrent validity of an automated algorithm for computing the center of pressure excursion index (CPEI). *Gait Posture* 59: 7–10. doi: 10.1016/j.gaitpost.2017.09.022. Epub 2017 Sep 22.Gait Posture. 2018.

[23] Vaughan C.L., Davis B.L. and O'Connor J.C. (1999). *Dynamics of Human Gait*, 2nd ed. Cape Town, South Africa: Kiboho Publishers.

[24] Nieuwenhuys A, De Cat J, Desloovere K, *et al.* (2016). Identification of joint patterns during gait in children with cerebral palsy: a Delphi consensus study. *Dev. Med. Child Neurol.* 58(3): 306–313. https://www.ncbi.nlm.nih.gov/pubmed/26330338.

[25] Bax M., Goldstein M., Rosenbaum P., *et al.* (2005). Proposed classification of cerebral palsy. *Dev. Med. Child Neurol.* 47(8): 571–576. doi: 10.1017/s001216220500112x.

[26] Palisano R., Rosenbaum P., Walter S., Russell D., Wood E. and Galuppi G. (1997). Development and reliability of a system to classify gross motor function in children with cerebral palsy. *Dev. Med. Child Neurol.* 39: 214–223.

[27] Ramakrishnan T., Lahiff C.-A. and Reed K.B. (2018). Comparing gait with multiple physical asymmetries using consolidated metrics. *Front Neurorobot.* 12: 2. doi:10.3389/fnbot.2018.00002.

[28] Rozumalski A. and Schwartz M.H. (2011). The GDI-Kinetic: a new index for quantifying kinetic deviations from normal gait. *Gait Posture.* 33(4): 730–732. doi:10.1016/j.gaitpost.2011.02.014.

[29] Baker R., McGinley J.L., Schwartz M.H., et al. (2009). The gait profile score and movement analysis profile. *Gait Posture* 30(3): 265–269. doi:10.1016/j.gaitpost.2009.05.020. Epub 2009 Jul 24.

[30] Graham H.K., Harvey A., Rodda J., Nattrass G.R. and Pirpiris M. (2004). The Functional Mobility Scale (FMS). *J. Pediatr. Orthop.*24(5): 514–520. doi:10.1097/00004694-200409000-00011.

[31] Perry J. and Burnfield J.M. (2010). Gait Analysis. *Normal and Pathological Function*, 2nd ed. Thorofare: SLACK Inc.

[32] Kendall F.P. and Kendall McCreary E. (1983). *Muscles: Testing and Function*, 3rd ed. Baltimore, MD: Williams & Wilkins.

[33] Meseguer-Henarejos A.B., Sánchez-Meca J., López-Pina J.A. and Carles-Hernández R. Inter- and intra-rater reliability of the Modified Ashworth Scale: a systematic review and meta-analysis. *Eur J Phys Rehabil Med.* 2018; 54(4): 576–590. doi: 10.23736/S1973-9087.17.04796-7.

Chapter 9
Gait analysis in rehabilitation

Man Sang Wong[1], *Babak Hassan Beygi*[1], *Hui Dong Wu*[1], *Nicola Eddison*[2], *Nachiappan Chockalingam*[2] *and Aoife Healy*[2]

9.1 Introduction

Human gait is considered as a complex movement where the evolutionary aspects of the musculoskeletal system including the upper limbs, lower limbs, and spine, synchronize to facilitate bipedal propulsion. Coordination of lower limbs on the movement of the body, with the assistance of the trunk, maintains the balance and interaction of this complex with the upper limbs. Increasing the efficiency of the movement makes human gait a very particular phenomenon [1].

The malfunction of the central and peripheral nervous system caused by diseases or aging may disrupt the motor and postural control and be reflected in the fluctuation of measured gait parameters in those individuals as compared to the counterpart healthy population [2].

Normal gait is usually explained in terms of a gait cycle as initial contact of one foot to the next initial contact of the same foot. The gait cycle is subdivided into stance and swing phases where during the stance phase, there will be periods of single or double limb support. During standing and walking, the support limb must have the adequate muscular strength to support the weight of the body to facilitate the progression of the contralateral limb and the whole body. The vector of the ground reaction force (GRF) should be well controlled during walking to increase the stability of the lower limbs and decrease the joint moments and muscle forces. Healthy adults generally use a self-preference walking speed of around 1.3–1.4 m/s and are able to accelerate or decelerate it within the range of 0.8–1.8 m/s. Knee flexion is a key maneuver to shorten the lower limb and facilitate the ground clearance of the trailing foot throughout the swing phase. Healthy people implement these mechanisms during walking to save mechanical energy facilitating forward progression with less metabolic cost [3].

[1]Department of Biomedical Engineering, The Hong Kong Polytechnic University, Hong Kong, People's Republic of China
[2]Centre for Biomechanics and Rehabilitation Technologies, Staffordshire University, Stoke on Trent, UK

Observational gait analysis is conducted with the individual wearing minimal clothing to provide a whole picture of lower limb movement in relation to the stability of the whole body. Initially, walking is observed barefoot to enable the clinician to capture a full understanding of the individual's gait, followed by observation of their gait in the prescribed orthosis or assistive device. This observational evaluation is further facilitated by utilizing the video recording of the gait and possibly force plates [3]. Gait analysis may include the physical examination (muscle strength, spasticity, passive and active joint range of motion, and rotational alignment of the lower extremities), video recording in the coronal and sagittal planes as well as 3-D analyses of kinematics and kinetics [4].

A pathologic gait pattern often occurs due to neuromuscular pathology, joint instability, pain, and other clinical conditions. Either excessive or diminished joint mobility may result in gait deviation patterns throughout the gait cycle which might be directly related to the pathologies or as a result of a compensatory maneuver. These abnormalities can be detected by eye alone, but due to the complexity and the speed at which the phases of the gait cycle change, particularly during the switching of moments around the joints of the lower limbs, only gross deviations can be observed, and the detail of the pathology is harder to detect. The use of 3D motion analysis is a more reliable way to observe gait. Commonly observed primary gait deviations include equinovarus, drop foot (weakness of dorsiflexor muscles leading to insufficient dorsiflexion in swing phase), knee hyperextension, knee instability, and genu varum and valgum. The most commonly seen compensatory deviations in a pathological gait include increased step width, vaulting (exaggerated plantar flexion of uninvolved ankle to provide the clearance of the affected limb), steppage gait (exaggerated hip and knee flexion during swing phase), hip hiking, and trunk lean in different directions (anterior, posterior, or lateral) [3,5]. In this chapter, neuromuscular/musculoskeletal impairments and their corresponding gait deviations and possible rehabilitation modalities are briefly described. The chapter will also discuss the gait of lower limb amputees and their prosthetic management.

9.2 Muscular dystrophies

Muscular dystrophies (MD) are inherited diseases of muscles caused by mutations in a group of genes [6]. Duchenne MD is the most common type of MD and mainly affects males. Other types include myotonic, facioscapulohumeral, Becker, congenital, Emery-Dreifuss, oculopharyngeal, and limb-girdle muscular dystrophies [7]. These disorders vary in age of onset, severity, and pattern of affected muscles, causing chronic and progressive muscle degeneration and weakness, balance disorder, limited range of movement, respiratory difficulty, etc., and as a result, people with MD may walk with an impaired gait or even become unable to walk.

9.2.1 Gait deviations

People with MD, especially with Duchenne MD, often experience impaired gait, including spatiotemporal (e.g., decreased step length and gait speed), kinematic (e.g., decreased dorsiflexion during the swing and increased knee range of motion), and kinetic (e.g., decreased maximal power generation at the hip, maximal knee extension torque, maximal dorsiflexion torque, and maximal power generation at the ankle) parameters [8]. These gait deviations can be associated with muscle weakness and joint contractures/stiffness in the lower extremities. Dorsiflexor weakness or plantar-flexor contractures may cause decreased dorsiflexion during the stance and swing phases. Decreased dorsiflexion during stance can result in the GRF moving anterior to the knee joint center, which may result in knee hyperextension, although people with MD often rely on knee hyperextension during stance phase to compensate for knee extensor weakness, giving them the stability required to continue walking. Dorsiflexor weakness can be the reason for the decreased maximal power generation at the ankle, and for the increased knee flexion during the swing promoting toe clearance. The decreased maximal power generation at the hip may result from the weakness of hip extensor muscles.

9.2.2 Rehabilitation

There is no specific treatment to stop or reverse the progression of MD. All available treatments mainly aim to minimize impairment and to prolong mobility, cardiac, and respiratory function [9]. An interdisciplinary approach to care is essential for the management of MD [10]. Physiotherapy aims to retain muscle strength and joint flexibility as much as possible, to improve the individuals' gait ability; nonetheless, effectiveness of muscular exercise is controversial. Although the active exercise and electrical stimulation might have a small effect on muscular strength, the meta-analysis of the literature showed that muscular exercise did not improve the muscle strength yet modest enhancement in walking endurance was observed in myotonic dystrophy. Respiratory exercises may improve endurance during the exercise period [11]. According to the physiotherapy guidelines, stretching, especially a home stretching program focusing on the ankle, knee, and hip is often recommended for the whole course of MD. At the non-ambulatory stage, individualized orthotic and occupational therapy can be applied to the upper extremities with orthotic and occupational therapy guidelines if necessary [10].

Orthoses aim to position joints, stretch muscles, and provide support during walking. Night-time orthoses combined with passive stretching can prevent contractures or slow the progression of joint contractures. Research has shown a positive effect of an ankle-foot orthosis (AFO) on gait, including an increase in peak dorsiflexion angle and dorsiflexor moment and a decrease in plantar flexion angle and ankle joint power generation, peak hip flexion angle, and hip power absorption [12]. However, there is no reliable evidence suggesting which type of AFOs are most effective in the management of ankle range of motion and/or in prolonging gait ability [13]. Also, there is a lack of evidence to support at which age, and level of joint flexibility, the orthotic treatment should be initiated [14]. Therefore, a regular multidisciplinary clinic is desirable for timely and appropriate intervention [13]. A knee orthosis can be considered for unstable knees, promoting gait ability [15]. Spinal orthoses may help delay scoliosis but may not be recommended in ambulatory and early non-ambulatory stages of MD. Assistive devices such as standing frames, walkers, and wheelchairs may preserve mobility as long as possible.

9.3 Multiple sclerosis

Multiple sclerosis (MS) is a chronic inflammatory disorder that damages the insulating covers of nerve cells in the brain and spinal cord [16]. The onset of MS is often aged 20–40 with unknown cause(s) [16]. The prevalence varies considerably across regions and over time (20/100,000–200/100,000) and is higher in females than males with a ratio of 3:1 [17]. Symptoms are predictable but often include vision problems, muscle weakness, stiffness and painful spasms, fatigue, balance, or coordination dysfunction [18–20] which may be associated with the location and degree of the sclerotic lesions. Many symptoms can impair mobility [21].

9.3.1 Gait deviations

Approximately, 75% of people with MS walk with impaired gait patterns [22] such as decreased gait speed [19,23,24], shorter stride length [19], lower cadence [21], shorter step length [21,23], wider base of support [21,23], and longer dual support-time [19,23]. During gait, the absence of heel-rise in terminal stance, poor clearance in the swing phase, and impaired plantar flexion of the ankle in mid-swing can be observed in some cases [25]. These abnormalities are associated with weakness and asymmetric strength of muscles in lower extremities, and disorders of balance or coordination [23,26]. Wider support-base and longer dual support-time may be strategies to adapt to their poor balance or coordination. Reduced propulsive force is a distinct kinetic change that is often associated with kinematic alternations of the knee and hip joints [25]. Abnormal movement in the head is also observed in some cases. People with MS tend to walk with their heads slightly down and view the environment close to the body, especially in an obstacle condition [27]. It is difficult to generalize the gait characteristics of people with MS since they correlate with the associated neurological disability [21] whose lesions vary greatly among individuals. Gait impairments can occur in both early and advanced stages of MS, even in minimally impaired MS (Expanded Disability Status Scale range: 0–2.5) [28].

9.3.2 Rehabilitation

Various interventions for the management of MS have been proposed yet no gold standard intervention is available. A recent summary of a comprehensive systematic review revealed moderate evidence which supported the efficacy of 8 weeks of weekly physiotherapy focusing on balance and gait improvement, and weaker evidence supporting the efficacy of other therapies, including comprehensive cross-specialty rehabilitation, individualized exercise programs, motor balance training, and breathing-enhanced arm exercises [29]. A systematic review supported the positive effect of treadmill training with or without body weight support (BWS) or robotic-assisted training on the improvement of gait speed and endurance; however, it remained unclear which type of treadmill training is most effective [30]. Also, BWS less than 30% was suggested for treadmill training [31]. Functional electrical stimulation (FES) is also recommended for facilitating functional activities [20]. A recent systematic review has shown a positive effect of FES on gait, especially gait speed [32], yet the literature is equivocal [33]. Therefore, more robust evidence is still needed to substantiate the effectiveness of FES. Peroneal FES is the most common site of stimulation [32]. Transcranial direct current stimulation can also be used to improve gait and mobility; however, its effect needs further investigation [34].

Orthoses may help improve deficits in gait and facilitate safe ambulation in people with neuromuscular diseases like MS [20]. AFOs were found effective in knee and ankle control during walking [35]. Various types of AFOs are available for different gait impairments resulting from MS: AFOs which restrict plantar flexion are beneficial for an isolated dorsiflexor weakness and will control the limb in the swing phase only. If the patient has both weak dorsi and plantar flexors but no coronal instability, an AFO with a form of energy return may be appropriate in an attempt to replace the lost power at push-off during terminal stance. A bespoke rigid AFO, with trimlines anterior to the malleoli will control the limb in the sagittal and coronal planes during the swing and stance phases. All AFOs should be biomechanically optimized to ensure optimum gait efficiency for the patient [36]. Stimulating insoles may benefit balance and gait, but, a recent systematic review revealed a non-evident effect [37]. If necessary, assistive devices such as canes, forearm crutches, and walkers can be used by people with balance impairment or for pain relief by distributing the loading on the lower extremities [20].

9.4 Osteoarthritis and rheumatoid arthritis

Arthritis is an umbrella condition that characterizes inflammation of one or more joints in the human body. Osteoarthritis (OA) and rheumatoid arthritis (RA) are very different forms of the same broader condition; OA is a degenerative disorder and RA is an autoimmune disorder, both result in a gradual breakdown of joint tissues over time [38,39]. These two diseases commonly affect joints including hands, fingertips, hips, knees, and spine, and typically at the neck or lower back, and almost share similar symptoms in joints including pain, tenderness, swelling, and stiffness. However, RA often presents symmetrically and in multiple joints at once, whereas OA can affect isolated joints depending on the mechanism of the disorder. When the condition gets severe, people may experience increased joint pain, decreased joint range of motion, and joint instability, along with muscle weakness, bone spur, and joint deformity which might affect the gait ability [38,39].

9.4.1 Gait deviations

People with OA or RA in the lower extremities may experience impaired gait. When bilateral hips are affected by OA, individuals may demonstrate abnormal gait characteristics, such as decreased cadence, ankle power generation, step width, hip extension and abduction angles, and hip abduction moment, along with a dropped pelvis during stance and an anterior pelvic tilt during the whole gait cycle [40]. For people with unilateral hip OA, an asymmetric gait can be observed, such as shorter step length and stance duration in the affected limb [41] aggravated by gait speed [40]. Knee OA commonly happens in the medial compartment, causing genu varum and increased adduction moment [40]. Individuals often walk at reduced gait speed along with the external rotation of the leg [40,42]. A significant increase of lateral trunk inclination toward the support limb along with decreased external hip adduction moment is also observed [42–44] but, alternations in the sagittal plane are not evident [43]. This may be a strategy to reduce the external knee adduction moment and loading on the medial compartment. Subsequently, this biomechanical adaptation may cause hip abductor weakness and lower back pain. When the ankle and foot are affected, decreased gait speed along with a reduction of power and range of motion of the ankle joint in the sagittal and transverse planes can be observed [40,45]. Similar to OA, people with foot and ankle RA often adopt an antalgic gait, such as reduced gait speed, cadence, and stride length, ankle power with increased peak plantar pressure at the forefoot, and increased stance duration, while they tend to avoid extreme positions of the foot/ankle complex [40,46,47]. Limiting the joint range of motion may be a strategy to accommodate the ankle and foot pain or deformity. Limiting dorsiflexion can contribute to an increase in knee and hip flexion [40]. Specific to forefoot RA, a late stance and reduced hallux dorsiflexion at the terminal stance and pre-swing can be found. For hindfoot RA, increased heel eversion, internal tibial rotation, and forefoot inversion and decreased navicular height can be observed [40]. Hindfoot RA tends to affect gait more than forefoot RA [40].

9.4.2 Rehabilitation

Physiotherapy is important to the management of arthritis, including pain relief and functional preservation. Exercises for arthritis may contain strengthening as well as active stretch exercises aiming to retain joint stability and mobility. Exercise is widely recommended for the management of arthritis [48]. A study reported exercise can ease the symptoms in some cases but did not decrease the joint moment of force, especially for knee OA [49]. Specific exercises for foot and ankle arthritis can include strengthening of intrinsic muscles, active stretching of plantar fascia, Achilles-tendon, and peroneal muscles, and active exercises to improve joint mobility [50]. Real-time biofeedback gait training was reported to be effective in the reduction of knee adduction moment and pain as well as in functional improvement of people with knee OA, however, stronger evidence is needed [51]. Transcutaneous electrical nerve stimulation is recommended for pain relief in knee OA, yet its effect has been inconclusive [52]. Acupuncture is also suggested for the management of chronic moderate to severe pain in knee OA. A recent meta-analysis reported that acupuncture (1–3 sessions per week, at least 4 points needled at least 20 min or more for each painful knee) had a positive effect on long- and short-term physical function but only short-term pain relief [53]. Also, education/self-management, weight management, and thermal modalities can be deployed, if necessary, in people with arthritis [48].

Orthoses may be helpful for symptomatic alleviation in certain joints. Offloading knee orthoses and lateral wedged insoles or footwear are often prescribed for knee OA to correct the varus alignment and reduce the knee adduction moment during gait. Studies demonstrated the positive effect of knee orthoses on pain reduction and gait improvement (joint range of motion, gait speed, and step length) [54,55]. Wedged footwear may be beneficial for decreasing knee adduction moment and increasing gait speed [56,57]. It was reported that wedged footwear combined with medial arch support may help minimize ankle eversion but may be associated with a decrease in the reduction of the knee adduction moment [57]. Hence, balancing the effect on both knee moment and ankle eversion is particularly important. It was found that a combination of knee orthoses and wedged footwear can provide a greater reduction in the knee adduction moment than either standalone application of knee orthosis or wedged footwear [58]. But their bulky appearance and skin irritation may result in compliance issues. Although the positive effect of knee orthoses and wedged footwear has been widely reported, the literature is equivocal [48]. Therefore, further research is required. For ankle and foot arthritis, customized footwear may be helpful for people with abnormal foot function, foot joint deformity, malalignment of the foot, and forefoot plantar hyperpressure to improve their gait [50,59,60]. A conclusive answer on the effectiveness of footwear for RA is still under discussion [61]. If people experience impaired balance or pain, assistive devices such as a cane and walker can be considered to prevent falls or relieve pain by distributing the loading in the lower extremities.

9.5 Spina bifida

Spina bifida is caused by the failure of a portion of the fetal spinal cord to close in the early stages of pregnancy, resulting in partial exposure of the spinal cord at birth [62]. The lower back is the most common location [62]. Myelomeningocele is the most severe type of spina bifida and can cause muscle paralysis and imbalance, resulting in impaired mobility.

9.5.1 Gait deviations

Children with myelomeningocele may present with an impaired gait, such as decreased speed, cadence, step length, maximum knee extension, knee flexion during swing, an increase in pelvic rotation/obliquity, hip abduction and flexion, pes valgus, and abnormal femur and tibial rotation [63–65]. Gait deviations are associated with the severity and level of the lesion [63]. An increased pelvic obliquity could necessitate hyper-mobility of the lumbar spine and late lumbar degenerative alterations and instability and lower back pain. Commonly, extensive weakness of plantar flexors with or without additional weakness of dorsiflexors can cause kinematic changes in trunk, pelvis, hip, and knee in the sagittal, coronal, and transverse planes [66]. In this case, increased extensor loading can be seen on the knee to adapt to the loss of ankle joint moment capacity [67]. People with weakness in knee extensors often show knee hyperextension during the stance with forward trunk lean in the early stance to move the gravity line anterior to the knee joint center, locking the knee in extension. By contrast, extensive weakness of the hip abductor can result in larger kinematic changes in all three anatomical planes, including increased trunk lateral sway, trunk and pelvic rotation, pelvic hike, and knee extension [66]. Also, the hip may present an abnormal adduction moment and replace the knee as the energy absorber in an early stance. When extensive weakness occurs in the hip extensors, apart from reduction of sagittal movement and posterior trunk tilt, people often adopt a similar gait strategy to those with weakness of hip abductors [66].

9.5.2 Rehabilitation

Physiotherapy is used to optimize muscle function and preserve physical function. Exercises (e.g., muscular endurance and strengthening) and passive/active stretching can be indicated in cases of neuromuscular imbalance. A program of a 12-week home-based progressive treadmill training for ambulatory children had a long-term effect on ambulation (e.g., gait speed) but only a short-term effect on aerobic fitness [68]. A neuromuscular physiotherapy program of 6-month home-based whole-body vibration with interval intensive therapy showed a positive effect on gait speed but only after the active training period [69]. Threshold night-time electrical stimulation resulted in small improvements in muscular strength but was deemed impractical due to the recommended schedule of 6 nights per week, which was reported to be too intensive by users [70]. Besides, a transition to adult care, psychological concerns and quality of life, education and employment, etc., should also be considered when making a rehabilitation plan [71].

When thoracic and high-lumbar levels are involved, hip-knee-ankle-foot orthosis (HKAFO) or reciprocating gait orthosis (RGO) can be used to support the upright weight-bearing and mobility [72]. Wheelchairs are often considered as an energy-efficient way for people with high-level lesions to achieve mobility [72]. However, the side effects of long-term use of a wheelchair such as diminished vital capacity, respiratory muscle endurance, and shoulder muscle strength, should be considered [73]. Hence, special rehabilitation programs such as stretching, resistance training, and aerobic exercise are suggested for long-term wheelchair users [71]. When lesions occur at lower-lumbar and sacral levels, below-knee muscles can be affected. A rigid bespoke AFO can compensate for weak or absent ankle plantar and dorsiflexors [67,72], improving gait (e.g., reduced knee flexion in the stance and internal knee-extension moment, increased stride length and gait speed) and decreasing the energy cost of walking [74]. Research has demonstrated that AFOs combined with twister cables can improve the control of rotational malalignment before reaching an appropriate age for surgical correction [72], although in practice, this may only be successful in a small portion of cases. Knee-ankle-foot orthoses (KAFO) with a free knee joint may help reduce excessive valgus stress at the knee joint caused by rotational malalignment [72]. Forearm crutches allow upper extremities to share the loading in lower extremities [72] and increase the base of support, decreasing the weight-bearing of the lower extremities and medial–lateral control demands. Hence, using a crutch may help improve gait (e.g., decreased deviations in pelvic and hip kinematics, and increased gait speed) and reduce the energy cost of walking [72,74,75]. When walking with an AFO, additional use of crutches may help lessen the trunk dynamic range of motion and obliquity [76].

9.6 Poliomyelitis

Poliomyelitis (polio) is an acute enterovirus-infectious disease that damages the anterior horn cells of the spinal cord and brainstem while disrupting motor units and subsequently leading to muscle weakness or complete paralysis [9]. The prevalence varied across countries from 165 in high-income to 425 in low-middle-income countries per 100,000 person-year [77]. People who suffer from polio may experience muscular weakness, fatigue, pain, and respiratory problems [52]. The muscles in the lower extremities are most often affected and usually present unilaterally resulting in a decline in gait [9]. The role of the dorsiflexor muscles is important during the gait cycle: the eccentric action from initial heel contact to foot flat to control the plantar flexion, contraction of the foot from foot flat to mid-stance to move the body over the supporting foot, and action in the swing phase to keep the foot in dorsiflexion position to facilitate ground clearance. Mild weakness of these muscles will present in the form of "foot slap" following the heel contact while the severe weakness will induce a "foot drop" which interferes with foot clearance during the swing phase and may result in the compensatory movement of the hip and knee flexion [78].

9.6.1 Gait deviations

People with polio often walk in an asymmetric gait pattern with a decreased gait speed, cadence, step length, and double support duration [9,79,80]. People with post-polio syndrome tend to present with a reduced gait speed than those without the syndromes [9]. Deviations in lower-extremity kinematics, such as ankle plantar flexion at initial contact, hip flexion at toe-off, and excessive hip flexion during swing phase were also observed [9]. The knee is mostly affected; when partial or complete paralysis of the knee extensor occurs, people usually extend the knee excessively during stance with anterior trunk lean in early stance, moving the line of gravity anterior to the knee center to stabilize the knee [81]. If the hip extensor is also affected, people often push the thigh backward or perform a sudden extension movement of the trunk to lock the knee in the stance phase [81]. External rotation of the lower extremity can also be a possible compensation to stabilize the knee joint when several groups of muscles are paralyzed [81]. With hip abductor paralysis, classic limps or swing of the shoulder on the contralateral limb can be seen [81]. When calf muscles become excessively weak, two dominant gait patterns can be found, one with non/mild-deviation in ankle angle, ankle moment, and knee angle, and one with strong deviation in ankle angle and non/mild-deviation in ankle moment and knee angle [82]. Foot slap, calcaneal gait, and flail foot can be observed when paralysis occurs in the dorsiflexors, plantar flexors, and all muscles of the ankle and foot, respectively [83]. Weakness in the lower-extremity muscles is the major contributor to the gait deviations as well as fatigability [84]. Other factors such as balance disorders and pain could also contribute to gait performance, especially over a longer distance [85].

9.6.2 Rehabilitation

Comprehensive interdisciplinary rehabilitation is important for the management of post-polio syndrome, and rehabilitation programs should address the individuals' treatment goals and emphasize individualized and goal-oriented objectives [86]. Muscular strengthening can be beneficial where deficiency and pain are not evident or where strength is in the phase of progressive loss [87,88]. If increasing fatigue of muscle or pain is reported, a period of rest or ceasing training altogether is suggested [87]. Stretching may help prevent contractures or slow progression of mild contractures of joints in neuromuscular disease [89]. Aerobic training programs conducted at submaximal limits have shown a positive effect on tolerance [87,88] but only for the short-term [87]. A training program of 6–8 weeks, 2–3-times per week in intervals of 20–40-min at moderate intensity (not more than 70% of the theoretical maximum heart rate) is suggested [87]. Aquatic physiotherapy can be an attractive tolerated rehabilitation therapy with a positive impact on muscle strength and gait capacity as well as on pain relief [87]. When people with polio experience increasing fatigue or face difficulty with activities of daily living, energy conservation strategies such as orthoses, assistive aids, and environmental and lifestyle adjustment should be considered [87].

Orthoses play an important role in the rehabilitative care of people with polio. A variety of AFOs, designed for the individual patients with partial or complete restriction of dorsiflexion and/or plantar flexion can be used in people with weakness of the dorsiflexors, plantar flexors, or all muscles in the ankle and foot [83]. AFOs have been shown to improve gait (speed, ankle dorsiflexion, and knee flexion in mid- and terminal-stance) [90]. KAFOs with posterior knee support may be advantageous for people with excessive knee extension, e.g., more than 10° [81,83]. People with polio can benefit from conventional KAFO designs (consisting of the drop-locked knee joints and leather/metal construction), and more advanced KAFOs using polypropylene or carbon-composite shells [91], and powered KAFOs [92,93]. Carbon-composite KAFOs have been reported to improve knee and ankle kinematics/kinetics (knee flexion, forward excursion of the center of pressure, peak ankle moment, and timing of peak ankle power) due to their lower energy-cost demands as compared to conventional KAFOs [91]. Powered KAFOs may facilitate a more natural knee flexion in swing [94] and decreased compensatory displacement of the pelvis [92] but reduced gait speed [92–94] and distance walked [93] were reported in comparison with the conventional drop-locked KAFO.

However, any KAFO design must take into account the individual presentation of the patient and hip muscle power, as it is a prerequisite for the prescription of more high-tech powered and stance-control KAFOs. When using a KAFO, additional heel-to-toe rocker sole adaption may provide a positive effect on gait [95]. It should be noted that walking on the real-life ground surface usually requires greater demands (e.g., higher shear forces in knee and external peak knee flexion moments) from people with polio and their KAFO for support and stability than walking on the smooth level ground [96]. HKAFO with a full caliper and pelvic band may be considered for people with the paralyzed hip abductors [83]. A positive impact of using a walking device on gait (increased gait cadence and shorter stride time) was demonstrated [79].

9.7 Spinal deformities

9.7.1 Kyphosis

Kyphosis is a physiologic posterior sagittal thoracic curve of the spine with an angle of 20°–40° [97]. Hypo-kyphosis, known as flat back, refers to a decrease of kyphosis while hyper-kyphosis, known as round back, refers to an increase of kyphosis, resulting in sagittal malalignment of the spine. Hyper-kyphosis is more common in the upper thoracic than in other locations. People with mild hypo- or hyper-kyphosis can be asymptomatic, but when the condition becomes severe, back pain associated with movement and postural imbalance may occur. In some severe cases especially in adults, a reduction of spinal motion can be found due to the presence of degenerative processes [98]. Severe sagittal malalignment mostly contributes to pain and disability in adults with spinal deformity [99] resulting in impaired gait performance which can deteriorate with an increase of hyper-kyphosis [98].

9.7.1.1 Gait deviations

People with spinal deformities, especially the elderlies with hyper-kyphosis may present with an abnormal gait (decreased maximum hip abduction moment in late stance phase and increased posterior tilt and pelvic rotation) [100]. When positive spinal balance (shifting of the C7 passing plump line more than 2 cm in front of the posterosuperior corner of the S1 vertebral body) occurs, these individuals may walk in a crouch gait (increased knee flexion at the mid terminal stance and decreased step length) [101].

9.7.1.2 Rehabilitation

The biomechanical base of conservative intervention is to reduce loading on anterior walls of the vertebral body and intervertebral disc [97]. Concerning physiotherapy for the management of hyper-kyphosis, the International Scientific Society on Scoliosis Orthopaedic and Rehabilitation Treatment (SOSORT) guideline makes specific recommendations: self-control of posture and self-elongation with use of proprio-ception are the most common physiotherapy exercises used for the management of hyper-kyphosis; other exercises, such as muscular endurance, strengthening, and breathing, can also be considered at the same time; they should be repeated on a daily basis for 20 min; they are mainly indicated for juvenile kyphosis and back pain especially when kyphosis is rigid; it is useful before orthotic intervention. The aims of physiotherapy and modalities remain controversial [97]. Recommendations of physiotherapy in hypo-kyphosis are rarely found.

Apart from reducing continuous loading on the anterior walls of the vertebral body and disc, orthoses may help realign muscular forces and give proper space for optimum development of vertebral bodies and discs. Specific recommendations for orthotic management of hyper-kyphosis given by SOSORT are as follows: the primary indications for early orthotic management are rigidity and severity of curve; orthosis wearing should start at the beginning of puberty and wean off at the end of growth but without waiting for bone maturity at Risser 5; for an adolescent thoracic hyper-kyphosis, rigid orthosis with 4- or 5-point force system is preferred and must be worn for the whole night and part of the day; for an adolescent thoracolumbar hyper-kyphosis, rigid orthosis with 4-point force system is preferred and must be worn during the sitting positions; and for a juvenile hyper-kyphosis, Milwaukee orthosis is recommended and must be worn part-time [97]. For adult hyper-kyphosis or hypo-kyphosis, few orthotic recommendations were found in the literature.

9.7.2 Scoliosis

Adolescent idiopathic scoliosis (AIS), the most prevalent type of scoliosis, is a growth deformity that affects the spine structure, trunk balance, and mobility with a modified locomotion. This condition is a 3D deformity of the spine and rib cage where the vertebral body shifts to the convexity of the curve as opposed to the spinous processes which displace toward the concave region of the curve [102]. The prevalence of this condition for curves greater than 10° in the coronal plane ranges from 1.5% to 3.0% [103].

9.7.2.1 Gait deviation

People with scoliosis demonstrate a distorted gait pattern. The resulting gait includes a decreased step length, diminished range of motion in extremities, asymmetry of trunk rotation, and GRF. Compared to healthy controls, the energy cost and oxygen consumption in the scoliotic population increases while the muscle efficiency shows a reduction trend [1].

Traditional gait parameters such as time, distance, vertical and anterior-posterior GRF, and the sagittal plane joint angles do not reveal a significant asymmetry between the left and right sides. Rotational alignment of the upper trunk in the transverse plane seems asymmetric in individuals with AIS as the orientation of the shoulder is asymmetric in response to the line of progression while symmetrical rotation of the pelvis on both sides is observed. Spinal orthosis was found to reduce the trunk and pelvic movements, but it had no effect on the correction of abnormal rotational trunk-pelvis alignment [104]. In another study, a decrease of 7% in step length of people with AIS was shown in comparison to normal participants [1].

The role of somatosensory, visual, and vestibular inputs in balance control is paramount. Studies show that there is a large asymmetry in recorded bilateral latency of somatosensory evoked potentials (SSEPs) in people with AIS which indicates the abnormal somatosensory function in these individuals [105]. The GRF acts as a vector representation of the acceleration of the center of mass (CoM). This disorder in sensory integration might prevent the spatial calibration of the center of pressure (CoP) in response to the CoM and that might be the reason why the balance impairment in people with AIS is observed [106].

Previous literature has contradictory results in terms of impairment in the postural balance of people with scoliosis as compared to their able-body counterparts. This discrepancy might be attributed to the characteristics of involved curves including the curve type and location and the interaction of these curves with postural balance though there are studies in which no relationship between the noted gait asymmetry and curve direction, magnitude, or vertebral rotation was observed [107]. This indicates the need for a comprehensive combined study of kinematics and kinetics to appreciate the neurological impairment in scoliosis cases. Considering kinematics, scoliotic cases represent asymmetrical gait patterns in the frontal and transverse planes [108]. There are asymmetries in the evaluation of kinetic gait parameters (symmetry index) too. There is an obvious difference between the mediolateral CoP displacement of the left and right sides in people with scoliosis [109,110]. Kinetic parameters and indices might be helpful indicators for a deeper understanding of the etiology of scoliosis [110].

The body mechanics and gait pattern of people with AIS might be represented by decreased knee range of motion in the sagittal plane, hip motion in the frontal and transverse plane, and asymmetry of trunk movement in the transverse plane [104,111].

The effect of scoliosis and its severity on kinematic and electromyography (EMG) gait parameters and spatio-temporal gait parameters has shown a significant increase compared to the control group [1].

9.7.2.2 Rehabilitation

The decision to proceed with treatment modalities in people with AIS (physical therapy, orthotic intervention, and various surgical procedures) depends on the severity of the scoliotic curve as well as the skeletal maturity. Spinal orthoses are the most common non-invasive modality to prevent curve progression. They are generally prescribed in cases with a Cobb angle of 20°–45° [111].

The effect of spinal orthoses in the postural performance of AIS cases is another controversial topic. In some studies, the orthosis negatively affected the balance of school children with AIS if the visual and proprioceptive inputs were coincidentally disturbed. A potential mechanism to explain the effect of an orthosis on balance can be linked to the mechanical limitation of the trunk motion entrapped between the thoracic and sacral spine with adverse effects on balance [112]. In contrast, in another study, the student with AIS who underwent a balance test while wearing an orthosis had a better balance outcome as compared to scoliotic cases who were not using an orthosis. Based on this study, it was suggested that the application of a spinal orthosis might have a long-term training effect to improve balance [113]. It was stated in another study that orthoses do not change the oxygen intake and the timing of the lumbopelvic muscle activity neither in healthy volunteers nor in AIS cases. People with AIS wearing the orthosis experienced more restriction in the frontal part of the hips, shoulder movement, and pelvis rotation compared to the healthy participants in the orthosis condition [114].

In another study, two types of spinal orthoses (a rigid and soft design) were found to induce a significant kinematic change in the hip joint and pelvis during walking. Both designs decreased the pelvic obliquity and hip abduction/adduction while neither had a significant effect on kinematics of the knee and ankle joints. To draw a conclusion, the importance of regular evaluation of orthotic intervention over a long period of time should be emphasized [115].

Versatile orthoses have been utilized within the last decades to prevent curve progression. The mechanism of orthoses is based on the force vector/corrective pads as well as enhancement of proprioception. They can be broadly categorized into flexible/rigid designs as well as the origin they were initially designed. The Milwaukee orthosis was used as a cervico thoracolumbosacral orthosis (CTLSO) for many years to treat the high thoracic curves of idiopathic scoliosis (T6 and above). Thoracolumbosacral orthoses (TLSOs) were developed to provide curve control to those who might refuse to wear the Milwaukee orthosis. The most common types of TLSOs developed in North America include the Boston Brace, Wilmington orthosis, Rosenberger orthosis, Charleston bending orthosis (night-time), and the Providence orthosis. The Boston brace is a prefabricated posterior opening rigid orthotic design that was originally prescribed for the curves with the apical vertebra at T8 or below. In contrast, the European orthoses include Cheneau and its derivates, Lyon orthosis, TriaC, and Sforzesco orthosis. The Cheneau orthosis is an anterior opening asymmetric rigid orthosis that is effective in correcting the translational forces to control scoliosis. On comparing rigid versus soft orthoses, it has been shown that the rigid design has superior results, while the significant problem with toileting was found in the soft design due to placement of the groin straps in SpineCor orthosis [103,116]. As different orthotic designs exist, the decision to prescribe and fit an appropriate orthosis should be individualized to each case and be considered according to the experience of treating physicians and orthotists [103].

Orthotic treatment is planned according to the age and degree of the curvature. Younger children with Cobb angle below 20° would be kept under observation. Orthoses are prescribed if the angle is within the range of 20°–45°. The mechanism of biomechanical action of spinal orthoses in the correction of the deformity originates in end point control and the transverse control. In end point control, the emphasis is to fix the orthosis on the base of the spine at the level between the greater trochanter and the iliac crest. Transverse control works according to the three-point pressure systems to apply the force to the apical vertebra opposed by two counterforces above and below the apex. This can be done by the spinal orthosis applying the corrective forces depending on the flexibility of the spine, the degree of curvature, progression of deformity, and the age of the patient [102].

9.8 Cerebral palsy

Cerebral palsy (CP) is the most common neuromotor disability of childhood with a persistent disorder of movement and posture in children due to a lesion in the immature brain [117]. This non-progressive condition caused by damage to the motor centers of the brain leads to poor motor control [118]. A prevalence of 1.5–2.5 per 1,000 live births is reported [117,119]. The birth of infants with CP is highly associated with increased prematurity, adverse events of the prenatal stage, and the frequent pregnancies of the mother [117]. Some of the early signs of preterm infants with CP are represented in the form of scarcity or excessive type of movement, tone abnormalities (hyper/hypotonia), hyperextension of the head and neck, and the presence of asymmetric tonic neck reflex (ATNR) [117]. CP can be classified based on different parameters including the pathology of brain injury (infective, traumatic, inflammatory, vascular) and timing of brain injury occurrence (prenatal, perinatal/postnatal). According to the motor impairment, the modified Swedish classification categorizes the CP in response to the muscle tone type (Spastic, Dyskinetic, Ataxic) and the number and extent of the affected limbs (tetraplegia "global impact on gray and white matter structures in forebrain," diplegia "damage of white matter of internal capsule," hemiplegia "caused by intracerebral hemorrhage or ischemia on one hemisphere") [117,120]. More than half of children with CP walk independently, while many present spastic hemiplegia or diplegia [121]. Children with spastic CP manifest as hypotonic in the first months after birth and gradually turn to hypertonic.

9.8.1 Gait deviation

3D instrumental gait analysis is a common option in the process of preoperative examination and decision-making in children with CP. It may modify the treatment plans of surgical proposals which were planned based on the clinical examination by orthopedic surgeons though according to some papers, a good agreement was found between the gait analysis recommendations and the subsequent surgical plans [4]. The summary of indices developed from the data may represent the quality of the gait pattern in a "single score" format. The most common summary measured scores are the Gait Profile Score (GPS) and the Gait Deviation Index (GDI). Both indices are calculated according to the distance between the patient's data and the reference dataset on different gait features in the kinematics of pelvis, hip, knee, and ankle [122]. Gait analysis may help in the detection of certain abnormalities that benefit from the intervention. It also may increase the agreement between the surgeons [122,123].

As a general rule, children with CP do not suffer from muscle weakness but the muscles do not contract in the correct timing of the gait cycle due to the lack of coordination and a co-contraction of antagonist muscles. Deficiency in equilibrium reactions, as well as loss of selective muscle control, is underlined in these individuals [5]. Since CP is not a homogenous disorder, different presentations of pathological gait within this patient group might be observed. Spastic equinus is a common presentation in which the gait pattern and the balance are completely disturbed. The heels do not contact the ground and due to the increased tone of the plantar flexors, the normal function of the ankles is prevented. In addition, the equinus position of the ankles may contribute to the development of knee hyperextension deformity [78].

Spastic valgus position of the foot is another gait manifestation in these children where it results in increased pressure in the medial side of the foot [78] and if left untreated, can lead to a mid-foot break, whereby the patient pushes off from the medial aspect of the foot as opposed to the forefoot, at terminal stance.

Weakness of antigravity muscles and spasticity in both lower limbs are commonly seen in children with spastic diplegic CP. The common type of gait in these children is the crouched gait. This pathological gait is characterized by an increased amount of internal rotation of the thigh and leg, excessive dorsiflexion at the ankle in combination with excessive flexion at the knee and hip. Compensatory movement including the hip hike, external rotation of the foot, and circumduction of the swing leg is manifested in this gait pattern. The GRF line falls behind the knee joint center during the stance phase (single-limb support). The scissor gait pattern is another common type of gait deviation in children with spastic diplegia which is depicted by spastic hip adductors and flexors [120]. Due to the deficiency of power generation in the ankle joint of people with CP, extra power should be generated in the hip as well as knee joints. EMG could be used together with kinematics in the evaluation of the muscles such as the rectus femoris. Equinovalgus with excessive pronation in a child with CP might be seen. Simultaneous activation of synergists and antagonists leads to an inappropriate sequence of muscle contractions in children with CP [5].

9.8.2 Rehabilitation

During the acute phase, spasticity limits the movement (active and passive) in the direction of agonist muscles. Later and toward the chronic phase, the spasticity and restriction of movements lead to the contracture of immobilized muscles and joint stiffness with subsequent restriction of movement. Therefore, the role of early management is crucial to prevent the subsequent impairments [3]. The application of orthotic devices should be seen as part of a holistic treatment plan in people with spasticity and not as a standalone modality [3]. In general, the three-point force system is applied to control the ankle in the corrected joint positions. AFOs may resist the plantar flexion and the tendency for the foot to migrate out of the orthosis which needs to be inhibited through the anterior ankle straps. Any shortening of the gastrocnemius muscle must be fully accommodated in the AFO by ensuring the angle of the ankle in the AFO matches the passive length of the muscle with the knee extended [36]. Gastrocnemius is a biarticular muscle and as such, if it does not have the required length for the resulting AFO angle with the knee extended, the result will be insufficient length at the knee which will prevent the required knee extension at mid-stance, terminal stance, and terminal swing. In addition, the muscles will not be able to produce power when the sarcomeres are stretched to their maximal length. All AFOs should be tuned, to ensure the optimum result in each patient. Ankle-foot orthosis-footwear combination (AFO-FC) tuning can be defined as the process whereby fine adjustments are made to the design of the AFO-FC to optimize its performance during a particular activity. The term biomechanical optimization is used to encompass the whole process of designing, aligning, and tuning the AFO-FC [36]. To optimize the entry and exit from midstance, external shoe adaptations including the cushion heels, point loading rockers, and rounded rockers can be added to compromise the restricted ankle functions in rigid AFOs [36,78].

CP is a disorder of walking movement originating in the weakness and abnormalities of motor control sequences and synergies. These individuals have inappropriate muscle activity which may lead to debilitating symptoms such as the clearance of the foot in the swing phase due to the overactivity of plantar flexor muscles. The gait should be carefully observed to evaluate the stability, foot clearance during the swing phase, foot positioning at the end of the swing phase and before initial contact, and stride length. To evaluate the abnormality, the observation of the child walking in barefoot condition is suggested. Musculoskeletal evaluation of these children is required for the potential exploration of the hip subluxation, scoliosis, hamstring contracture, and tendoachilles. Equinus deformity is the most common type of deformity in people with CP which is induced by spastic contracture of the gastrocnemius muscle; it might be manifested in the form of tip-toe or toe-heel gait. If the muscle shortening turns to a fixed deformity, surgical intervention such as tendoachilles lengthening might be required [117], but careful consideration should be exercised in order to prevent over lengthening and a resultant crouch gait.

Previous studies suggest that gait analysis may alter the surgical plans in more than 50% of people with CP [124]. The agreement was more on the lengthening of the gastrocnemius and hamstring. In some cases, the preoperative gait analysis may change the policy from stage surgical operation to multilevel surgical intervention. In addition to the surgical operation, physiotherapy and botulinum toxin injection should also be considered.

The limitation of clinical examination may affect surgical decision-making. For example, the evaluation of a child walking on their toes based on visual gait analysis and clinical examination may bring the misconception of the existence of real equinus; however, the sagittal plane kinematics in the same case may show that the ankle has a normal range of dorsiflexion but the existence of excessive flexion of hip and knee throughout the stance phase is depicted [4]. Gait analysis in preoperative decision-making might be more important for people with diplegia than for those with hemiplegia [4].

Treatment of gait problems in CP has been changed gradually in recent decades. Injection of botulinum toxin to the gastrocnemius and soleus muscles may postpone and decrease the need for surgery [125]. Multidisciplinary approaches involve versatile treatments including physical therapy, orthotics, chemical neurolysis, or orthopedic surgery with the target to improve the gait in children with CP [126].

The goal of orthotic prescription in these cases might include tone reduction, control of contracture, and facilitatory approach on neuromotor recovery; they are defined based on two theories of orthotic application in the treatment of spasticity either through the neurophysiologic approach or the biomechanical perspective. It depends on whether the orthosis is planned to influence the muscle tone by facilitation/inhibition or it is going to address the contracture through the stretching of muscles and according to the biomechanical principles. Therefore, an orthosis may not only rely on the mechanical characteristics to obtain the treatment goal but also it may address the sensation of wearing an orthosis in contact with the skin to change the movement patterns in an intended direction which may serve as a kinesthetic reminder [3].

People with mild to moderate spastic diplegic or hemiplegic CP as well as hypotonic cases might be prescribed with dynamic AFOs and supramalleolar orthoses [120]. People with spastic CP often require AFOs, assistive devices, or wheelchairs for locomotion. Those with significant involvement of the upper limb and trunk may require equipment for feeding and activities of daily living. The abnormal tone might be managed by resting splints to decrease the risk of musculoskeletal deformities or they might be used in the post-surgical stage to correct the deformity. The stability of children with dyskinetic CP could be achieved by external support during locomotion. They may need seating systems and upper or lower extremity orthoses. In the case of ataxic CP, orthotists may implement AFOs to provide external support and positioning [120].

Due to the complexity of spasticity, management should include multidisciplinary approach including medical management (such as nerve block or neurotoxin injection) followed by rehabilitation intervention. The target of rehabilitation in these cases should include biomechanical restoration, prevention of additional musculoskeletal impairments, and improvement of motor control and limb function [3].

9.9 Cerebrovascular accident

A cerebrovascular accident (CVA) is the impairment in the cerebral circulation which might lead to stroke or onset of a neurological dysfunction [127]. The acute stage of this condition is usually characterized by flaccidity or lack of muscular tone. Due to the muscle weakness, joint malalignment and loss of activity in the musculoskeletal structure are commonly seen. This lack of activity will later lead to pain and contracture [118]. Stroke is also known as a brain insult. The prevalence of stroke is 8.1% in the US population aged over 65 years [128]. The most common type is ischemic stroke (88%) in two forms of thrombus and embolism as compared to hemorrhagic stroke (12%) [120]. While uncontrolled hypertension and ruptured aneurysm may lead to hemorrhagic stroke, hypertension, obesity, smoking, atrial fibrillation, and other parameters might be considered as risk factors of ischemic stroke. The prognosis of stroke for the ischemic type relies on the severity of the site of occlusion. The risk of recurrence of embolic cases is higher than the ones with thrombosis. The hemorrhagic type has the highest morbidity and mortality rate. In terms of muscle tone, these individuals initially present hypotonia due to neurologic shock. While some remain hypotonic, most cases develop different extents of hypertonus following the stroke [120].

9.9.1 Gait deviation

To evaluate the gait of these patients, it should be checked whether the post-CVA hemiplegic patient has some extent of control over specific muscles rather than abnormal mass activation of flexion or extension. The ability to dorsi-flex the ankle during the progression of the tibia in mid-stance might be counteracted by contraction of plantar flexors and the equinus position of the foot. The extension synergy pattern brings the lower limb to excessive extension at the hip and knee and the foot in equinovarus. Due to the limited amount of knee flexion and ankle dorsiflexion in swing, compensatory strategies such as circumduction may facilitate the swing phase clearance [120].

Common gait deviations include equinus or equinovarus, lateral instability in the ankle, and knee hyperextension in the stance phase. The walking speed is generally reduced while the step length on the impaired side is decreased. The timing for double stance shows an increase while single stance on affected side decreases. Backward leaning of the trunk is shown in these individuals, as well [127].

Spastic equinovarus might be seen in cerebral injuries including individuals with CVA. Equinovarus position of the foot hinders the clearance of the foot during the swing phase and may lead to the initial contact of the lateral border of the foot and the toes. The treatment of spastic equinovarus should be based on two sets of three-force systems to prevent both the plantar flexion and the supination though the priority should be given to maximum correction of the varus deformity. Regarding the knee extensor spasm, this deviation is mainly associated with an equinovarus deformity of the foot and ankle and a lack of complete extension in the hips. It manifests during the stance phase in the shape of exaggerated knee extension as a result of abnormal loading of the forefoot and the corresponding knee extension moment. Due to the stretch of the posterior components of the knee joint, further deformity of the joint and pain gradually emerge. In addition, a lack of normal knee flexion and ankle dorsiflexion could affect the foot clearance during the swing phase [78].

The upper extremities often follow a flexion pattern while the lower extremity is prone to extension. Force generation is also impaired. In terms of mobility and coordination, asymmetry in using the trunk and limbs during the activities is shown and the abnormal synergy of flexion in the upper extremity and extension of the lower extremity is also observed [120].

9.9.2 Rehabilitation

Regarding rehabilitation, people with hemiplegia often have an altered tone, change in joint flexibility, and change in motor strength. Improvement of gait quality relies on the natural history of pathology and the role of gait training strategies in the rehabilitation program. After a few weeks of rehabilitation, more than half of the stroke cases might be able to walk unaided if they wear an appropriate AFO and use assistive devices. After 6 months, more than 80% of these individuals can be ambulatory independent [120].

The target of orthotic treatment for people with post-stroke hypertonicity of the lower extremity is to control ankle motion and facilitate tibial advancement. Two types of orthotic strategies include the prescription of a rigid AFO which controls the foot/ankle in all three planes, or range permitting to incorporate an articulated AFO with the possibility of slight ankle motion toward plantar and dorsiflexion. It is worth noting that limited or blocked ankle in AFO designs enhances the stability of the stance phase, but the step length might decrease as a result of affected forward progression of tibia. In this case, application of a rocker bottom sole to the shoe may facilitate to simulate the second rocker of gait. Furthermore, considerations of footwear adaptations should form part of the routine AFO-FC tuning process. It should be mentioned that the orthosis prescribed in the early phase of rehabilitation, might become inappropriate at later stages once the patient gradually recovers [118]. Thus, the orthotist needs to maintain a regular review of the suitability of the AFO design throughout the patient's rehabilitation.

Following a stroke, AFOs have been shown to improve an individual's balance and walking abilities, playing a substantial role to decrease the fear of falling, and enhancing the kinetic and kinematic parameters of gait while attempting to decrease the energy consumption of walking [129–131]. The AFOs must be tuned to the most appropriate shank to vertical angle, to ensure optimum gait is achieved. Application of AFOs may improve the step length in both affected and non-paretic sides and improve the symmetry of gait during the swing phase. Walking speed and cadence might also be improved [127]. A KAFO might be prescribed to counteract the large extension moment of the knee. As it was indicated in the AFO designs, a similar principle should be implemented in the design of KAFOs to facilitate the correction of accompanying equinovarus deformity [78].

9.10 Aging and balance disorders

Gait disorder and falls are considered among the main reasons for chronic disability in elderly people leading to a lower level of quality of life and may intensify the risk of mortality [132]. The fall may cause a range of complications from a simple bruise/injury to complicated fractures or head injury. It also reduces mobility and the level of independence. It later deteriorates as the elderly population might face the fear of falling. In addition, geriatric gait disorders might indicate the risk of cardiovascular disease and the emergence of dementia [132,133]. Knowing the fact that 20% of elderly individuals maintain the ability of normal walking, it implies that those elderly with gait disturbance might be considered as early signs of underlying diseases linked to the brain (white matter) or visual/vestibular system. The safety of walking in the elderly is a critical parameter to decrease the risk of falling [2].

9.10.1 Gait deviation

Different gait deviations with a wide spectrum might be observed in elderly people. Antalgic gait which is the reaction to pain represents a reduced stance phase on the involved limb often resulting in limping. The paretic gait may manifest as high steppage, drop foot, and waddling (or duck-shape walking through the sway of the trunk from one side to the other). Spastic gait may present with circumduction or scissoring (bilateral form) [132]. Hypokinetic-rigid gait impairment which is the disease of the brain mainly manifests with a shuffling pattern of gait with a reduced step height accompanying a shorter stride length and consequently very slow walking with reduced arm swing. The presence of the shuffle is due to the continuous propulsive movement of the foot at the time of the initial contact. The trunk rotation in the majority of these people occurs in phase with the pelvis rather than the opposite twisting pattern [5]. The base of support is normal in Parkinson's disease (PD) while it increases in atypical parkinsonism. Festination is observed in the advanced phase of disease where the elderly use rapid small steps to keep the feet under the propulsive trunk [132,134]. Cautious gait is presented by slow and wide base walking and decreased movement of the trunk. This is a common gait in elderly people who suffer from the fear of falling. Dyskinetic gait disorder is linked to the presence of involuntary movement or posture. This gait impairment may lead to falls associated with exaggerated trunk movement which is beyond the boundary of stability [132].

9.10.2 Rehabilitation

The use of wearable technology in the documentation and monitoring of gait disorders is becoming more accepted and can be implemented in people with PD as well. Such inertial sensors can facilitate the evaluation of people in a non-lab environment [134,135]. In addition, the application of exergame (a term used for video games that are also a form of exercise) in the rehabilitation of people with PD is feasible and even more effective than regular exercise. Exergames facilitate the performance of home rehabilitation which can be remotely monitored by therapists. This is an advantageous feature that is not applicable in traditional rehabilitation settings [136].

The deteriorated sensation of cutaneous pressure is part of the natural aging process, and it has been associated with postural instability in the geriatric population. Foot orthoses have been studied to enhance postural stability which is particularly crucial to the elderly [137]. To decrease the risk of falling, different insoles [138,139] and AFOs [129] showed positive results. Footwear management is usually recommended as a non-invasive modality to reduce the risk of falling in elderly people. They might be suggested to wear a thin, hard-soled shoe with a high collar to decrease the risk of falling. Vibratory and magnetic insoles have been effective in balancing the control [137]. Wearing thick socks may reduce the plantar pressure sensitivity and tactile input and subsequently intensify the postural sway with an increased risk of falling [140].

The sensitivity of the plantar surface of the foot in people with PD is significantly decreased. Foot orthoses with texture may regulate the dynamic stability and improve the turning performance of these individuals [141].

9.11 Biomechanical optimization of ankle-foot orthoses and footwear combinations

As outlined above, attempting to normalize gait in people with movement disorders is often achieved using an AFO. However, the footwear that is worn with an AFO is integral in determining the overall biomechanical control provided, so the AFO and footwear have been termed AFO-FC [142]. The effects of footwear on gait when wearing AFOs have been documented [143,144]. More recently, AFO-FC tuning has been recognized as an essential aspect of clinical practice when prescribing AFOs [145–148]. Issuing a sub-optimal AFO-FC may have an immediate detrimental effect on function and in the longer term, it may contribute to the deterioration [36,148]. AFO-FC tuning can be defined as the process whereby fine adjustments are made to the design of the AFO-FC to optimize its performance during a particular activity. The term biomechanical optimization is used to encompass the whole process of designing, aligning, and tuning the AFO-FC [149].

9.12 Amputation and prosthetic management

The alignment of a lower limb prosthesis facilitates the optimization of the gait in people with amputation. A prosthesis with improper alignment will negatively affect the walking of an amputee and may induce complications in the interface between the residual limb and the socket [150].

The gait of lower limb amputees highly depends on the level of the amputation and as a general rule, individuals with shorter residual limbs may face more challenges in terms of ambulation parameters (kinematics and kinetics) as well as the level of energy consumption. Though microprocessor-controlled prosthetic knees and ankles have improved the kinematic parameters of walking as compared to the standard prosthetic components, the walking pattern does not approximate the missing intact limb.

The motion of the ankle in the sagittal plane is traditionally thought to consist of three "rockers"; the first rocker from the initial contact to foot flat with controlled ankle plantar flexion, the second rocker with the tibial advancement over the neutral positioned foot towards dorsiflexion, and the third rocker where the ankle moves to a plantar flexion position in late stance. However, in pre-swing, the metatarsal-phalangeal joints (MTPJs) continue to extend combined with a change in the position of the ankle from a dorsi-flexed position at the end of terminal stance to a plantar-flexed position at the end of pre-swing. Thus, it is suggested that dividing the forefoot rocker into a third and a fourth rocker recognizes the difference in ankle kinematics in terminal stance compared to pre-swing. Prosthetic feet can simulate the first two rockers while they could not address the third and fourth rockers as they have no active push-off force in late stance (except the designs with powered prosthetic feet). Therefore, it mainly depends on the amputee's ability to use the spring characteristics of the prosthetic foot to mimic the push-off action of the sound limb in late stance [151].

Prosthetic components are mainly prescribed based on the amputee's functional level where the amputees with lower functional levels are prescribed with less sophisticated components as contrary to the state-of-the-art units in higher functional level cases.

The target of walking in amputees is to bring back the kinematic and kinetic parameters of the gait as much as possible close to the pre-amputation stage though some asymmetry in walking of the amputees will be perceived due to the structural asymmetries of the amputation.

Gait deviations cause improper loading of the forces on the residual limb while inducing an increased level of weight-bearing on the contralateral side and an increase in energy consumption. In the long term, it might lead to orthopedic complications. Therefore, detection and addressing these deviations might be helpful to decrease the potential detrimental impairments to the amputee.

It has been shown that an angular alignment change of less than 6° and translational alignment changes of smaller than 2 mm are less likely to be sensed by amputees. It is suggested to resolve the gait deviation in the sagittal plane first before dealing with the coronal plane alignment. The alignment of the components in a prosthesis is defined based on the linear and angular categories in the coronal and sagittal planes. Internal or external rotation in response to the line of progression is considered for the definition of the transverse plane orientation [151,152].

In the following sections, the transtibial and transfemoral prosthetic gait, their corresponding gait deviations, and prosthetic management options will be discussed.

9.12.1 Transtibial gait analysis

The assessment of prosthetic gait requires a complete understanding of normal gait. Prosthetic gait is attributed to different parameters including the length of the residual limb, the inherent mechanical capabilities of prosthetic components, the strength of the remaining muscles in the amputated side as well as the range of motion of the hip or hip and knee on the same side. Hip and knee flexion contractures larger than 10° may negatively affect gait and increase energy consumption.

From weight acceptance to foot flat, the knee flexes around 6°–10° as compared to 18° in normal gait. The heel of the prosthetic foot might be compressed slightly depending on the weight of the amputee and the characteristics of each prosthetic foot. Knee flexion in this phase is regulated by eccentric contracture of the knee extensors.

During mid-stance, the total weight is applied to the prosthesis. In the anterior-posterior view, there is minimum lateral socket displacement observed while the pylon which is the connecting component between the socket and the prosthetic foot should remain perpendicular to the floor and the foot should be kept flat on the floor. In the lateral view, the limb continues to move forward over the foot depending on the amount of dorsiflexion the foot design allows.

From mid-stance to toe-off, the prosthetic foot unloads and weight transfers to the sound side.

During the swing phase, the knee flexes easily and allows the prosthetic toes to clear the ground. The socket should remain securely on the residual limb [153].

9.12.1.1 Transtibial gait deviations

Gait deviations might be attributed to a poor fitting socket, improper prosthetic alignment, pain in the residual limb, lack of training, and the walking habits of an amputee. Gait deviations will increase energy consumption and the amputee's discomfort. If the gait deviation could be realized earlier, it may prevent it from becoming a permanent problem of the amputee's gait. It is clinically accepted to observe a mild lateral movement of the socket at the mid-stance phase. This is attributed to the length of the residual limb, the suspension system as well as adduction of the femur [153].

Most of the transtibial amputees have normal knee extensor and flexor muscles. The resultant moments at the knee are determined by the selection of the prosthetic foot component. In general, the knee kinematics of the prosthetic side is almost similar to the intact knee.

Although the amputation of ankle disarticulation is advantageous to the weight bearing of the amputee, such surgical intervention might limit the prosthetic component options. 18–25 cm of distance between the end of the residual limb to the floor warrants the selection of different foot components and implementation of proper strategies for the functional performance targets [151].

Initial contact to foot flat

Excessive knee extension: The amputee may feel a sense of walking up a hill. The extended knee may lead to skin abrasion and pain in the anterior distal part of the flexed socket. The main reasons for observing excessive knee extension are a large toe lever arm (posterior placement of the socket over the prosthetic foot), insufficient provision of socket flexion (not being set in 5°–8° of flexion), or a soft plantar flexion bumper in the prosthetic foot.

Knee instability: Knee instability can be seen at the initial or terminal phases of stance. The major reasons for knee instability include a very short toe lever arm, alignment of the socket in the excessive flexion, or a hard heel plantar flexion [153].

Mid-stance

Excessive vertical movement of the hip on the prosthetic side: It can be related to the prosthetic length such that if the prosthetic length is longer than the intact leg, the amputee adopts a compensatory maneuver to raise the center of gravity over the support joint during the stance phase. In contrast, a very short prosthetic leg may lead the hip and shoulder on the prosthetic side to drop significantly [153].

Frontal plane knee instability: During mid-stance, a mild knee varus moment might be observed which is considered clinically acceptable. Deviation from this alignment should be kept into consideration as it might be associated with excessive medial/lateral dimension of the socket or outset/inset alignment of the foot in relation to the socket [151].

Wide-based gait: This deviation imposes additional pressure to the medial distal end of the residual limb as well as the proximal lateral wall of the socket. This wide-based gait is mainly induced due to two factors: the outset alignment of the prosthetic foot or medial leaning of the pylon where the top of the pylon is medial to the bottom section [151].

Narrow-based gait and thrust of the prosthesis: To keep the pelvis level, the amputee may lean away from the prosthetic side during stance or allow the pelvis to drop and lean toward the prosthetic side. Regardless of the direction, this increased leaning may induce exaggerated movement of the hip and knee, toward or away from the prosthesis. This gait deviation looks similar to the representation of gait deviation with poor gluteus medius muscle. This problem is caused by two factors: lateral leaning of pylon or medial placement of the foot in response to the alignment line [153].

Terminal stance

Knee extension: It is mainly due to a long toe lever arm where the knee joint remains in extension within the last part of the stance phase [153].

Premature heel rise: It mostly occurs if the prosthetic toe lever is too rigid to allow the prosthetic advancement over the foot during the single-limb support.

Intact-side vaulting: It is often caused by uncertainty on the prosthesis to clear the ground, in which the amputee may lengthen the contralateral limb by a sudden contraction of the calf muscles on the sound side to facilitate prosthetic limb clearance [151].

Swing phase

Pistoning (loss of suspension): If the suspension is inadequate, the socket will pull out vertically once the foot clears the floor. A certain amount of vertical movement could be considered reasonable depending on the suspension mechanism adopted. It might also be observed if the volume of the residual limb has changed dramatically, and the socket no longer matches the volume of the residual limb.

Unequal step length: This step length asymmetry might be altered to a habit of amputees where they may increase or decrease the step length in the contralateral side though a short step on the contralateral side is more often observed in which, the length of the prosthetic side would be longer. It might be attributed to insufficient socket flexion as the above-knee (AK) amputees use lumbar lordosis to propel the intact limb [120,151,153].

9.12.2 Transfemoral gait analysis

An AK or transfemoral amputee uses the femoral residual limb (gluteal muscle and iliopsoas) to achieve hip movement and control the prosthetic knee components.

Knee components generally lack the flexion required during early stance. It should be noted that even the microprocessor-controlled knee units fail to present a normal kinematic knee pattern which might be due to the lack of confidence in the amputee or be linked to the poor alignment of the knee component in the sagittal plane.

Knee hyperflexion (flexion beyond 60°) in the swing phase can be controlled and adjusted in mechanical or hydraulic joints to prevent knee-buckling during the swing phase and decrease the risk of falls [151].

All parameters previously mentioned regarding below-knee prostheses, including the socket fit, leg length, suspension, and static alignment, should also be considered for the prosthetic management at the transfemoral level. In addition, knee stability is a key factor to be considered at this level. All types of knee mechanisms should be aligned slightly posterior to a line joining the trochanter, anatomical knee, and ankle axis. This will create an extension moment during initial contact. Mechanical swing phase knee mechanisms also rely on the amputee to extend the residual limb against the posterior wall of the socket to maintain knee extension from heel strike to mid-stance. The microprocessor knees' hydraulic components include a mechanism that controls knee stability even if the knee cannot be fully extended at initial contact and in turn, they approximate the normal gait by facilitating an initial knee flexion [153]. Except for the microprocessor knees, other prosthetic knee designs are kept in extension from heel contact to mid-stance. During mid-stance, the amputee should show good medial–lateral stability and balance on the prosthetic side where the contralateral foot can clear the floor for the swing phase. There might be a slight pelvic drop of less than 5° and mild lateral trunk bending according to the length of the residual limb. From mid-stance to toe-off, a smooth heel-rise is depicted as the weight is transferred forward over the prosthetic forefoot. The hip extends without lumbar lordosis. During the swing phase, the foot clears the ground, and the prosthetic knee bends smoothly. The shank swings smoothly and prepositions in extension prior to the next initial contact [151,153].

9.12.2.1 Transfemoral gait deviations

Initial contact to mid-stance

Foot slap: If the forefoot descends fast, a slapping sound will be created as it hits the floor. It might be caused by soft plantar flexion resistance or a short heel lever arm.

Knee instability: The knee instability might be caused by the following factors: the anterior placement of knee axis to trochanteric knee ankle line, a lack of adequate socket flexion, or a hard heel bumper. A severe hip flexion contracture not being accommodated in the socket alignment or a prosthetic foot set in excessive dorsiflexion may also contribute to this gait deviation.

Foot external rotation: From initial contact to foot flat, the prosthetic foot should be kept in the line of progression. If the socket is poorly suspended or being donned improperly, an external rotation in the prosthetic foot might be observed. In addition, the placement of the foot in external rotation alignment may contribute to this gait deviation as well [120,151,153].

Mid-stance

Lateral trunk bending: Individuals actively bend the trunk over the prosthesis during the stance phase which might be helpful in elevation of the pelvis in the contralateral side and positioning of the center of mass closer to the prosthetic foot. Excessive bending might be due to different causes: a socket not set in proper adduction alignment, the high medial brim of the socket, a short prosthesis, short lever arm (residual limb), and/or a weak hip abductor on the prosthetic side.

Abducted gait: This deviation might be due to the following reasons: a very long prosthesis, a high medial wall, abduction contracture.

Excessive trunk extension: It might be caused by the following issues: insufficient initial socket flexion, flexion contracture, weakness of abdominal muscles, and hip extensors [151,153].

Terminal stance (pre-swing)

Drop-off: Sudden downward movement of the trunk due to the short toe lever arm may induce the drop-off.

Inadequate heel off: If the subject does not feel secure on weight shift forward over the toe of the prosthesis, the heel may not come off the floor until the whole trunk brought forward [153].

Swing phase

Circumducted gait: The causes of circumduction include the length of the prosthesis and lack of confidence in the amputee to bend the prosthetic knee. The circumduction should be addressed in the early stage of the rehabilitation to prevent it from being an established habitual compensatory strategy.

Vaulting: Similar to circumduction, vaulting is another strategy to ensure the prevention of foot contact with the floor during the swing phase as they basically share similar underlying causes. In this gait deviation, the amputee rises on the toe of the intact foot to facilitate the prosthesis swing with little knee flexion. It might be caused by a very long prosthesis, limitation in knee flexion, and inadequate socket suspension.

Medial or lateral whips: Whips are observed by the medial or lateral shift of the heel at the beginning of the swing phase and upon the occurrence of initial flexion. They might be related to excessive external and internal rotation of the prosthetic knee, respectively. In addition, an improperly donned socket or a socket that is too tight may induce similar results.

Excessive terminal impact: In general, amputees prefer to have a terminal impact at the end of the swing phase to increase their confidence in spatial positioning and stability of the prosthetic knee beneath their trunks. Though the level of terminal impact chosen by amputees is a subjective preference, the adjustment could facilitate the balance between the knee flexion in early swing and knee extension in late swing [151–153].

Uneven steps and uneven arm swing throughout the gait cycle

These two deviations occur together. The arm on the prosthetic side is kept closer to the body and the amputee puts a short stance phase on the prosthetic side. It is mainly due to the weight-bearing issue and the lack of gait training for amputees [153].

9.12.3 Prosthetic prescription

9.12.3.1 Distal amputations

As compared to higher levels of amputation, distal levels facilitate the reduction of energy cost, improve function, and contribute to a more normal gait. In a ray amputation which is the removal of both the toe and the metatarsal bone, a custom-molded foot prosthesis with a filler is prescribed. In a transmetatarsal amputation, a high level of load is applied to the skin closure during the late stance phase. Prostheses are basically either an insert being fitted into a shoe or a socket fitted to the patient. The socket can be made of silicone, polyurethane, resins, leather, or a combination of materials. The sockets should adhere to a semi-rigid foot plate with a toe filler and the completed design should enable the amputee to have a partially normal gait. Chopart amputation which is the disarticulation at midtarsal joint level leaves the limb in an equinus deformity which may necessitate an Achilles tendon lengthening. The shoe insert attached to a footplate and toe filler, as well as the shoe sole on the sound side, are the options used in Lisfranc and Chopart amputations. It is essential to note that the intact central ray is angled at 40° cephalad from the long axis of the foot centered at the base of the third metatarsal (midfoot), where the loss of the midfoot in a Chopart amputation results in the loss of the supporting structures, causing the calcaneus to move into a plantar-flexed position, resulting in a bulbous residual limb and a significant leg length discrepancy. The amputees should avoid barefoot walking. In Syme's amputation which is the amputation through the distal tibia and fibula bone, the heel fat pad is preserved which facilitates the end bearing. The patient can walk without any prosthesis for a short distance. The length difference between the amputated and sound sides provides an opportunity to utilize different prosthetic foot components including the SACH (Solid Ankle Cushion Heel) foot or the energy-storing carbon foot to enhance the gait efficiency. As for most of the individuals, the proximal trimline of the socket can be located below the tibial tuberosity. However, if the weight-bearing is limited, the socket can be further extended to make the socket similar to the transtibial socket designs in an attempt to distribute the weight on extended areas [151,152].

9.12.3.2 Transtibial amputation

A transtibial prosthesis includes a custom-made socket, a pylon, a prosthetic foot, and additional suspension components (in some cases). The socket is the most important component as it is in contact with the amputee's skin. The total surface contact sockets allow the distribution of force over a residual limb. Therefore, during walking, the muscles locate into the channels in the socket to provide the function. Amputees with shorter residual limbs might be fitted with a socket design with higher medial-lateral and anterior trimline called supracondylar/suprapatellar. Such design facilitates self-suspension of the socket on the residual limb. Many people use elastic sleeves or silicone liners as an interface rolled over the prosthesis to improve the suspension. In traditional systems, cuff straps wrapped around the distal thigh or thigh corsets with metal side joints were utilized. Regarding the distal components, pylons and the foot and ankle units are considered in a transtibial prosthetic system. Endoskeletal modular pylons may incorporate a shock-absorbing mechanism to absorb the vertical shock in an early stance. A variety of foot and ankle units are available in the market. They are made of materials such as wood, rubber, urethane, titanium, and graphite. Based on the mechanism, prosthetic feet are categorized as SACH foot, single and multi-axis feet, and dynamic response units. For those who would like to change their shoes on different occasions (with different heels), adjustable ankles are the available options [120,152].

9.12.3.3 Transfemoral amputation

Different socket designs have been used for decades in the transfemoral amputation level; the quadrilateral socket, the ischial containment, and the dynamic design (incorporation of the flexible socket and carbon fiber reinforced frame) are different options in the socket design process. In terms of the suspension, it may not be common for all amputees in AK level to tolerate the suction suspension. Suspension belts can be auxiliary components to hold the socket in the correct alignment/contact with the residual limb. The design of the prosthetic knee joints ranges from a simple single-axis to polycentric to sophisticated microprocessor-controlled knee units. A microprocessor automatically adjusts the response of the knee joint to the walking terrains. The knee should be adjusted by a computer software interface and according to an individual's gait pattern. The general concept for the shank and prosthetic ankle and foot are in line with the principles of transtibial amputees as previously discussed [151–153].

References

[1] Mahaudens P, Banse X, Mousny M and Detrembleur C. Gait in adolescent idiopathic scoliosis: kinematics and electromyographic analysis. *European spine Journal: Official Publication of the European Spine Society, the European Spinal Deformity Society, and the European Section of the Cervical Spine Research Society*. 2009;18(4):512–21.

[2] Callisaya ML, Blizzard L, Schmidt MD, McGinley JL and Srikanth VK. Ageing and gait variability – a population-based study of older people. *Age Ageing*. 2010;39(2):191–7.

[3] Coppard BM. In Lohman H, Shultz-Johnson K, Coppard BM, editors. Introduction to splinting: a clinical-reasoning & problem-solving approach. 2nd ed. St. Louis, MO: Mosby; 2001.

[4] Lofterød B, Terjesen T, Skaaret I, Huse A-B and Jahnsen R. Preoperative gait analysis has a substantial effect on orthopedic decision making in children with cerebral palsy: comparison between clinical evaluation and gait analysis in 60 patients. *Acta Orthopaedica.* 2007;78(1):74–80.

[5] Levine D, Richards J and Whittle M. Whittle's Gait Analysis. London: Churchill Livingstone/Elsevier; 2012.

[6] Mercuri E and Muntoni F. Muscular dystrophies. *The Lancet.* 2013;381 (9869):845–60.

[7] Mercuri E, Bönnemann CG and Muntoni F. Muscular dystrophies. *Lancet (London, England).* 2019;394(10213):2025–38.

[8] Goudriaan M, Van den Hauwe M, Dekeerle J, *et al.* Gait deviations in Duchenne muscular dystrophy – Part 1. A systematic review. *Gait & Posture.* 2018;62:247–61.

[9] Menant JC and Gandevia SC. Poliomyelitis. *Handbook of Clinical Neurology.* 2018;159:337–44.

[10] Birnkrant DJ, Bushby K, Bann CM, *et al.* Diagnosis and management of Duchenne muscular dystrophy. Part 2: respiratory, cardiac, bone health, and orthopaedic management. *The Lancet Neurology.* 2018;17(4):347–61.

[11] Gianola S, Castellini G, Pecoraro V, Monticone M, Banfi G and Moja L. Effect of muscular exercise on patients with muscular dystrophy: a systematic review and meta-analysis of the literature. *Frontiers in Neurology.* 2020;11:958.

[12] de Souza MA, Figueiredo MM, de Baptista CR, Aldaves RD and Mattiello-Sverzut AC. Beneficial effects of ankle-foot orthosis daytime use on the gait of Duchenne muscular dystrophy patients. *Clinical Biomechanics (Bristol, Avon).* 2016;35:102–10.

[13] Gupta A, Nalini A, Arya SP, *et al.* Ankle-foot orthosis in Duchenne muscular dystrophy: a 4 year experience in a multidisciplinary neuromuscular disorders clinic. *Indian Journal of Pediatrics.* 2017;84(3):211–5.

[14] Kern V, Wicklund M, Haulman A, *et al.* Ankle bracing practices in ambulatory, corticosteroid-naive boys with Duchenne muscular dystrophy. *Muscle Nerve.* 2020;61(1):52–7.

[15] O'Connor J, McCaughan D, McDaid C, *et al.* Orthotic management of instability of the knee related to neuromuscular and central nervous system disorders: systematic review, qualitative study, survey and costing analysis. *Health Technology Assessment (Winchester, England).* 2016;20(55):1–262.

[16] Neuman RM, Shearin SM, McCain KJ and Fey NP. Biomechanical analysis of an unpowered hip flexion orthosis on individuals with and without multiple sclerosis. *Journal of Neuroengineering and Rehabilitation.* 2021;18(1):104.

[17] Kingwell E, Marriott JJ, Jetté N, *et al.* Incidence and prevalence of multiple sclerosis in Europe: a systematic review. *BMC Neurology.* 2013;13:128.

[18] Compston A and Coles A. Multiple sclerosis. *Lancet (London, England).* 2008;372:1502–17.

[19] Martin C, Phillips B, Kilpatrick T, *et al.* Gait and balance impairment in early multiple sclerosis in the absence of clinical disability. *Multiple Sclerosis.* 2006;12:620–8.

[20] Stevens V, Goodman K, Rough K and Kraft GH. Gait impairment and optimizing mobility in multiple sclerosis. *Physical Medicine and Rehabilitation Clinics of North America.* 2013;24(4):573–92.

[21] Givon U, Zeilig G and Achiron A. Gait analysis in multiple sclerosis: characterization of temporal-spatial parameters using GAITRite functional ambulation system. *Gait & Posture.* 2009;29(1):138–42.

[22] Sutliff MH. Contribution of impaired mobility to patient burden in multiple sclerosis. *Current Medical Research and Opinion.* 2010;26(1):109–19.

[23] Sandroff BM, Sosnoff JJ and Motl RW. Physical fitness, walking performance, and gait in multiple sclerosis. *Journal of the Neurological Sciences.* 2013;328(1–2):70–6.

[24] Pau M, Coghe G, Atzeni C, *et al.* Novel characterization of gait impairments in people with multiple sclerosis by means of the gait profile score. *Journal of the Neurological Sciences.* 2014;345(1–2):159–63.

[25] Kempen JCE, Doorenbosch CAM, Knol DL, Groot Vd and Beckerman H. Newly identified gait patterns in patients with multiple sclerosis may be related to push-off quality. *Physical Therapy.* 2016;96(11):1744–52.

[26] Remelius JG, Hamill J, KentUBraun J and Van Emmerik REA. Gait initiation in multiple sclerosis. *Motor Control.* 2008;12(2):93U108.

[27] Remelius JG. Adaptations to stride Patterns and Head Movements During Walking in Persons with and without Multiple Sclerosis. Amherst: University of Massachusetts; 2012.

[28] Kurtzke JF. Rating neurologic impairment in multiple sclerosis: an expanded disability status scale (EDSS). *Neurology.* 1983;33:1444–52.

[29] Haselkorn JK, Hughes C, Rae-Grant A, *et al.* Summary of comprehensive systematic review: rehabilitation in multiple sclerosis. *American Academy of Neurology.* 2015;85:1896–903.

[30] Swinnen E, Beckwee D, Pinte D, Meeusen R, Baeyens JP and Kerckhofs E. Treadmill training in multiple sclerosis: can body weight support or robot assistance provide added value? A systematic review. *Multiple Sclerosis International.* 2012;2012:240274.

[31] Swinnen E, Baeyens JP, Pintens S, *et al.* Trunk kinematics during walking in persons with multiple sclerosis: the influence of body weight support. *NeuroRehabilitation.* 2014;34(4):731–40.

[32] Springer S and Khamis S. Effects of functional electrical stimulation on gait in people with multiple sclerosis – a systematic review. *Multiple Sclerosis and Related Disorders.* 2017;13:4–12.

[33] Miller L, Rafferty D, Paul L and Mattison P. The impact of walking speed on the effects of functional electrical stimulation for foot drop in people with multiple sclerosis. *Disability and Rehabilitation: Assistive Technology.* 2015;11(6):478–83.

[34] Pilloni G, Choi C, Coghe G, *et al.* Gait and functional mobility in multiple sclerosis: immediate effects of transcranial direct current stimulation (tDCS) paired with aerobic exercise. *Frontiers in Neurology*. 2020;11:310.

[35] Feng Y and Song Y. The categories of AFO and its effect on patients with foot impair: a systemic review. *Physical Activity and Health*. 2017;1(1): 8–16.

[36] Eddison N, Healy A, Needham R and Chockalingam N. The effect of tuning ankle foot orthoses-footwear combinations on gait kinematics of children with cerebral palsy: a case series. *Foot (Edinburgh, Scotland)*. 2020;43:101660.

[37] Alfuth M. Textured and stimulating insoles for balance and gait impairments in patients with multiple sclerosis and Parkinson's disease: a systematic review and meta-analysis. *Gait & posture*. 2017;51:132–41.

[38] Diseases NIoAaMaS. Osteoarthritis: National Institutes of Health; 2019. Available from: https://www.niams.nih.gov/health-topics/osteoarthritis#tab-causes.

[39] Diseases NIoAaMaS. Rheumatoid Arthritis: National Institutes of Health; 2019. Available from: https://www.niams.nih.gov/health-topics/rheumatoid-arthritis#tab-causes.

[40] Brostrom EW, Esbjornsson AC, von Heideken J and Iversen MD. Gait deviations in individuals with inflammatory joint diseases and osteoarthritis and the usage of three-dimensional gait analysis. *Best Practice & Research: Clinical Rheumatology*. 2012;26(3):409–22.

[41] Constantinou M, Barrett R, Brown M and Mills P. Spatial-temporal gait characteristics in individuals with hip osteoarthritis: a systematic literature review and meta-analysis. *Journal of Orthopaedic & Sports Physical Therapy*. 2014;44(4):291–B7.

[42] Magalhães CMBd and Kirkwood RN. Strategies to reduce joint load in the medial compartment of the knee during gait in individuals with osteoarthritis: a review of the literature. *Fisioterapia em Movimento*. 2016;29(4): 831–42.

[43] Iijima H, Shimoura K, Ono T, Aoyama T and Takahashi M. Proximal gait adaptations in individuals with knee osteoarthritis: a systematic review and meta-analysis. *Journal of Biomechanics*. 2019;87:127–41.

[44] Turcot K, Armand S, Lubbeke A, Fritschy D, Hoffmeyer P and Suva D. Does knee alignment influence gait in patients with severe knee osteoarthritis? *Clinical Biomechanics (Bristol, Avon)*. 2013;28(1):34–9.

[45] Valderrabano V, Nigg BM, von Tscharner V, Stefanyshyn DJ, Goepfert B and Hintermann B. Gait analysis in ankle osteoarthritis and total ankle replacement. *Clinical Biomechanics (Bristol, Avon)*. 2007;22(8):894–904.

[46] Carroll M, Parmar P, Dalbeth N, Boocock M and Rome K. Gait characteristics associated with the foot and ankle in inflammatory arthritis: a systematic review and meta-analysis. *BMC Musculoskeletal Disorders*. 2015;16:134.

[47] Baan H, Dubbeldam R, Nene AV and van de Laar MA. Gait analysis of the lower limb in patients with rheumatoid arthritis: a systematic review. *Seminars in Arthritis and Rheumatism.* 2012;41(6):768–88 e8.

[48] Nelson AE, Allen KD, Golightly YM, Goode AP and Jordan JM. A systematic review of recommendations and guidelines for the management of osteoarthritis: the chronic osteoarthritis management initiative of the U.S. bone and joint initiative. *Seminars in Arthritis and Rheumatism.* 2014;43(6): 701–12.

[49] Lewinson RT and Stefanyshyn DJ. Wedged insoles and gait in patients with knee osteoarthritis: a biomechanical review. *Annals of Biomedical Engineering.* 2016;44(11):3173–85.

[50] Tenten-Diepenmaat M, van der Leeden M, Vliet Vlieland TPM and Dekker J, Group RAFE. Multidisciplinary recommendations for diagnosis and treatment of foot problems in people with rheumatoid arthritis. *Journal of Foot and Ankle Research.* 2018;11:37.

[51] Richards R, van den Noort JC, Dekker J and Harlaar J. Gait retraining with real-time biofeedback to reduce knee adduction moment: systematic review of effects and methods used. *Archives of Physical Medicine and Rehabilitation.* 2017;98(1):137–50.

[52] Frontera WR, Silver JK and Rizzo TD. *Essentials of Physical Medicine and Rehabilitation – Musculoskeletal Disorders, Pain, and Rehabilitation.* 4th ed. New York, NY: Elsevier; 2020.

[53] Lin X, Huang K, Zhu G, Huang Z, Qin A and Fan S. The effects of acupuncture on chronic knee pain due to osteoarthritis: a meta-analysis. *Journal of bone and joint surgery. American volume.* 2016;98(18):1578–85.

[54] Thoumie P, Marty M, Avouac B, *et al.* Effect of unloading brace treatment on pain and function in patients with symptomatic knee osteoarthritis: the ROTOR randomized clinical trial. *Scientific Reports.* 2018;8(1):10519.

[55] Maleki M, Arazpour M, Joghtaei M, Hutchins SW, Aboutorabi A and Pouyan A. The effect of knee orthoses on gait parameters in medial knee compartment osteoarthritis: a literature review. *Prosthetics and Orthotics International.* 2016;40(2):193–201.

[56] Erhart JC, Mundermann A, Elspas B, Giori NJ and Andriacchi TP. A variable-stiffness shoe lowers the knee adduction moment in subjects with symptoms of medial compartment knee osteoarthritis. *Journal of Biomechanics.* 2008;41(12):2720–5.

[57] Shaw KE, Charlton JM, Perry CKL, *et al.* The effects of shoe-worn insoles on gait biomechanics in people with knee osteoarthritis: a systematic review and meta-analysis. *British Journal of Sports Medicine.* 2018;52(4):238–53.

[58] Moyer RF, Birmingham TB, Dombroski CE, *et al.* Combined effects of a valgus knee brace and lateral wedge foot orthotic on the external knee adduction moment in patients with varus gonarthrosis. *Archives of Physical Medicine and Rehabilitation.* 2013;94(1):103–12.

[59] Tenten-Diepenmaat M, der MLv, Vlieland TV, Roorda LD and Dekker J. THU0750-HPR The effectiveness of therapeutic footwear in patients with

rheumatoid arthritis: a systematic review and meta-analysis. *Poster Presentations.* 2017. p. 1486.1.

[60] Frecklington M, Dalbeth N, McNair P, *et al.* Footwear interventions for foot pain, function, impairment and disability for people with foot and ankle arthritis: a literature review. *Seminars in Arthritis and Rheumatism.* 2018;47(6):814–24.

[61] Healy A, Farmer S, Pandyan A and Chockalingam N. A systematic review of randomised controlled trials assessing effectiveness of prosthetic and orthotic interventions. *PloS One.* 2018;13(3):e0192094.

[62] Spina bifida Wikipedia: Wikipedia; 24 August 2020. Available from: https://en.wikipedia.org/wiki/Spina_bifida#.

[63] Duffy CM, Hill AE, Cosgrove AP, Corry IS, Mollan RAB and Graham HK. Three-dimensional gait analysis in spina bifida. *Journal of Pediatric Orthopaedics.* 1996;16(6):786–91.

[64] Mueske NM, Ounpuu S, Ryan DD, *et al.* Impact of gait analysis on pathology identification and surgical recommendations in children with spina bifida. *Gait & Posture.* 2019;67:128–32.

[65] Borish CN, Mueske NM and Wren TAL. A comparison of three methods of measuring tibial torsion in children with myelomeningocele and normally developing children. *Clinical Anatomy.* 2017;30(8):1043–8.

[66] Gutierrez EM, Bartonek Å, Haglund-Åkerlind Y and Saraste H. Characteristic gait kinematics in persons with lumbosacral myelomeningocele. *Gait & Posture.* 2003;18(3):170–7.

[67] Gutierrez EM, Bartonek A, Haglund-Akerlind Y and Saraste H. Kinetics of compensatory gait in persons with myelomeningocele. *Gait & Posture.* 2005;21(1):12–23.

[68] de Groot JF, Takken T, van Brussel M, *et al.* Randomized controlled study of home-based treadmill training for ambulatory children with spina bifida. *Neurorehabilitation and Neural Repair.* 2011;25(7):597–606.

[69] Stark C, Hoyer-Kuhn HK, Semler O, *et al.* Neuromuscular training based on whole body vibration in children with spina bifida: a retrospective analysis of a new physiotherapy treatment program. *Child's Nervous System.* 2015;31(2):301–9.

[70] Walker JL, Ryan SW and Coburn TR. Does threshold nighttime electrical stimulation benefit children with spina bifida? A pilot study. *Clinical Orthopaedics and Related Research.* 2011;469(5):1297–301.

[71] Dicianno BE, Kurowski BG, Yang JM, *et al.* Rehabilitation and medical management of the adult with spina bifida. *American Journal of Physical Medicine and Rehabilitation.* 2008;87(12):1027–50.

[72] Swaroop VT and Dias L. Orthopedic management of spina bifida. Part I: hip, knee, and rotational deformities. *Journal of Children's Orthopaedics.* 2009;3(6):441–9.

[73] Martins EJ, Gastaldi AC, Davoli GBQ, Leonardi-Figueiredo MM and Mattiello-Sverzut AC. Decreased respiratory performance of children and adolescents with myelomeningocele who use a wheelchair – preliminary

data. *Brazilian Journal of Medical and Biological Research.* 2019;52(8): e8671.

[74] Ivanyi B, Schoenmakers M, van Veen N, Maathuis K, Nollet F and Nederhand M. The effects of orthoses, footwear, and walking aids on the walking ability of children and adolescents with spina bifida: a systematic review using International Classification of Functioning, Disability and Health for Children and Youth (ICF-CY) as a reference framework. *Prosthetics and Orthotics International.* 2015;39(6):437–43.

[75] Vankoski S, Moore G, Statler KD, Sarwark JF and Dias L. The influence of forearm crutches on pelvic and hip kinematics in children with myelomeningocele: don't throw away the crutches. *Developmental Medicine and Child Neurology.* 2008;39(9):614–9.

[76] Gupta RT, Vankoski S, Novak RA and Dias LS. Motion amplitudes of the upper body segments are related to the degree of muscle weakness of the lower limbs. *Journal of Pediatric Orthopedics.* 2005;25(1):85–94.

[77] Jones KM, Balalla S, Theadom A, Jackman G and Feigin VL. A systematic review of the worldwide prevalence of survivors of poliomyelitis reported in 31 studies. *BMJ Open.* 2017;7(7):e015470.

[78] Condie DN. *Atlas Lower Limb Orthotic Prac.* New York, NY: Taylor & Francis; 1998.

[79] Portnoy S and Schwartz I. Gait characteristics of post-poliomyelitis patients: standardization of quantitative data reporting. *Annals of Physical and Rehabilitation Medicine.* 2013;56(7–8):527–41.

[80] Vreede KS, Henriksson J, Borg K and Henriksson M. Gait characteristics and influence of fatigue during the 6-minute walk test in patients with post-polio syndrome. *Journal of Rehabilitation Medicine.* 2013;45(9):924–8.

[81] Genet F, Schnitzler A, Mathieu S, *et al.* Orthotic devices and gait in polio patients. *Annals of Physical and Rehabilitation Medicine.* 2010;53(1):51–9.

[82] Ploeger HE, Bus SA, Nollet F and Brehm MA. Gait patterns in association with underlying impairments in polio survivors with calf muscle weakness. *Gait & posture.* 2017;58:146–53.

[83] Faraj AA. Poliomyelitis: orthopaedic management. *Current Orthopaedics.* 2006;20(1):41–6.

[84] Flansbjer UB, Brogardh C and Lexell J. Muscle strength is only a weak to moderate predictor of gait performance in persons with late effects of polio. *NeuroRehabilitation.* 2013;33(3):457–64.

[85] Gylfadottir S and Dallimore M, E D. The relation between walking capacity and clinical correlates in survivors of chronic spinal poliomyelitis. *Archives of Physical Medicine and Rehabilitation.* 2006;87:944–52.

[86] Larsson Lund M and Lexell J. A positive turning point in life – how persons with late effects of polio experience the influence of an interdisciplinary rehabilitation programme. *Journal of Rehabilitation Medicine.* 2010;42 (6):559–65.

[87] Tiffreau V, Rapin A, Serafi R, *et al.* Post-polio syndrome and rehabilitation. *Annals of Physical and Rehabilitation Medicine.* 2010;53(1):42–50.

[88] Orsini M, de Souza JA, Araujo Leite MA, *et al.* Previous acute polio and post-polio syndrome: recognizing the pathophysiology for the establishment of rehabilitation programs. *Neurology International.* 2015;7(1):5452.

[89] Lo JK and Robinson LR. Post-polio syndrome and the late effects of poliomyelitis: Part 2. treatment, management, and prognosis. *Muscle Nerve.* 2018;58(6):760–9.

[90] Ploeger HE, Bus SA, Brehm MA and Nollet F. Ankle-foot orthoses that restrict dorsiflexion improve walking in polio survivors with calf muscle weakness. *Gait & Posture.* 2014;40(3):391–8.

[91] Brehm MA, Beelen A, Doorenbosch CA, Harlaar J and Nollet F. Effect of carbon-composite knee-ankle-foot orthoses on walking efficiency and gait in former polio patients. *Journal of Rehabilitation Medicine.* 2007;39(8):651–7.

[92] Arazpour M, Chitsazan A, Bani MA, Rouhi G, Ghomshe FT and Hutchins SW. The effect of a knee ankle foot orthosis incorporating an active knee mechanism on gait of a person with poliomyelitis. *Prosthetics and Orthotics International.* 2013;37(5):411–4.

[93] Arazpour M, Ahmadi Bani M, Samadian M, *et al.* The physiological cost index of walking with a powered knee-ankle-foot orthosis in subjects with poliomyelitis: a pilot study. *Prosthetics and Orthotics International.* 2016;40(4):454–9.

[94] Arazpour M, Moradi A, Samadian M, *et al.* The influence of a powered knee-ankle-foot orthosis on walking in poliomyelitis subjects: a pilot study. *Prosthetics and Orthotics International.* 2016;40(3):377–83.

[95] Mojaver A, Arazpour M, Aminian G, *et al.* The effect of footwear adapted with a multi-curved rocker sole in conjunction with knee-ankle-foot orthoses on walking in poliomyelitis subjects: a pilot study. *Disability and Rehabilitation Assistive Technology.* 2017;12(7):747–51.

[96] Andrysek J, Klejman S and Kooy J. Forces and moments in knee-ankle-foot orthoses while walking on irregular surfaces: a case series study. *Prosthetics and Orthotics International.* 2014;38(2):104–13.

[97] Mauroy Jd, Weiss H, Aulisa A, *et al.* 7th SOSORT consensus paper: conservative treatment of idiopathic & Scheuermann's kyphosis. *Scoliosis.* 2010;5:9.

[98] Sangtarash F, Manshadi FD and Sadeghi A. The relationship of thoracic kyphosis to gait performance and quality of life in women with osteoporosis. *Osteoporos International.* 2015;26(8):2203–8.

[99] Glassman SD, Bridwell K, Dimar JR, *et al.* The impact of positive sagittal balance in adult spinal deformity. *Spine.* 2005;30(18):2024–9.

[100] Kurokawa TA, Katuhira JB and Maruyama HB. Biomechanical analysis of elderly people with kyphosis during gait. *Rigaku Ryoho Kagaku.* 2010; 25(4):589–94.

[101] Gottipati P, Fatone S, Koski T, Sugrue PA and Ganju A. Crouch gait in persons with positive sagittal spine alignment resolves with surgery. *Gait & Posture.* 2014;39(1):372–7.

[102] Jain S and Jain G. *Spine and Spinal Orthoses.* New Delhi: Jaypee Brothers Medical Publishers Pvt. Limited; 2015.

[103] Fisk J. Atlas of Spinal Orthotics; 2017. Available from: https://www.exceed-worldwide.org/atlas-of-spinal-orthotics.
[104] Kramers-de Quervain IA, Müller R, Stacoff A, Grob D and Stüssi E. Gait analysis in patients with idiopathic scoliosis. *European Spine Journal.* 2004;13(5):449–56.
[105] Guo X, Chau WW, Hui-Chan CW, Cheung CS, Tsang WW and Cheng JC. Balance control in adolescents with idiopathic scoliosis and disturbed somatosensory function. *Spine.* 2006;31(14):E437–40.
[106] Beaulieu M, Toulotte C, Gatto L, *et al.* Postural imbalance in non-treated adolescent idiopathic scoliosis at different periods of progression. *European Spine Journal: Official Publication of the European Spine Society, the European Spinal Deformity Society, and the European Section of the Cervical Spine Research Society.* 2009;18(1):38–44.
[107] Schizas CG, Kramers-de Quervain IA, Stüssi E and Grob D. Gait asymmetries in patients with idiopathic scoliosis using vertical forces measurement only. *European Spine Journal: Official Publication of the European Spine Society, the European Spinal Deformity Society, and the European Section of the Cervical Spine Research Society.* 1998;7(2):95–8.
[108] Yang JH, Suh SW, Sung PS and Park WH. Asymmetrical gait in adolescents with idiopathic scoliosis. *European Spine Journal: Official Publication of the European Spine Society, the European Spinal Deformity Society, and the European Section of the Cervical Spine Research Society.* 2013;22(11):2407–13.
[109] Chockalingam N, Bandi S, Rahmatalla A, Dangerfield PH and Ahmed E-N. Assessment of the centre of pressure pattern and moments about S2 in scoliotic subjects during normal walking. *Scoliosis.* 2008;3(1):10.
[110] Chockalingam N, Dangerfield PH, Rahmatalla A, Ahmed el N and Cochrane T. Assessment of ground reaction force during scoliotic gait. *European Spine Journal: Official Publication of the European Spine Society, the European Spinal Deformity Society, and the European Section of the Cervical Spine Research Society.* 2004;13(8):750–4.
[111] Karimi MT, Kavyani M and Kamali M. Balance and gait performance of scoliotic subjects: a review of the literature. *Journal of Back & Musculoskeletal Rehabilitation.* 2016;29(3):403–15.
[112] Chow DH, Leung DS and Holmes AD. The effects of load carriage and bracing on the balance of schoolgirls with adolescent idiopathic scoliosis. *European Spine Journal: Official Publication of the European Spine Society, the European Spinal Deformity Society, and the European Section of the Cervical Spine Research Society.* 2007;16(9):1351–8.
[113] Adler N, Bleck EE, Rinsky LA and Young W. Balance reactions and eye-hand coordination in idiopathic scoliosis. *Journal of Orthopaedic Research: Official Publication of the Orthopaedic Research Society.* 1986; 4(1):102–7.
[114] Mahaudens P, Banse X, Mousny M, Raison M and Detrembleur C. Very short-term effect of brace wearing on gait in adolescent idiopathic scoliosis

girls. *European Spine Journal: Official Publication of the European Spine Society, the European Spinal Deformity Society, and the European Section of the Cervical Spine Research Society*. 2013;22(11):2399–406.

[115] Wong MS, Cheng CY, Ng BK, *et al.* The effect of rigid versus flexible spinal orthosis on the gait pattern of patients with adolescent idiopathic scoliosis. *Gait & Posture*. 2008;27(2):189–95.

[116] Wong MS, Cheng JC, Lam TP, *et al.* The effect of rigid versus flexible spinal orthosis on the clinical efficacy and acceptance of the patients with adolescent idiopathic scoliosis. *Spine*. 2008;33(12):1360–5.

[117] Aneja S. Evaluation of a child with cerebral palsy. *Indian Journal of Pediatrics*. 2004;71(7):627–34.

[118] Jacobs M and Austin N. *Splinting the Hand and Upper Extremity: Principles and Process*. Baltimore, MD: Lippincott Williams & Wilkins; 2003.

[119] Yeargin-Allsopp M, Van Naarden Braun K, Doernberg NS, Benedict RE, Kirby RS and Durkin MS. Prevalence of cerebral palsy in 8-year-old children in three areas of the United States in 2002: a multisite collaboration. *Pediatrics*. 2008;121(3):547–54.

[120] Lusardi MM. In Jorge M, Nielsen CC, editors. Orthotics and Prosthetics in Rehabilitation. 3rd ed. *St. Louis, MO*: Elsevier Saunders; 2012.

[121] Brégou Bourgeois A, Mariani B, Aminian K, Zambelli PY and Newman CJ. Spatio-temporal gait analysis in children with cerebral palsy using, foot-worn inertial sensors. *Gait & Posture*. 2014;39(1):436–42.

[122] Rasmussen HM, Nielsen DB, Pedersen NW, Overgaard S and Holsgaard-Larsen A. Gait deviation index, gait profile score and gait variable score in children with spastic cerebral palsy: intra-rater reliability and agreement across two repeated sessions. *Gait & Posture*. 2015;42(2):133–7.

[123] Fuller DA, Keenan MA, Esquenazi A, Whyte J, Mayer NH and Fidler-Sheppard R. The impact of instrumented gait analysis on surgical planning: treatment of spastic equinovarus deformity of the foot and ankle. *Foot & Ankle International*. 2002;23(8):738–43.

[124] Cook RE, Schneider I, Hazlewood ME, Hillman SJ and Robb JE. Gait analysis alters decision-making in cerebral palsy. *Journal of Pediatric Orthopedics*. 2003;23(3):292–5.

[125] Hägglund G, Andersson S, Düppe H, Lauge-Pedersen H, Nordmark E and Westbom L. Prevention of severe contractures might replace multilevel surgery in cerebral palsy: results of a population-based health care programme and new techniques to reduce spasticity. *Journal of Pediatric Orthopedics Part B*. 2005;14(4):269–73.

[126] Damiano DL, Alter KE and Chambers H. New clinical and research trends in lower extremity management for ambulatory children with cerebral palsy. *Physical Medicine and Rehabilitation Clinics of North America*. 2009;20(3):469–91.

[127] Seymour R. *Prosthetics and Orthotics: Lower Limb and Spinal*. Philadelphia, PA; Lippincott Williams & Wilkins; 2002.

[128] Labarthe D. *Epidemiology and Prevention of Cardiovascular Diseases: A Global Challenge*. Sudbury, MA: Jones & Bartlett Learning; 2011.

[129] Wang C, Goel R, Zhang Q, Lepow B and Najafi B. Daily use of bilateral custom-made ankle-foot orthoses for fall prevention in older adults: a randomized controlled trial. *Journal of the American Geriatrics Society*. 2019; 67(8):1656–61.

[130] Tyson SF and Kent RM. Effects of an ankle-foot orthosis on balance and walking after stroke: a systematic review and pooled meta-analysis. *Archives of Physical Medicine and Rehabilitation*. 2013;94(7): 1377–85.

[131] Bouchalová V, Houben E, Tancsik D, *et al.* The influence of an ankle-foot orthosis on the spatiotemporal gait parameters and functional balance in chronic stroke patients. *Journal of Physical Therapy Science*. 2016;28(5):1621–8.

[132] Snijders AH, van de Warrenburg BP, Giladi N and Bloem BR. Neurological gait disorders in elderly people: clinical approach and classification. *The Lancet Neurology*. 2007;6(1):63–74.

[133] Marquis S, Milar Moore M, Howieson DB, *et al.* Independent predictors of cognitive decline in healthy elderly persons. *Archives of Neurology*. 2002; 59(4):601–6.

[134] Shah VV, McNames J, Harker G, *et al.* Effect of bout length on gait measures in people with and without Parkinson's disease during daily life. *Sensors (Basel)*. 2020;20(20):5769.

[135] Hale LA, Pal J and Becker I. Measuring free-living physical activity in adults with and without neurologic dysfunction with a triaxial accelerometer. *Archives of Physical Medicine and Rehabilitation*. 2008; 89(9): 1765–71.

[136] Garcia-Agundez A, Folkerts A-K, Konrad R, *et al.* Recent advances in rehabilitation for Parkinson's disease with exergames: a systematic review. *Journal of NeuroEngineering and Rehabilitation*. 2019;16(1):17.

[137] Aboutorabi A, Bahramizadeh M, Arazpour M, *et al.* A systematic review of the effect of foot orthoses and shoe characteristics on balance in healthy older subjects. *Prosthetics and Orthotics International*. 2015;40(2): 170–81.

[138] Otis MJD, Ayena JC, Tremblay LE, Fortin PE and Ménélas B-AJ. Use of an enactive insole for reducing the risk of falling on different types of soil using vibrotactile cueing for the elderly. *PloS One*. 2016;11(9).

[139] Najafi B, de Bruin ED, Reeves ND, Armstrong DG and Menz HB. The role of podiatry in the prevention of falls in older people: a JAPMA special issue. *Journal of the American Podiatric Medical Association*. 2013;103(6): 452–6.

[140] Ma C, Wong D, Lam W, Wan A and Lee W. Balance improvement effects of biofeedback systems with state-of-the-art wearable sensors: a systematic review. *Sensors*. 2016;16:434.

[141] Robb KA and Perry SD. Textured foot orthotics on dynamic stability and turning performance in Parkinson's disease. *Journal of Motor Behavior.* 2020;52(4):396–403.

[142] Owen E. Shank angle to floor measures and tuning of ankle-foot orthosis footwear combinations for children with cerebral palsy, spina bifida and other conditions. *MSc Thesis*. Glasgow: University of Strathclyde; 2004.

[143] Cook TM and Cozzens B. The effects of heel height and ankle-foot orthosis configuration on weight line location: a demonstration of principles. *Prosthetics and Orthotics International.* 1976;30(4):43–6.

[144] Churchill AJ, Halligan PW and Wade DT. Relative contribution of footwear to the efficacy of ankle-foot orthoses. *Clinical Rehabilitation.* 2003; 17(5):553–7.

[145] Owen E. Tuning of ankle foot orthosis footwear combinations for children with cerebral palsy, spina bifida and other conditions. In *Proceedings of the European Society of Movement Analysis in Adults and Children (ESMAC) Seminars*; Warsaw, Poland; 2004; pp. 23–25.

[146] Bowers R and Ross K. Development of a best practice statement on the use of ankle-foot orthoses following stroke in Scotland. *Prosthetics and Orthotics International.* 2010;34:245–53.

[147] Bowers R and Ross K, editors. A Review of the Effectiveness of Lower Limb Orthoses Used in Cerebral Palsy. Recent Developments in Healthcare for Cerebral Palsy: Implications and Opportunities for Orthotics; Oxford: Wolfson College; 2009.

[148] Owen E. Tuning of AFO-footwear combinations for neurological disorders. *11th World Congress of the International Society for Prosthetics and Orthotics, Hong Kong*; 2004.

[149] Owen E. Segmental approach to orthotic management: ankle-foot orthosis/ footwear combination. In: Rahlin M, editor. *Physical Therapy for Children with Cerebral Palsy; An Evidence-Based Approach.* Chicago, IL: Slack Incorporated; 2016; pp. 341–70.

[150] Chow DH, Holmes AD, Lee CK and Sin SW. The effect of prosthesis alignment on the symmetry of gait in subjects with unilateral transtibial amputation. *Prosthetics and Orthotics International.* 2006; 30(2):114–28.

[151] Krajbich JI, Pinzur MS, Potter BK and Stevens PM. *Atlas of Amputations and Limb Deficiencies: Surgical, Prosthetic, and Rehabilitation Principles*. Rosemont, IL: American Academy of Orthopaedic Surgeons; 2016.

[152] Carroll K and Edelstein JE. *Prosthetics and Patient Management: A Comprehensive Clinical Approach.* Chicago, IL: SLACK Incorporated; 2006.

[153] May BJ. In Lockard MA, editor. *Prosthetics & Orthotics in Clinical Practice: A Case Study Approach*. Philadelphia, PA: F.A. Davis; 2011.

Chapter 10
Forensic gait analysis – Is there a case?

Yasushi Makihara[1], *Daigo Muramatsu*[2] *and Yasushi Yagi*[3]

We describe techniques of video-based gait analysis and their application to forensics in this chapter. First, we describe a brief history of the development of video-based gait analysis techniques including early human perception-based one, model-based and appearance-based gait representations, machine learning techniques, and recent deep learning approaches, as well as its use cases in forensic purposes in the world. We also describe our gait verification system for criminal investigation, which is composed of subject enrollment, silhouette extraction, and verification with feature extraction, where a user, i.e., a criminal investigator, confirms the correctness at each stage step by step. The gait verification system copes with a variety of circumstances including view angles, image size, frame-rate, and region to be masked out of matching. We then elaborate on a situation of gait forensics with the developed gait verification system in Japan and conclude this chapter with a summary and future direction of the gait forensics.

10.1 Introduction

People might have ever identified their family/friend walking at a distance, where his/her face is not clearly seen. One of the cues to identify a pedestrian (i.e., a walking person) at a distance is a walking pattern, that is, gait. Gait has been considered as one of the behavioral biometrics, and a technique to identify a person from gait has been called gait recognition. Gait recognition has drawn much attention from a variety of application sides such as surveillance, access control, and forensics [1–3], thanks to a couple of advantages over the other biometrics (e.g., DNA, fingerprint, finger vein, palmprint, iris, face, etc.): identification at a distance from a camera; identification of an unconscious/unconstrained target; the difficulty to disguise.

After the possibility of gait recognition was shown in the biological motion research community in the 1970s [4], studies on vision-based gait recognition/analysis got started in the 1990s [5,6]. In the 2000s, more and more researchers

[1]Institute for Advanced Co-Creation Studies, Osaka University, Osaka, Japan
[2]Faculty of Science and Technology, Seikei University, Tokyo, Japan
[3]SANKEN, Osaka University, Osaka, Japan

participated in studies on gait recognition in the world, and gait recognition has been established as a solid research area in research communities of biometrics, computer vision, and pattern recognition as a result [7–16].

One of the major projects in this area is DARPA HumanID at a Distance program [17]. In this project, research on the face and gait recognition at a distance was conducted. As one of the outcomes, a major publicly available gait database named USF gait database, and a baseline algorithm for gait recognition based on silhouette-sequence matching were proposed [18], and they have been contributed to the development of subsequent studies on gait recognition whose accuracies were compared/evaluated with the baseline algorithm/database [19–24]. Some other major publicly available small-/middle-sized gait databases were also released in the 2000s and early 2010s [25–30].

One of the major topics in gait recognition in the 2000s and early 2010s is gait representation. The gait representation mainly falls into two families: model-based representation and appearance-based representation.

Since gait is the pattern of movement of the limbs of animals during locomotion over a solid substrate [31], it was considered to be natural to model the motion using an articulate human model, and in fact, many of the gait recognition studies focused on the model-based representation in the early 2020s, e.g., using leg motion [8,10], stride [32], a latent motion with an ellipsoidal human model [33]. The model-based representations had not been dominant for a long time due to several difficulties: erroneous human model fitting; high computational cost; and requirement of relatively high image resolution despite that one of the advantages of gait recognition is recognition even at a distance (i.e., with a low-resolution image).

The gait recognition community therefore geared to appearance-based representation, particularly, silhouette-based representation. A nice property with the silhouette representation is that it is not affected by the clothes' color and texture, which should be essentially irrelevant with gait biometrics, although they are still useful in person re-identification problems where a target person may wear the same clothes. One of the most straightforward ways to use the silhouettes is a silhouette sequence matching with frame synchronization as proposed in a baseline algorithm [18] or other studies [23,34–36]. The direct sequence matching is, however, sensitive to frame-by-frame silhouette noise (e.g., over-/under-segmentation), and hence gait representations which are aggregated over the temporal dimension, became more popular since the temporal aggregation statistically reduce the frame-by-frame silhouette noise.

One of the simplest yet effective representations is the gait energy image (GEI) [21] a.k.a. an averaged silhouette [37], which is obtained by averaging the silhouettes over one gait period. Most of the appearance-based gait representation employ the temporal aggregation (e.g., frequency-domain feature [38], motion silhouette image [39], gait entropy image [16], gait motion descriptor [40], gait flow image [39], chrono-gait image [20], gradient histogram energy image [41], Gabor GEI [42]).

On the other hand, gait recognition needs to cope with a variety of covariates such as view angle, clothing, carried objects, walking speed, which are mainly caused by unconstrained targets (e.g., a target may change his/her walking speed and direction, clothes, and carried objects depending on situations). Particularly, the appearance-based gait representations are directly affected by the above-mentioned covariates.

One of the effective countermeasures for this is to develop a more robust gait feature against the covariate. For example, single-support GEI [43] and higher-order shape configuration via Procrustes shape analysis [44] were proposed for speed invariance, while masked GEI [45] and gait energy response function [46] were proposed for clothing and carrying-status invariance.

Another countermeasure is to employ machine learning techniques to make the gait recognition robust against the covariates. Because the temporally aggregated gait representation is not sequential data but actually a single or multiple image template (s) (e.g., GEI [21] and Gabor GEI [47]), most of the standard metric learning techniques for images can be applied. Typically adopted machine learning techniques in gait recognition include but are not limited to principal component analysis and linear discriminant analysis [21], singular value decomposition [38], uncorrelated discriminant simplex analysis [48], kernel discriminant analysis [49], random subspace method [50], linear discriminant analysis with tensor representation [51], analysis of variance (ANOVA) [52], multi-view discriminant analysis [53], support vector regression [54], mutual subspace method [55], cluster-pairwise discriminant analysis [56], and ranking support vector machine [57].

As a deep learning era opens in the late 2010s, the gait recognition community also geared to deep learning approaches with the help of significantly large-scale gait databases [58–60]. Early work on deep learning-based gait recognition still adopted the temporally aggregated gait representations as input for the deep neural networks [61–70], some other work achieved the state-of-the-art accuracy using a set of silhouette frames [71], a sequence of silhouettes [72] or RGB images [73]. Moreover, thanks to the recent progress of deep learning-based approaches to pose estimation (i.e., human model fitting), the previous concerns on the model-based gait representation have been getting solved and hence studies on the model-based gait recognition using pose sequences have recently resumed [74–76].

Gait recognition technologies have been put into practical use along with the progress of the above-mentioned gait recognition. At the early stage, human perception-based forensic gait analysis has been started with the Aalsgaarde case [77], where some similarities were found in the gait pattern between suspect and perpetrator (e.g., hyperextension in the knee joints). Subsequently, it was also utilized in the Noerager case of a perpetrator who robbed a bank, in Denmark in 2004. Forensic experts from the University of Copenhagen, listed up several gait characteristics such as long/short steps and outward rotation of feet/ankle joints and judged whether they match between the perpetrator and a suspect based on human perception.

Thereafter, the first use case of automatic gait recognition took place in the UK in 2008 [1,78,79]. Gait recognition researchers from the University of Southampton tested their algorithm on a burglar caught on closed-circuit television (CCTV), and the result has been admitted as evidence in UK courts for the first time and the burglar was convicted thanks to his gait. In addition, a start-up company in China released a commercially available gait recognition system for surveillance and forensic purpose [80].

In this chapter, we will elaborate on our gait verification system for forensics developed in Osaka University [81] in Section 10.2. We also introduced situations of deploying our gait forensics techniques in Japan in Section 10.3.

10.2 Gait verification system

We assume that a pair of image sequences are given for the gait verification system. For example, one is an image sequence of a perpetrator walking captured at a crime scene, and the other is an image sequence of a suspect walking. In cases there are multiple suspects, multiple pairs are generated from that of the crime scene and that of each suspect, and each pair is processed. And an expected output of a gait verification system for forensic is not a binary classification result such as the same or different, but an objective score that shows how the given pair originates from the same subject. Therefore, the role of the gait verification system for forensic is calculating an objective score from a pair of gait image sequences. Our gait verification system [81] computes a posterior probability that extracted perpetrator's gait features and suspect's gait features originate from the same subject. In this system, we focus on a silhouette-based approach and incorporate GEI as a gait feature. Because GEI is affected by several covariates such as view angle and frame-rate, image size, and clothes, impacts associated with these covariates must be taken into account for a posterior probability calculation. We, therefore, consider the circumstance that is defined by covariates associated with the extracted gait feature and compute the circumstance-dependent posterior probability. In order to realize this computation from image sequences, the following processes are necessary: (1) a target person selection from the image sequence, (2) silhouette image creation from the target person's image sequences, and (3) gait feature extraction and circumstance-dependent posterior probability computation. Each process is implemented as a separated module in our gait verification system, and therefore operators can execute gait verification step by step. From the viewpoint of efficiency, a full-automatic system is preferable, but a full-automatic system is inappropriate for the forensic system. An operator must check if each step is correctly processed and justify the processed data in order to exclude incorrect verification, and in case they find inappropriate processed data such as the noisy silhouette, they need to apply an appropriate process to correct the data. So, visualization function and modification function are indispensable for the system. A gait verification system for forensics must be designed by taking these properties into account. Now, we explain each module step by step.

10.2.1 Target person selection module

The first step of the gait verification is to select the regions of the target person for verification from the image sequences. For this selection, two points should be cared: the spatial point and the temporal point. Spatial point is that the target person must be included in the selected region in each frame, and the temporal point is that the target person is walking naturally in the selected sequences. The spatial point is necessary because the system cannot understand which person is the target, and the temporal point is necessary because the current gait verification algorithm implemented in the system assumes that the person is walking naturally. The operator boots the target person selection module, and reads an image sequence such as CCTV footage, and finds the image sequences that include the target person. For the selection, an operator selects and cuts off the target image sequence that satisfies the temporal point, and then sets a bounding box (BB) to the region of the target person in each frame as shown in Figure 10.1 (a). In this figure, the set BB in each frame is shown in yellow lines. To complete the BB setting, the operator is requested to check the BB; and in case the BB is reasonable, the operator accepts the BB by checking the "Rect entry" as shown in Figure 10.1(b). Then, the accepted BB is shown in white lines as shown in Figure 10.1(c). This additional process is very important to exclude the mistakes that lead to incorrect verification. After the confirmation, the operator registers the BB together with information associated with the target image sequences, such as frame-rate (fps) of the image sequences and view angle of the target person. In the current system, fps of 30, 15, 10, 7, 5, 6, and 5 are acceptable, and view angles are selected from 24 candidate views.

10.2.2 Silhouette creation module

The next step is to create a silhouette image sequence of the target person from the selected image sequences. There are several approaches to create a silhouette image sequence. The simplest and direct approach is to use a mouse-based manual paint function and extract a silhouette from a scratch by painting the foreground region (person body). Through this approach, a reasonable silhouette can be created. But as there are many images to be processed, creating the silhouette from a scratch is very time consuming. Moreover, in several low-quality images such as dark ones, the creation of silhouettes with consistent quality becomes difficult. We, therefore, implemented a semi-automatic silhouette creation module. In this module, the system utilizes the background image sequences of the target scenes. These background image sequences are to be cut off from the original image sequences, and selection of the background image sequence is realized using functions in the target person selection module. Background subtraction and graph-cut-based segmentation functions are implemented together with several image processing functions such as area filtering, dilation, and erosion. Using these functions and manual painting, the operator creates a high-quality silhouette image frame by frame. Figure 10.2 shows an example of the silhouette creation process using the module. The silhouette of the target person is a region contoured by a red line. For the gait verification, high-quality silhouette quality is necessary, so, several auxiliary functions are also devised to help silhouette creation more efficiently. For example, the operator can easily switch the displayed contents by just pushing one key: raw image with

(a) (b) (c)

Figure 10.1 BB setting. (a) BB is set by pointing upper left and bottom right of the target region. (b) In order to accept the BB, the operator is requested to do an additional confirmation process. (c) Once BB is accepted, the color of BB change from yellow to white.

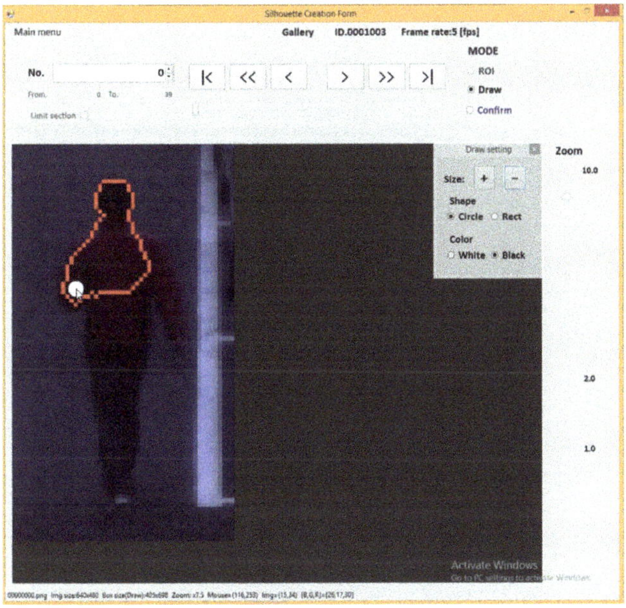

Figure 10.2 Silhouette creation on the draw mode. The contour of the silhouette can be drawn by mouse operation. Multiple image processing functions are also executable in this mode.

contour, raw image, and binary silhouette as shown in Figure 10.3. When a raw image with contour is displayed, the operator can change the transparency of the contour line to improve visibility. These auxiliary functions are additionally implemented according to the criminal investigator's request. This silhouette creation is processed in the silhouette draw mode. After creating silhouettes, silhouette images are saved in the system. Created silhouettes saving is possible only in the confirm mode as shown in Figure 10.4. In this mode, the created silhouette is shown together with the height information, and vertical and horizontal lines shown in the left, right, upper, and bottom (these parts should be checked in the system). On the confirm mode, the operator can easily check the quality of the created silhouettes, and in case the operator finds an inappropriate silhouette, he modifies the silhouette on draw mode. Confirm mode seems to be redundant, but confirm mode enables to find an inappropriate silhouette and can contribute to avoiding inappropriate silhouette saving.

10.2.3 Feature extraction and posterior probability calculation module

This module is mainly composed of three processes: (1) gait feature generation, (2) dissimilarity score calculation, and (3) circumstance-dependent posterior probability calculation.

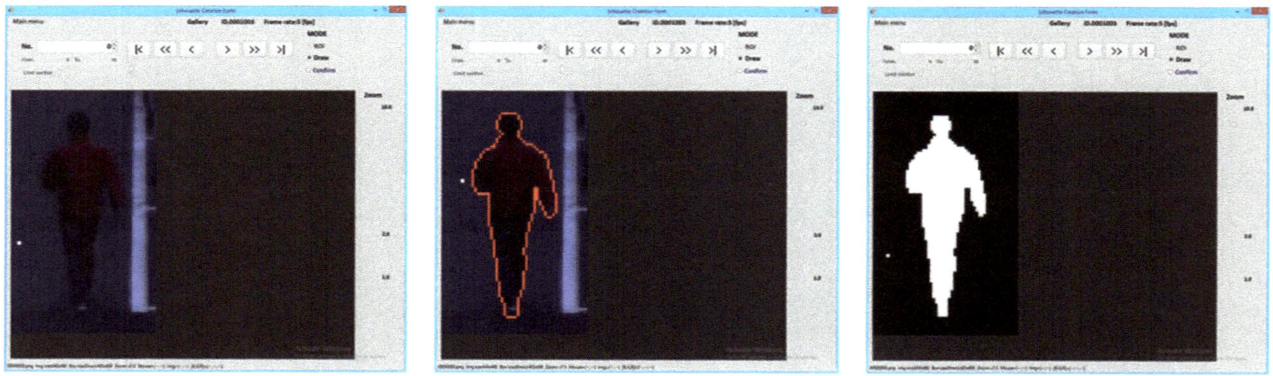

Figure 10.3 Three different types of contents are displayed. (Left) Raw image, (middle) raw image with contour, and (Right) binary silhouette

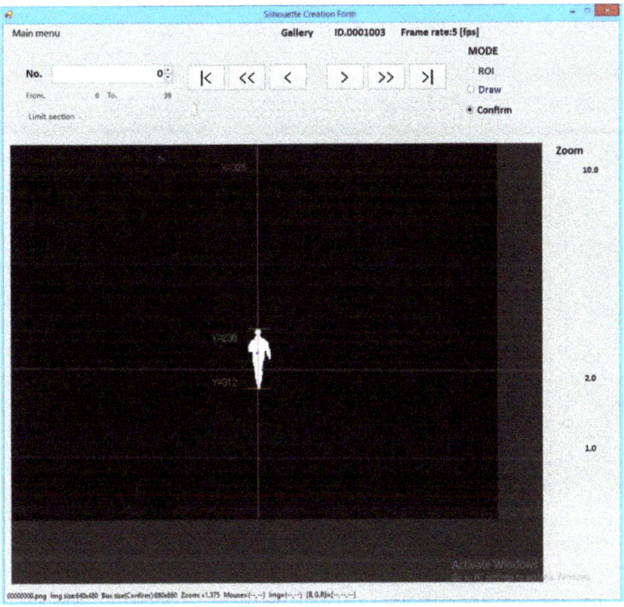

Figure 10.4 Silhouette saving on the confirm mode

10.2.3.1 Gait feature generation

GEIs [21] are generated from the silhouette image sequences. For the generation, an operator simply selects the target subject ID* and pushes the create button, then the system automatically calculates the gait period of the target subject, and creates a set of GEIs.

10.2.3.2 Dissimilarity score calculation

A dissimilarity score is calculated from two GEI sets. In this system, we calculate dissimilarity between GEIs registered as probe and GEIs registered as gallery. So, we assume that through the gait feature generation processes, two sets of GEIs are generated. In Figure 10.6 (upper left), the probe is a GEI set of subject ID "0001000," and the gallery is a GEI set of subject ID "0001003." Basically, a dissimilarity score can be computed by summing up the difference of each corresponding pixel of the whole gait features. This type of score is reasonable in the case where the view of the gallery and the probe are the same, and there is no outlier in the GEIs, but it cannot work well with view differences or outliers. Here, outliers include large differences in clothes, wearing a hat, carrying belongings, etc. For the view difference issue, our gait verification system applies the view transformation technique [38] to the gait features before dissimilarity calculation if the

*Subject's data in this system is managed by the subject ID input by operator in the target person selection module.

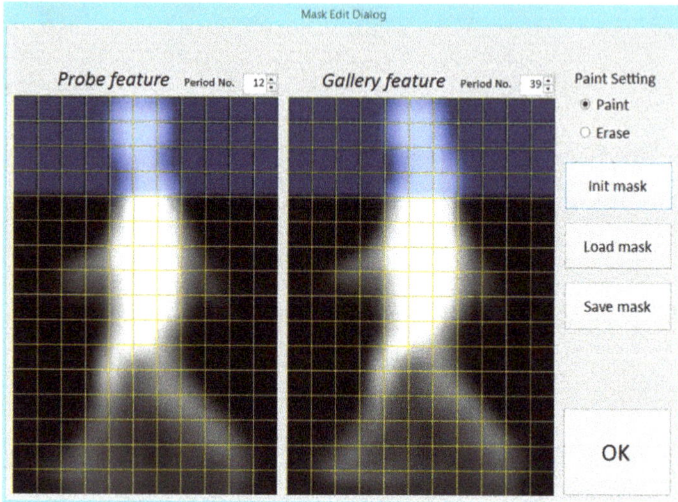

Figure 10.5 Block-wise masking. The masked region is selected by manual painting. Pixels included in the painted blocks are not used for dissimilarity calculation.

view of the gallery and the probe are different. By applying the view transformation technique, GEIs from both views (the probe view and gallery view) are generated, and we can calculate a dissimilarity score from GEIs with the same view. Because view information is registered in the target person selection module, the gait verification system can automatically decide if view transformation is necessary. And for the outlier issue, our system equips block-wise masking function as shown in Figure 10.5. By painting the blocks associated with outliers, we can eliminate the painted block from the dissimilarity calculation. In the example of Figure 10.5, a person in the probe wears the helmet, but a person in the gallery does not; so the head region must be excluded from the dissimilarity calculation. By this masking, the gait verification system can mitigate the impact of outliers.

10.2.3.3 Circumstance-dependent posterior probability calculation

A posterior probability that the gallery gait features and the probe gait features originate from the same subject is computed from the calculated dissimilarity score. For the computation, dissimilarity score distribution of the same subjects and dissimilarity score distribution of different subjects are employed, and a dissimilarity score calculated from the gallery and the probe is transformed to a posterior probability. Let x_g and x_p be the gait features of the gallery and the probe. And let χ be a label whose value is the "same (S)" if the gallery and the probe originate from the same subject, and is "different (D)" otherwise. Using the x_g and x_p, we would like to calculate the probability $P(\chi = \mathrm{S}\ x_g, x_p)$. In place of the direct calculation of $P(\chi = \mathrm{S}\ x_g, x_p)$, our gait system calculates the $P(\chi = \mathrm{S}\ d(x_g, x_p))$,

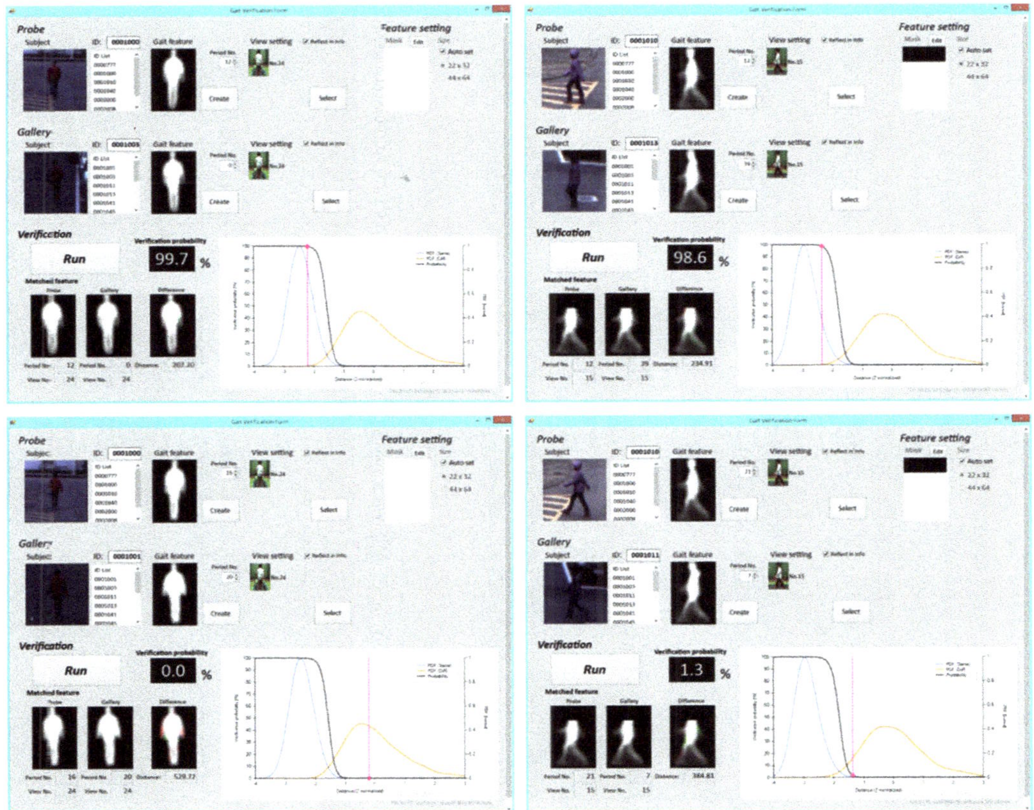

Figure 10.6 *Results of the gait verification system. In this figure, four results are displayed. (Upper left) Same subject case from frontal view. (Lower left) Different subject cases from frontal view. (Upper right) Same subject case from side view with mask. (Lower right) Different subject case from side view with mask*

where $d(x_g, x_p)$ is a dissimilarity score calculated from x_g and x_p. This probability can be described by:

$$P(\chi = S|d(xg,xp)) = \frac{p(d(xg,xp)|\chi = S)P(S)}{p(d(xg,xp)|\chi = S)P(S) + p(d(xg,xp)|\chi = D)P(D)} \tag{10.1}$$

Here, $p(d(x_g, x_p)\ \chi = \mathrm{S})$ and $p(d(x_g, x_p)\ \chi = \mathrm{S})$ are probability density functions (PDF) of the dissimilarity score associated with gait features matching of the same subjects and PDF of the dissimilarity score associated with gait features matching of the different subjects, respectively. Because these PDFs are heavily dependent on the covariates, mask, and usage of view transformation, we consider a circumstance-dependent framework. We focus on the following factors that impact dissimilarity score distributions:

- image size of the target subjects (spatial resolutions (SR))
- frame-rates (temporal resolutions (TR))
- view angles (view)
- mask region (mask)

and we define circumstance by the combination of each factor's status.

Let ξ_{SR}, $\xi_{T\ R}$, ξ_{view}, and ξ_{mask} be status associated with the gallery and the probe of SP, TR, view, and mask, respectively. And let $\Xi = (\xi_{SR}, \xi_{T\ R}, \xi_{view}, \xi_{mask})$ be a symbol that indicates the circumstance. In the circumstance-dependent framework, a posterior probability is calculated by taking the circumstance Ξ into account:

$$P(\chi = S|d(xg,xp);\Xi) = \frac{p(d(xg,xp)|\chi = S;\Xi)P(S)}{p(d(xg,xp)|\chi = S;\Xi)P(S) + p(d(xg,xp)|\chi = D;\Xi)P(D)} \tag{10.2}$$

In order to realize this, PDF of the same subjects and that of different subjects are computed dependent on Ξ.

Figure 10.6 shows examples of verification results. In this figure, four results are displayed; the first column shows the results of frontal view matching, and the second column shows the results of side view matching with a mask. In both columns, the first row is the verification results of the same subject cases, and the second row is the verification results of different subject cases. In each result, generated gait features, information for dissimilarity score, information associated with circumstance, and PDFs are displayed on the upper left, lower left, upper right, and lower right, respectively, and the posterior probability is shown in a little left and below the middle. From these, you can see that high posterior probabilities are computed for the same subject case while low posterior probabilities are computed for the different subject cases. And also you can observe that PDFs of the right column and left column are different. This is because, PDFs are computed depending on the circumstance, and the circumstances of the left column case and the right column case are different.

10.2.4 Future direction of the gait verification system

Recently, image processing techniques have been drastically improved thanks to deep learning and a large scale of training data. Among these, person detection algorithms such as [82], and semantic segmentation algorithms such as [83,84] can be implemented in order to improve the efficiency of the system. But it should be noted that even though the state-of-the-art algorithms are implemented, a full-automatic system is not preferable for forensics. Moreover, many sophisticated gait verification algorithms have been developed. These algorithms are candidates for further improvement. In case we introduce the new algorithms, we need to carefully design how to compute an objective score for forensics.

10.3 Use cases of gait forensics in Japan

In this section, we will describe situations of our gait forensic techniques in Japan.

The automatic gait recognition technology developed in Osaka University has been applied to a case of a firer in 2008 for the first time in Japan. Nara Prefectural Police (NPP) brought a pair of videos capturing a perpetrator (i.e., the firer) and a suspect to us and asked us to examine whether they were the same or not by gait. In one of the videos, the perpetrator was walking in front of the scene of the crime (i.e., a garage) and was caught on a private surveillance camera installed at the garage. On the other hand, the suspect was confidentially captured by a police inspector from NPP. Because the perpetrator's upper body (i.e., a head and shoulders) was partially out of an image frame (i.e., unobserved), we masked out the unobserved body parts both for the perpetrator's and suspect's videos. Also, the suspect carried a bag with his hand while walking, we masked out arm swing regions from matching since the carried object may affect the arm swing. As a result of matching using a frequency-domain feature of size-normalized silhouettes [38], we obtained 90% posterior probability of the same subject and submitted an expert evidence to report the result of gait forensics to NPP.

NPP counted the evidence as one of the clues and then decided to charge the crime and arrest the suspect. The evidence was also submitted to Nara District Court (NDC) to judge the case. A gait recognition researcher from Osaka University, who was in charge of the gait forensics for this case, was invited to NDC as an expert witness and gave evidence on the gait forensics for this case. Although gait forensics has not been admitted as evidence since it was not regarded as a matured technology at that time, the presiding judge gave an exceptional positive comment on the expectation of utilization of gait forensics in the future.

We then started a research project (2010–14) on criminal investigation supported by the Research and Development Program for Implementation of Anti-Crime and Anti-Terrorism Technologies for a Safe and Secure Society, in part by the Funds for Integrated Promotion of Social System Reform and Research and Development through the Ministry of Education, Culture, Sports, Science, and Technology. The project aimed at developing criminal investigation technologies from CCTV footage, more specifically, an ultra-fast person retrieval system based on face and/or full-body cues including gait, height, and clothes. The project involved six research groups: four from universities; one from a private company; and the other one from a national research institute, i.e., National Research Institute of Police Science (NRIPS) in Japan.

We developed our gait verification system for criminal investigation described in the previous section, during the project. NRIPS has started a trial use of the gait verification system since 2013 after we gave a detailed lecture on how to use the gait verification system to researchers of NRIPS. The gait verification system has been also introduced as a novel technique of video-based forensics in the Police White Paper of 2014.

In addition, in order to prevail the gait verification system all over the prefectures in Japan, we gave a whole day lecture on video-based gait analysis and hands-on tutorial on the gait verification system to criminal investigators once or twice per year, which have been held as a part of a series of lectures on image-based forensic techniques at NRIPS. More specifically, eight criminal investigators from a variety of prefectures in Japan joined for each lecture, and a total of 104 criminal investigators have already experienced the gait verification system as of 2020.

After the above-mentioned activity to prevail the gait forensics in Japan, the expert evidence of the gait forensics, which was made by researchers from NRIPS with the gait verification system, has been finally admitted in the Japanese court for the first time in 2016.

10.4 Conclusion and future prospects

In this chapter, we described a brief history of the development of video-based gait analysis techniques including early human perception-based one, model-based and appearance-based gait representations, machine learning techniques, and recent deep learning approaches, as well as its applications (or use cases) to forensics in the world. We also described our gait verification system for criminal investigation. Finally, we showed use cases of gait forensics, particularly, the gait verification system, in Japan.

While the gait verification system was developed before the deep learning era, i.e., was built upon hand-crafted gait features and classical machine learning techniques, the gait recognition community has already geared to deep learning approaches similar to the other research field of computer vision, pattern recognition, and biometrics. We are therefore considering making a major upgrade of the gait verification system by taking the deep learning techniques into account. Specifically, current manual tedious operations on registering (or cropping) a target person and silhouette extraction could be largely reduced by state-of-the-art deep learning approaches to pedestrian detection and semantic/instance segmentation [82–85].

Also, the state-of-the-art deep learning approaches to gait recognition [69–72] would significantly improve the accuracy.

While we focused on a single biometric modality, i.e., the gait, in this chapter, the use of multi-modal biometrics is worth exploring since it is well-known that multi-modal biometric fusion improves accuracy in the biometric research field [86]. In fact, a fusion of face and gait has been extensively studied [14,87–99]. Since face recognition from a low-resolution image [100–103] is also a hot research topic in computer vision, the fusion of face and gait from CCTV footage would become more popular. We have also already investigated the fusion of three modalities: gait, face (texture of a head region), and height in [104–106] and developed a prototype of the multi-modal verification system. We are therefore considering the extension to the multi-modal biometrics for CCTV footage-based forensics too.

References

[1] Bouchrika I, Goffredo M, Carter J, *et al.* On using gait in forensic biometrics. *Journal of Forensic Sciences.* 2011; 56(4): pp. 882–889.

[2] Iwama H, Makihara Y and Yagi Y. Foreground and shadow segmentation based on a homography-correspondence pair. In: *Proceedings of the 10th Asian Conference on Computer Vision*, Queenstown, New Zealand; 2010. pp. 2790–2802.

[3] Lynnerup N and Larsen P. Gait as evidence. *IET Biometrics.* 2014; 3(2): pp. 47–54.

[4] Johnson A. Spin-images: A Representation for 3D Surface Matching. PhD Thesis. Carnegie Mellon University; 1997.

[5] Niyogi S and Adelson E. Analyzing and recognizing walking figures in xyt. In: *Proceedings of the 7th IEEE Conference on Computer Vision and Pattern Recognition*; 1994. pp. 469–474.

[6] Stevenage SV, Nixon MS and Vince K. Visual analysis of gait as a cue to identity. *Applied Cognitive Psychology.* 1999;13(6): pp. 513–526.

[7] Bobick AF and Johnson AY. Gait recognition using static activity-specific parameters. In: *Proceedings of the 14th IEEE Conference on Computer Vision and Pattern Recognition*, Vol. 1; 2001. pp. 423–430.

[8] Yam CY, Nixon MS and Carter JN. Extended model based automatic gait recognition of walking and running. In: *Proceedings of the 3rd International Conference on Audio and Video-based Person Authentication*, Halmstad, Sweden; 2001. pp. 278–283.

[9] Phillips PJ, Sarkar S, Robledo I, *et al.* Baseline results for the challenge problem of HumanID using gait analysis. In: *Proceedings of Fifth IEEE International Conference on Automatic Face Gesture Recognition*; 2002. pp. 137–142.

[10] Cunado D, Nixon MS and Carter JN. Automatic extraction and description of human gait models for recognition purposes. *Computer Vision and Image Understanding.* 2003; 90(1): pp. 1–41.

[11] Cuntoor N, Kale A and Chellappa R. Combining multiple evidences for gait recognition. In: *Proceedings of the 28th IEEE International Conference on Acoustics, Speech, and Signal Processing*, Vol. 3; 2003. pp. 33–36.

[12] Wang L, Tan T, Ning H, *et al.* Silhouette analysis-based gait recognition for human identification. *IEEE Transactions on Pattern Analysis and Machine Intelligence*. 2003; 25(12): pp. 1505–1518.

[13] Ohara Y, Sagawa R, Echigo T, *et al.* Walking person identification dealing with resolution and appearance changes. In: *Proceedings of Korea-Japan Joint Workshop on Frontiers of Computer Vision*; 2004. pp. 89–94.

[14] Geng X, Wang L, Li M, *et al.* Distance-driven fusion of gait and face for human identification in video. In: *Proceedings of Image and Vision Computing*; 2007. pp. 19–24.

[15] Lu J, Zhang E, Zhang Z, *et al.* Gait recognition using independent component analysis. In: Wang J, Liao XF, Yi Z, editors. *Advances in Neural Networks – ISNN 2005*. Berlin, Heidelberg: Springer Berlin Heidelberg; 2005. pp. 183–188.

[16] Bashir K, Xiang T and Gong S. Feature selection on gait energy image for human identification. In: *2008 IEEE International Conference on Acoustics, Speech and Signal Processing*; 2008. pp. 985–988.

[17] Phillips PJ. Human identification technical challenges. In: *Proceedings of International Conference on Image Processing*, Vol. 1; 2002. pp. 49–52.

[18] Sarkar S, Phillips JP, Liu Z, *et al.* The HumanID gait challenge problem: data sets, performance, and analysis. *IEEE Transactions on Pattern Analysis and Machine Intelligence.* 2005; 27(2): pp. 162–177.

[19] Sagawa R, Makihara Y, Echigo T, *et al.* Matching gait image sequences in the frequency domain for tracking people at a distance. In: *Proceedings of the 7th Asian Conference on Computer Vision*, Vol. 2; 2006. pp. 141–150.

[20] Wang C, Zhang J, Wang L, *et al.* Human identification using temporal information preserving gait template. *IEEE Transactions on Pattern Analysis and Machine Intelligence.* 2012; 34(11): pp. 2164–2176.

[21] Han J and Bhanu B. Individual recognition using gait energy image. *IEEE Transactions on Pattern Analysis and Machine Intelligence.* 2006; 28(2): pp. 316–322.

[22] Liu Z and Sarkar S. Effect of silhouette quality on hard problems in gait recognition. *IEEE Transactions on Systems, Man, and Cybernetics Part B: Cybernetics*. 2005; 35(2): pp. 170–183.

[23] Liu Z and Sarkar S. Improved gait recognition by gait dynamics normalization. *IEEE Transactions on Pattern Analysis and Machine Intelligence.* 2006; 28(6): pp. 863–876.

[24] Lam THW, Cheung KH and Liu JNK. Gait flow image: a silhouette-based gait representation for human identification. *Pattern Recognition.* 2011; 44: pp. 973–987. Available from: http://dx.doi.org/10.1016/j.patcog.2010.10.011.

[25] Gross R and Shi J. The CMU Motion of Body (MoBo) Database. CMT; 2001.

[26] Shutler JD, Grant MG, Nixon M.S, *et al.* On a large sequence-based human gait database. In: *Proceedings of the 4th International Conference on Recent Advances in Soft Computing*, Nottingham, UK; 2002. pp. 66–71.

[27] Yu S, Tan D and Tan T. A framework for evaluating the effect of view angle, clothing and carrying condition on gait recognition. In: *Proceedings of the 18th International Conference on Pattern Recognition*, Hong Kong, China, Vol. 4; 2006. pp. 441– 444.

[28] Makihara Y, Mannami H, Tsuji A, *et al.* The OU-ISIR gait database comprising the treadmill dataset. *IPSJ Transactions on Computer Vision and Applications*. 2012; 4: pp. 53–62.

[29] Iwama H, Okumura M, Makihara Y, *et al.* The OU-ISIR gait database comprising the large population dataset and performance evaluation of gait recognition. *IEEE Transactions on Information Forensics and Security*. 2012; 7(5): pp. 1511–1521.

[30] Hofmann M, Geiger J, Bachmann S, *et al.* The TUM gait from audio, image and depth (GAID) database: multimodal recognition of subjects and traits. *Journal of Visual Communication and Image Representation.* 2014; 25(1): pp. 195–206. Available from: http://dx.doi.org/10.1016/j.jvcir.2013.02.006.

[31] gait. https://en.wikipedia.org/wiki/Gait.

[32] Wang L, Ning H, Tan T, *et al.* Fusion of static and dynamic body biometrics for gait recognition. In: *Proceedings of the 9th International Conference on Computer Vision*, Vol. 2; 2003. pp. 1449–1454.

[33] Urtasun R and Fua P. 3D tracking for gait characterization and recognition. In: *Proceedings of the 6th IEEE International Conference on Automatic Face and Gesture Recognition*; 2004. pp. 17–22.

[34] Murase H and Sakai R. Moving object recognition in eigenspace representation: gait analysis and lip reading. *Pattern Recognition Letters*. 1996; 17: pp. 155–162.

[35] Mori A, Makihara Y and Yagi Y. Gait recognition using period-based phase synchronization for low frame-rate videos. In: *Proceedings of the 20th International Conference on Pattern Recognition*, Istanbul, Turkey; 2010. pp. 2194–2197.

[36] Mowbray SD and Nixon MS. Automatic gait recognition via Fourier descriptors of deformable objects. In: *Proceedings of the 1st IEEE International Conference on Advanced Video and Signal Based Surveillance*; 2003. pp. 566–573.

[37] Liu Z and Sarkar S. Simplest representation yet for gait recognition: averaged silhouette. In: *Proceedings of the 17th International Conference on Pattern Recognition,* Vol. 1; 2004. pp. 211–214.

[38] Makihara Y, Sagawa R, Mukaigawa Y, *et al.* Gait recognition using a view transformation model in the frequency domain. In: *Proceedings of the 9th European Conference on Computer Vision*, Graz, Austria; 2006. pp. 151–163.

[39] Lam THW and Lee RST. A new representation for human gait recognition: motion silhouettes image (MSI). In: *Proceedings of the IAPR International Conference on Biometrics*; 2006. pp. 612–618.

[40] Bashir K, Xiang T and Gong S. Gait representation using flow fields. In: *Proceedings of the 20th British Machine Vision Conference*, London, UK; 2009. pp. 1–11.

[41] Hofmann M and Rigoll G. Improved gait recognition using gradient histogram energy image. In: *IEEE 2012 International Conference on Image Processing*, Orlando, FL, USA; 2012. pp. 1389–1392.

[42] Tao D, Li X, Wu X, *et al.* General tensor discriminant analysis and Gabor features for gait recognition. *IEEE Transactions on Pattern Analysis and Machine Intelligence.* 2007; 29(10): pp. 1700–1715.

[43] Xu C, Makihara Y, Li X, *et al.* Speed invariance vs. stability: cross-speed gait recognition using single-support gait energy image. In: *Proceedings of the 13th Asian Conference on Computer Vision (ACCV 2016),* Taipei, Taiwan; 2016. pp. 52–67.

[44] Kusakunniran W, Wu Q, Zhang J, *et al.* Gait recognition across various walking speeds using higher order shape configuration based on a differential composition model. *IEEE Transactions on Systems, Man, and Cybernetics, Part B: Cybernetics.* 2012; 42(6): pp. 1654–1668.

[45] Bashir K, Xiang T and Gong S. Gait recognition without subject cooperation. *Pattern Recognition Letters.* 2010; 31(13): pp. 2052–2060.

[46] Wei Li RZ and Wang X. Human reidentification with transferred metric learning. In: *The 11th Asian Conference on Computer Vision*; 2012.

[47] Tao D, Li X, Wu X, *et al.* General tensor discriminant analysis and gabor features for gait recognition. *IEEE Transactions on Pattern Analysis and Machine Intelligence.* 2007; 29(10): pp. 1700–1715. Available from: http://dx.doi.org/10.1109/TPAMI.2007.1096.

[48] Lu H, Plataniotis KN and Venetsanopoulos AN. A survey of multilinear subspace learning for tensor data. *Pattern Recognition.* 2011; 44(7): pp. 1540–1551.

[49] Yang J, Tang H, Ma Y, *et al.* Image super-resolution as sparse representation of raw image patches. In: *Conference on Computer Vision and Pattern Recognition*; 2008. pp. 1–8.

[50] Guan Y, Li CT and Roli F. on reducing the effect of covariate factors in gait recognition: a classifier ensemble method. *IEEE Transactions on Pattern Analysis and Machine Intelligence.* 2015; 37(7): pp. 1521–1528.

[51] Yan S, Xu D, Yang Q, *et al.* Discriminant analysis with tensor representation. In: *Proceedings of the IEEE Computer Society Conference on Computer Vision and Pattern Recognition*; 2005. pp. 526–532.

[52] Veres G, Gordon L, Carter J.N, *et al.* What Image Information Is Important in Silhouette-based gait recognition? In: *Proc*eedings *of IEEE Conf*erence *on Computer Vision and Pattern Recognition*, Vol. 2; 2004. pp. 776–782.

[53] Mansur A, Makihara Y, Muramatsu D, *et al.* Cross-view gait recognition using view-dependent discriminative analysis. In: *The 2nd IEEE International Joint Conference on Biometrics (IJCB 2014)*; 2014. pp. 1–8.

[54] Kusakunniran W, Wu Q, Zhang J, *et al.* Support vector regression for multi-view gait recognition based on local motion feature selection. In: *Proceedings of IEEE Computer Society Conference on Computer Vision and Pattern Recognition* 2010, San Francisco, CA, USA; 2010. pp. 1–8.

[55] Iwashita Y, Sakano H and Kurazume R. Gait recognition robust to speed transition using mutual subspace method. In: Murino V, Puppo E, editors.

Image Analysis and Processing—ICIAP 2015. Cham: Springer International Publishing; 2015. pp. 141–149.

[56] Makihara Y and Yagi Y. Cluster-pairwise discriminant analysis. In: *Proceedings of the 20th International Conference on Pattern Recognition*, Istanbul, Turkey; 2010. pp. 577–580.

[57] Martin-Felez R and Xiang T. Uncooperative gait recognition by learning to rank. *Pattern Recognition.* 2014; 47(12): pp. 3793–3806. Available from: http://www.sciencedirect.com/science/article/pii/S0031320314002325.

[58] Xu C, Makihara Y, Ogi G, *et al.* The OU-ISIR gait database comprising the large population dataset with age and performance evaluation of age estimation. *IPSJ Transactions on Computer Vision and Applications.* 2017; 9 (1): p. 24. Available from: https://doi.org/10.1186/s41074-017-0035-2.

[59] Takemura N, Makihara Y, Muramatsu D, *et al.* Multi-view large population gait dataset and its performance evaluation for cross-view gait recognition. *IPSJ Transactions on Computer Vision and Applications.* 2018; 10(1): p. 4. Available from: https://doi.org/10.1186/s41074-018-0039-6.

[60] Uddin MZ, Ngo TT, Makihara Y, *et al.* The OU-ISIR Large Population Gait Database with real-life carried object and its performance evaluation. *IPSJ Transactions on Computer Vision and Applications.* 2018; 10(1): p. 5. Available from: https://doi.org/10.1186/s41074-018-0041-z.

[61] Shiraga K, Makihara Y, Muramatsu D, *et al.* GEINet: view-invariant gait recognition using a convolutional neural network. In: *Proceedings of the 8th IAPR International Conference on Biometrics* (ICB 2016). O19, Halmstad, Sweden; 2016. pp. 1–8.

[62] Wu Z, Huang Y, Wang L, *et al.* A comprehensive study on cross-view gait based human identification with deep CNNs. *IEEE Transactions on Pattern Analysis and Machine Intelligence.* 2016; 39(2): pp. 209–226.

[63] Yu S, Chen H, Reyes E.B.G, *et al.* GaitGAN: invariant gait feature extraction using generative adversarial networks. In: *2017 IEEE Conference on Computer Vision and Pattern Recognition Workshops (CVPRW)*; 2017. pp. 532–539.

[64] Yu S, Liao R, An W, *et al.* GaitGANv2: invariant gait feature extraction using generative adversarial networks. *Pattern Recognition.* 2019; 87: pp. 179–189. Available from: http://www.sciencedirect.com/science/article/pii/ S0031320318303649.

[65] Takemura N, Makihara Y, Muramatsu D, *et al.* On input/output architectures for convolutional neural network-based cross-view gait recognition. *IEEE Transactions on Circuits and Systems for Video Technology.* 2019; 29(9): pp. 2708–2719.

[66] Yu S, Chen H, Wang Q, *et al.* *I*nvariant feature extraction for gait recognition using only one uniform model. *Neurocomputing.* 2017; 239(C): pp. 81–93. Available from: https://doi.org/10.1016/j.neucom.2017.02.006.

[67] Li X, Makihara Y, Xu C, *et al.* Joint intensity transformer network for gait recognition robust against clothing and carrying status. *IEEE Transactions on Information Forensics and Security.* 2019; 14(12): pp. 3102–3115.

[68] Li X, Makihara Y, Xu C, *et al.* Gait recognition invariant to carried objects using alpha blending generative adversarial networks. *Pattern Recognition.* 2020; 105:107376. Available from: http://www.sciencedirect.com/science/article/pii/S0031320320301795.

[69] Xu C, Makihara Y, Li X, *et al.* Cross-view gait recognition using pairwise spatial transformer networks. *IEEE Transactions on Circuits and Systems for Video Technology*. 2020; pp. 1–15.

[70] Li X, Makihara Y, Xu C, *et al.* Gait recognition via semi-supervised disentangled representation learning to identity and covariate features. In: *The IEEE/CVF Conference on Computer Vision and Pattern Recognition (CVPR)*; 2020.

[71] Chao H, He Y, Zhang J, *et al.* GaitSet: regarding gait as a set for cross-view gait recognition. In: *Proceedings of the 33rd AAAI Conference on Artificial Intelligence (AAAI 2019)*; 2019.

[72] Fan C, Peng Y, Cao C, *et al.* GaitPart: temporal part-based model for gait recognition. In: *The IEEE/CVF Conference on Computer Vision and Pattern Recognition (CVPR)*; 2020.

[73] Zhang Z, Tran L, Yin X, *et al.* Gait recognition via disentangled representation learning. In: *Proceeding of IEEE Computer Vision and Pattern Recognition*, Long Beach, CA; 2019.

[74] Liao R, Cao C, Garcia EB, *et al.* Pose-based temporal-spatial network (PTSN) for gait recognition with carrying and clothing variations. In: Zhou J, Wang Y, Sun Z, *et al.*, editors. Prof. of Chinese Conference on Biometric Recognition 2017. Cham: Springer International Publishing; 2017. pp. 474–483.

[75] An W, Liao R, Yu S, *et al.* Improving gait recognition with 3D pose estimation. In: Zhou J, Wang Y, Sun Z, *et al.*, editors. Prof. of Chinese Conference on Biometric Recognition 2018. Cham: Springer International Publishing; 2018. pp. 137–147.

[76] An W, Yu S, Makihara Y, *et al.* Performance evaluation of model-based gait on multi-view very large population database with pose sequences. *IEEE Transactions on Biometrics, Behavior, and Identity Science.* 2020; 7: pp. 1–11.

[77] Lynnerup N and Vedel J. Person identification by gait analysis and photogrammetry. *Journal of Forensic Sciences.* 2005; 50: pp. 112–8.

[78] How biometrics could change security; 2008. http://news.bbc.co.uk/2/hi/programmes/click online/7702065.stm.

[79] Nixon MS, Bouchrika I, Arbab-Zavar B, *et al.* On use of biometrics in forensics: gait and ear. In: *2010 18th European Signal Processing Conference*; 2010. pp. 1655–1659.

[80] Chinese gait recognition of tech IDs people by how they walk; 2018. https://apnews.com/bf75dd1c26c947b7826d270a16e2658a.

[81] Iwama H, Muramatsu D, Makihara Y, *et al.* Gait verification system for criminal investigation. *IPSJ Transactions on Computer Vision and Applications.* 2013; 5: pp. 163–175.

[82] Redmon J and Farhadi A. YOLO9000: better, faster, stronger. In: *2017 IEEE Conference on Computer Vision and Pattern Recognition (CVPR)*; 2017. pp. 6517–6525.

[83] He K, Gkioxari G, Dollar P, *et al.* Mask R-CNN. In: *2017 IEEE International Conference on Computer Vision (ICCV)*; 2017. pp. 2980–2988.
[84] Zhang S, Li R, Dong X, *et al.* Pose2Seg: detection free human instance segmentation. In: *2019 IEEE/CVF Conference on Computer Vision and Pattern Recognition (CVPR)*; 2019. pp. 889–898.
[85] Lin G, Milan A, Shen C, *et al.* RefineNet: Multi-path refinement networks for high-resolution semantic segmentation. In: *2017 IEEE Conference on Computer Vision and Pattern Recognition (CVPR)*; 2017. pp. 5168–5177.
[86] Ross AA, Nandakumar K and Jain AK. Handbook of Multibiometrics. *International Series on Biometrics*. Secaucus, NJ: Springer-Verlag New York, Inc.; 2006.
[87] Liu Z and Sarkar S. Outdoor recognition at a distance by fusing gait and face. *Image and Vision Computing.* 2007; 25(6): pp. 817–832. Available from: http://dx.doi.org/10.1016/j.imavis.2006.05.022.
[88] Hofmann M, Schmidt SM, Rajagopalan A, *et al.* Combined face and gait recognition using alpha matte preprocessing. In: *Proceedings of the 5th IAPR International Conference on Biometrics*, New Delhi, India; 2012. pp. 1–8.
[89] Zhou X and Bhanu B. Feature fusion of side face and gait for video-based human identification. *Pattern Recognition.* 2008; 41(3): pp. 778–795.
[90] Zhou X and Bhanu B. Integrating face and gait for human recognition at a distance in video. *IEEE Transactions on Systems, Man, and Cybernetics, Part B: Cybernetics.* 2007; 37(5): pp. 1119–1137.
[91] Zhou X and Bhanu B. Integrating face and gait for human recognition. In: *Proceedings of the IEEE Computer Society Workshop on Biometrics* 2006, New York, NY, USA; 2006.
[92] Zhou X, Bhanu B and Han J. Human recognition at a distance in video by integrating face profile and gait. In: *Proceedings of the 5th international conference on Audio- and Video-Based Biometric Person Authentication. AVBPA'05.* Berlin, Heidelberg: Springer-Verlag; 2005. pp. 533–543. Avail- able from: http://dx.doi.org/10.1007/11527923 55.
[93] Zhang T, Li X, Tao D, *et al.* Multimodal biometrics using geometry preserving projections. *Pattern Recognition.* 2008; 41(3): pp. 805–813.
[94] Kale A, Roy-Chowdhury A and Chellappa R. Fusion of gait and face for human identification. In: *Proceedings of the IEEE International Conference on Acoustics, Speech, and Signal Processing* 2004 (ICASSP'04), Vol. 5; 2004. pp. 901–904.
[95] Shakhnarovich G and Darrell T. On probabilistic combination of face and gait cues for identification. In: *Pro*ceedings of the Fifth IEEE International Conference on *Automatic Face Gesture Recognition*, Vol. 5; 2002. pp. 169–174.
[96] Zhou X and Bhanu B. Feature fusion of face and gait for human recognition at a distance in video. In: *Proceedings of the 18th International Conference on Pattern Recognition,* Hong Kong, China, Vol. 4; 2006. pp. 529–532.

[97] Lee TKM, Ranganath S and Sanei S. Fusion of chaotic measure into a new hybrid face-gait system for human recognition. In: *Proceedings of the 18th International Conference on Pattern Recognition,* Hong Kong, China, Vol. 4; 2006. pp. 541–544.

[98] Geng X, Smith-Miles K, Wang L, *et al.* Context-aware fusion: a case study on fusion of gait and face for human identification in video. *Pattern Recognition*. 2010; 43(10): pp. 3660–3673. Available from: http://dx.doi.org/10.1016/j. patcog.2010.04.012.

[99] Geng X, Wang L, Li M, *et al.* Adaptive fusion of gait and face for human identification in video. In: *Proceedings of the 2008 IEEE Workshop on Applications of Computer Vision*; 2008. pp. 1–6.

[100] Zou WWW and Yuen PC. Very low resolution face recognition problem. *IEEE Transactions on Image Processing.* 2012. 21(1): pp. 327–340.

[101] Marciniak T, Dabrowski A, Chmielewska A, *et al.* Face recognition from low resolution images. In: *Multimedia Communications, Services and Security*, Vol. 287. Berlin: Springer Berlin Heidelberg; 2012. pp. 220–229. Available from: http://dx.doi.org/10.1007/978-3-642-30721-822.

[102] Xu X, Liu W and Li L. Low resolution face recognition in surveillance systems. *Journal of Computer and Communications.* 2014; 2: pp. 70–77.

[103] Hennings-Yeomans PH, Baker S and Kumar BVKV. Recognition of low-resolution faces using multiple still images and multiple cameras. In: *IEEE Second International Conference on Biometrics: Theory, Applications and Systems*; 2008. pp. 1–6.

[104] Muramatsu D, Iwama H, Makihara Y, *et al.* Multi-view multi-modal person authentication from a single walking image sequence. In: *Proceedings of the 6th IAPR International Conference on Biometrics*; 2013. pp. 1–8.

[105] Kimura T, Makihara Y, Muramatsu D, *et al.* Quality-dependent score-level fusion of face, gait, and the height biometrics. *IPSJ Transactions on Computer Vision and Applications.* 2014; 6: pp. 543–547.

[106] Kimura T, Makihara Y, Muramatsu D, *et al.* Single sensor-based multi-quality multi-modal biometric score database and its performance evaluation. In: *Proceedings of the 7th IAPR International Conference on Biometrics (ICB 2015)*. Paper ID 61, Phuket, Thailand; 2015. pp. 1–8.

Chapter 11
Future of gait analysis

Tom Shannon[1]

11.1 Introduction

The engineering problem to reliably capture and track points in space using marker-based optical motion capture systems has long been solved but research continues to better understand the association between the precise measurements of the location of bony landmarks on the skin surface and the actual location of the underlying skeletal structures during typical and pathologic locomotion. The disparity in the measures is commonly described as the *soft tissue artefact* so elegantly described by Dr David Sutherland [1] as trying *'...to measure the movements of a broomstick within a gunny sack'*. Many researchers have proposed techniques to identify joint centres and axes functionally through fitting the data to an idealised joint model that also incorporates some form of soft tissue artefact compensation. This work continues and like the earlier innovations is rapidly gaining acceptance through close collaborative efforts between clinicians, scientists and engineers in laboratories, academia and industry across the world.

Since the release of the film *Titanic* in 1997, there has been an explosion in the use of motion capture to enhance films, games, virtual reality environments, music videos and television that have brought many stories to life, thought almost impossible before the ready availability of the technology to the entertainment sector (Figure 11.1). With it has brought many challenges including the need to capture the motion of multiple characters simultaneously, in large volumes and with high camera counts. The sector has also demanded improved data resolution and capture rates to improve subtlety in the measurement of the motion of hands, feet and facial expressions with a secondary need to provide real-time reconstruction of the animated characters who may or may not be humanoid. The demands of life sciences including clinical gait analysis and sport, engineering including ergonomics and the entertainment sector have resulted in a synergy that has driven and improved the core technologies for the benefit of all. In just one example, the need in the life sciences to continuously improve skeletal modelling has benefited

[1]Centre for Biomechanics and Rehabilitation Technologies, Staffordshire University, Stoke on Trent, UK

Figure 11.1 *Teenage Mutant Ninja Turtles. Courtesy: Paramount Pictures*

animators to create ever-more realistic characters, whilst their requirements have directly improved how the technologies are applied and data are presented within the clinical gait and biomechanics communities. In the coming years, there are great opportunities to enhance the performance of apparatus used in the measurement of pathologic gait through the application of improved biomechanical models and algorithms by drawing upon the application of commensurate and emerging technologies and methodologies.

11.2 Application of functional calibration to the routine analysis of human gait

The repeatability of early kinematic and kinetic models was limited by the capability to accurately palpate anatomical landmarks in typically developed subjects or when in the presence of significant pathology. The common aspiration was to define actual joint centres and anatomical coordinate systems. The size of applied retro-reflective markers, video camera resolution limitations and low video frame rates further confounded the early attempts of providing useful routine clinical data.

Since the 1980s, similar biomechanical models were published based on the collaborative work of groups based within the Helen Hayes Hospital in New York and the Newington Children's Hospital in Hartford, Connecticut. The model was shared with six other laboratories [2] using early generation video-converter (VICON) systems (Oxford Metrics Ltd, Oxford, UK) that offered the opportunity to widen the experience and validations in a first multi-centre trial of a conventional gait model (CGM). During this period, the management of Oxford Metrics Ltd took the corporate decision to focus available development resources towards the emerging and growing clinical gait analysis market sector. In 1991, they released the first version of their 'Vicon Clinical Manager' which was created using a version of the CGM, in consultation with staff at both the Newington Children's and Helen Hayes Hospitals. Over the following years, the root model was improved and validated against the implementations managed by the two core hospitals. In 1998, the final code in the commercial model was exported into their VICON 'Plug-in-Gait (PiG)' offering.

Similarly, the Motion Analysis Corporation (Santa Rosa, CA) developed proprietary software to support a different model based on the Cleveland Clinic marker set [1]. The future growth and wider acceptance of the outcomes of clinical gait analysis, particularly when acquired using passive or active optical marker-based systems, remains dependent on the development of ever more elegant and independently validated biomechanical models. In response, contemporary use of clinical gait analysis has advanced through the application of numerical methods to define joint centres and axes of rotation independent of variations in operator palpation skill. The purpose of a number of different methods, grouped under a common description of functional calibration, has been to estimate hip joint centres and apply a mean helical axis to define knee joint flexion/extension axes, without relying on accurately locating anatomical landmarks.

The accurate interpretation of gait data and the clinical validity of any model are highly dependent on reliably describing the location of the hip joint centres, complicated by being unable to use palpation. Stagni *et al.* [3] studied the impacts of joint mislocation on kinematics and kinetics, demonstrating that a 30-mm error could result in approximately a 50% difference in mean flexion–extension hip moment and a 26% delay in the cross-over from extensor to flexor moment reporting over a given gait cycle. In 2011, Sangeux *et al.* [4] addressed the importance of accurately locating the hip joint centres as an essential element when modelling the lower limbs during kinematic and kinetic gait analyses. The hip joint is described biomechanically as a ball and socket arrangement with the femoral head located coincident with the inner surface of the acetabulum. Two predictive methods have been applied in routine clinical practice, based on widely used CGMs [5,6], or the application of regression equations [7], both based on using anthropometry to estimate the centre within the pelvis. More recently, inferring the position from the motion of the joint during functional calibration trials has been gaining traction in the community. Camomilla *et al.* [8] compared the methods against measurements taken from results obtained using magnetic resonance imaging and three-dimensional ultrasound captured from 19 typically developed adults. They determined that a functional calibration using a geometric sphere fitting technique defined by the motion of six markers on the thigh achieved mean distance errors of 15 mm (85% within 20 mm) with a regression equation approach [7] achieving very similar results. CGMs [6] applied to the acquired data during their study exhibited twice the mean distance errors. They further established that *in vivo* outcomes were an order of magnitude worse than estimations drawn from synthetic data implying that soft tissue artefact [9] remained a dominating error factor. Sangeux *et al.* [10] re-visited the question in 2014 with the advent in routine clinical use of the EOS^{TM} X-ray machine (EOS Imaging, SA, Paris, France). The ultra-low radiation dose technique facilitated the simultaneous capture of spatially calibrated coronal and sagittal plane images of whole upright load-bearing bodies. The resultant three-dimensional reconstruction of the skeleton provided scaled insight and visualisation of clinical parameters relevant to contemporary orthopaedic practice. Using the imagery, the researchers were able to establish the relative positions of the physical hip joint centre of 17 typically developed adults with 31 retro-reflective skin markers placed on the pelvis, thighs and shanks. They confirmed previous study results that the functional sphere fitting method achieved the best outcome through localising the hip centre to within and an average of 11 mm of the reference X-ray image, 100% to within 30 mm. Comparatively, Harrington *et al.*'s [7] approach achieved an average of 17 mm to within 97% of a 30-mm boundary. They concluded that accurate localisation of the centre was possible in routine gait analysis subject to any adaptations to match the clinical population under measurement. Where functional calibration motion is possible, they recommended using the geometrical sphere fitting method with Harrington *et al.*'s approach employed when defining patient hip joint motion is not clinically practical.

Equally crucial is the interpretation of hip rotation, defined as the angle between the medio-lateral axes of the femur and the pelvis in the transverse plane of the femur [11]. Schache *et al.* [12] suggested that the reliability of hip rotational kinematics was dependent on an accurate estimation of the medio-lateral axis during a gait analysis. In 2016, Sauret *et al.* [13] again used the EOS™ bi-planar X-ray method to simultaneously model the bones of the lower limbs and locate the retro-reflective markers in three-dimensional space from 13 healthy subjects in the standing posture. They achieved marker centre reconstruction to within ± 0.35 mm, anatomical landmarks to ± 1.0 mm, resulting in a condylar of the femur axis orientation precision of 1°. They further achieved a $\pm 3°$ registration of the measure with gait analysis data at a 95% confidence interval. The researchers compared the CGM [14] results which employed two markers placed by palpation over the medial and lateral epicondyles of the knee with three functional calibration methods [15–17] using additional markers on the thighs and patella with one case also applying a knee alignment device (KAD) (Figure 11.2). Results from the CGM demonstrated an average difference of 15° (standard deviation (SD) 6°), influenced by localisation errors in marker placement of the two anatomical landmarks which can typically be approximately 100 mm apart. The functional method mean using a KAD achieved alignment to the anatomical condylar axes of the subjects to within 1° (SD 5°).

Knee kinematics are also affected by the orientation of the medio-lateral axis during gait [12]. Functional calibration of the knee has a reduced reliance on palpating bony landmarks to determine knee angle using the relative motions of the thigh and shank segments to derive the medio-lateral axis of the femur. Most functional methods model the knee as a simple hinge joint with a single axis of rotation. More complex models allow the knee to exhibit both flexion–extension and internal–external rotations but in all cases, accuracy in tracking the movement of body segments will still be adversely affected by the presence of soft tissue artefact and by extension of the estimation of the medio-lateral axis of the femur. Sangeux *et al.* [18] captured the motion of the knee joint of 17 post-operative, full knee arthroplasty patients, using bi-planar fluoroscopy, undergoing treadmill walking. They simultaneously acquired the movement of shank and thigh marker

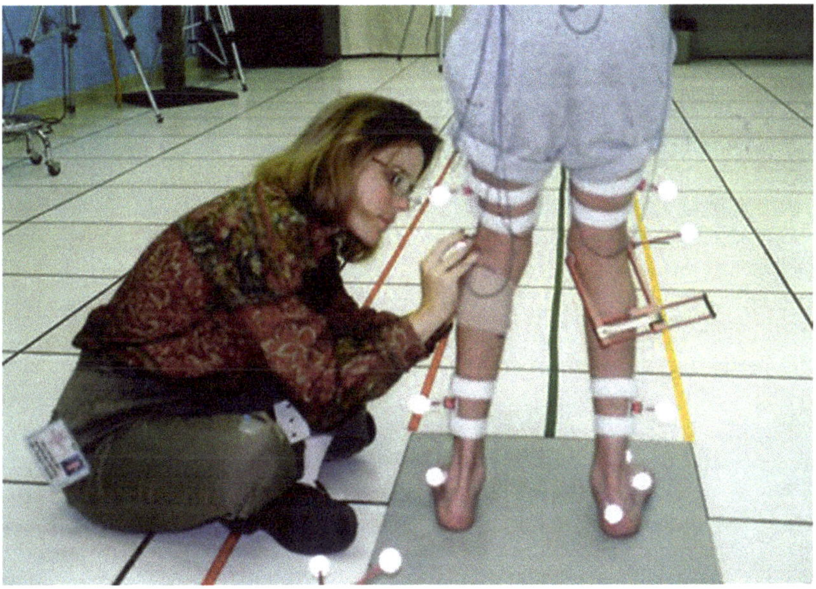

Figure 11.2 Applying the Helen Hayes marker set and KAD (Vicon Motion Systems Ltd)

clusters to quantify the effects of the soft tissue artefact against the reference imagery. The researchers applied three functional calibration methods to the data and concluded that all exhibited an improved tolerance (< 2° SD 5.2°) to the effects of soft tissue artefact when compared against the reference axis.

Firooz *et al.* [19] built on the understanding of lower limb biomechanics by measuring the ankle joint parameters of 23 typically developed adults to establish the location and axis of the ankle joint centre using a functional methodology. The goal of the research was to confirm or dispute the CGM assumption that the ankle joint centre lay at the centre point of the malleoli. Their results showed significant differences for the joint parameters in all axes relative to the malleoli line. Using a functional calibration approach, they concluded that there was a translation of 7% of the anterior foot length to the mid-point of the malleoli which may have an impact on future clinical gait analysis protocols.

Leboeuf *et al.* [20] have drawn on the experience of earlier CGMs and contemporary functional calibration research efforts to develop a series of 'open-source' biomechanical models that can be routinely applied in clinical gait analysis laboratories with the resultant data of great value in multi-centre validation of the methodology.

In 1995, Aurelio Cappozzo proposed the 'Calibration of Anatomical System Technique (CAST)' [21], commercialised by BTS Bioengineering SpA (Milan, Italy) which has been widely referenced in subsequent academic studies. C-Motion Inc. (Germantown, MD) later implemented the technique as six degrees of freedom models (6DoF) within their 'Visual3D' modelling and presentation software.

The method tracks lower limb segments independently without applying any joint constraints either using traditional single markers or rigid arrays of marker clusters as are becoming more common in routine practice today. Clusters reduce the impact of single marker soft tissue artefacts but can introduce errors related to their position and orientation with respect to the underlying bones [22]. Further improvements have been made to the CAST model at the Istituto Orthopedico Rizzoli in Bologna, Italy [23,24] resulting in a widely validated alternate offering to the CGM.

Global optimisation models or inverse kinematics have become popular since it was incorporated within 'OpenSim' open-source software [25] using a defined rigid body that is fitted to the observed physical markers locations using a weighted least-squares fit cost function. The approach can be prone to errors introduced in model definitions, anatomical landmark palpation or soft tissue artefacts but will likely be a contributor within the basic framework of future developments.

Traditional clinical and biomechanical models have assumed that the foot can be described as a single rigid segment. Similarly, contemporary kinematic protocols based on tracking three markers to describe the motion of an anatomical structure consisting of 26 bones and several joints are no longer considered acceptable clinically. Continuing improvements in camera imager resolution have translated to enhanced accuracy and precision performance offered by the major optical motion capture system manufacturers. A secondary improvement has been the wide adoption and reliable acquisition of the location of ever-smaller retro-reflective markers within larger motion capture volumes so facilitating the capability to achieve ever more complex gait cycle analyses of longer durations. The ready availability of appropriate and useful technology has spawned a significant number of proposals for multi-segment foot models. Leardini and Caravaggi [26] reviewed 30 different techniques and observed that only a limited number were validated for use in a clinical context including the Milwaukee foot model (1996) [27], Oxford foot model (2001) [28], Heidelberg foot measurement method (2006) [29], Rao *et al.* (2006) [30] and the Rizzoli foot model (2007 and 2014) [31,32].

11.3 Clinical gait analysis standardisation

The intended purpose of a clinical gait analysis system, as a classified medical device, can be described as providing information that can be used to take decisions for therapeutic purposes and to aid in the understating of the aetiology of gait abnormalities resulting from a diagnosis of cerebral palsy or Duchenne muscular dystrophy among presenting children and adolescent patients. The interpretation of resulting kinematic and independently acquired kinetic and electromyographic data remains in the hands of interdisciplinary teams of healthcare professionals and bio-mechanists with substantial knowledge of normal and pathologic human gait where characteristics are measured, abnormalities are identified, causes are postulated and treatments are proposed. The same technology is routinely applied to establish the efficacy of orthoses and prostheses, and during rehabilitation among the wider population. The purpose is to improve the efficiency of patient ambulatory locomotion.

Baker *et al.* [33] looked at the future directions of gait analysis recognising the growing importance of evidence-based practices and clinical governance within healthcare bodies [34]. They proposed that contemporary focus translated into a need for multi-centre standardisation, facilitated through the provision of independently validated procedures and processes. Bodies such as the European Society for Movement Analysis in Adults and Children (ESMAC), Clinical Movement Analysis Society (CMAS, UK and Ireland), Gait and Clinical Movement Analysis Society (GCMAS, USA) and the Commission for Motion Laboratory Accreditation Inc. (CMLA, USA) are attempting to ensure minimum standards are universally implemented as the first step in validation. The repeal of the European Union (EU) Medical Devices Directive 93/42/EEC as amended by EU Council Directive 2007/47/EC of 5 September 2007, by a more robust legislative framework within the Medical Devices Regulation (EU)2017/745 that came into force in 2017 will impact many manufactures during the transition period that ended on 26 May 2021. Gait analysis systems that were classified under the Directive as CLASS I with a measuring function can still continue to be supplied into the EU until 27 May 2024, under Annex V, as long as the manufacturer meets the requirements within Article 120 of the Regulation. The medical device classification rules described within Clause 6.3 of Annex VIII of the Regulation defines software intended to provide information which is used to take decisions with diagnosis or therapeutic purposes is to be classified as medium risk CLASS IIa, except if such decisions have an impact that may cause a serious deterioration of a person's state of health or a surgical intervention may be classified as the higher risk CLASS IIb. As all optical clinical gait analysis systems are reliant on supporting software. All manufacturers supplying equipment into the European Union will have had to re-classify their medical devices to meet the new Regulation either by the 27 May 2024 if currently classified CLASS I(m) and meeting the requirements of Article 120(2) and 120(3). Any systems not meeting Article 120 will have been re-classified before 26 May 2021.

The requirements for CE marking (Conformité Européenne) of all applicable devices will continue to be documented within legally published Declarations of Conformity under EU Notified Body oversight. As the Date of Applicability of the Regulation was set at a date post the transition period, the UK government will not adopt it into Great Britain Law (different for Northern Ireland) and will remain reliant on the existing Medical Devices Regulations 2002/(SI 2002 No 618, as amended). The UK government will cease to recognise the CE mark after 30 June 2023 for Great Britain and will be moving to the requirement for all affected devices to exhibit a UK Conformity Assessed (UKCA) mark. UK Notified Body oversight will be required for all devices classified as CLASS I with a measuring function. The US Food and Drugs Administration (FDA) classifies such systems under their 510(k) approvals process as CLASS 2 (Code LXJ). Similar recognition methodologies and protocols exist in many other countries.

Derived clinically relevant outputs such as joint angles from biomechanical models generated within commercial gait analysis software will have to be verified and validated as part of a certification process as predicted by Bake *et al.* [33] in 2017.

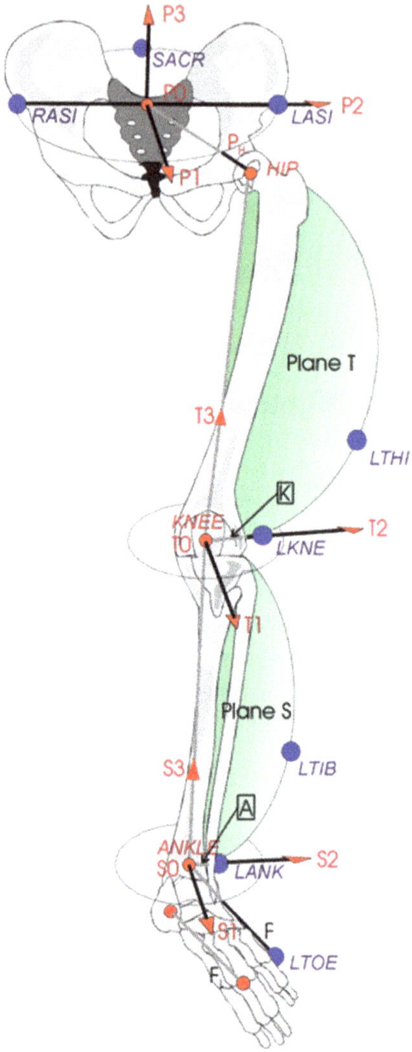

Figure 11.3 The Helen Hayes marker set (Vicon Motion Systems Ltd)

The decision by the EU to define software as a medical device is a sensible one and good practice. Manufacturers could be expected to take a similar view within their product risk management oversight and should be implementing enhanced validation processes ahead of the mandatory requirements for their product offerings.

Baker *et al.* also observed the reality that motion capture system applications are no longer restricted to clinical gait analysis laboratory environments. The commensurate ready availability of affordable personal computers with a wide acceptance and familiarity with user-friendly software applications has resulted in the technology becoming more ubiquitous in large and smaller healthcare and more generalist academic research organisations. The emphasis in many generic clinical gait analysis laboratories is to focus on patient data interpretation through implementing standardised and validated models rather than employing bio-mechanists to create bespoke versions. A secondary benefit using standardised models in multiple centres affords the opportunity for validating in-house repeatability and consistency against published benchmark data [35]. Baker *et al.* [33] accepted that scope remains to introduce further improvements in the CGM but recognised the considerable challenge in attempting to ensure where possible backward compatibility with earlier analysed patient databases.

The precision and three-dimensional reconstruction accuracy offered by marker-based motion capture systems, continuous improvements in the biomechanical modelling of the low limb joints and increased regulatory oversight in the UK, EU and the rest of the world, will ensure that the technology will continue to have a place in clinical gait analysis for the foreseeable future.

11.4 Animation and computer graphics

Perception by humans can best be described as organising, identifying and interpreting chemical and physical nervous stimuli from the senses to understand their environment that is also shaped by a recipient's expectations, cognition and attention to translate the information into high-level cerebral functions such as recognition and situational awareness. Researchers drawing on archaeological findings during the course of the Pleistocene age have started investigating social evolution and the perception of familiarity, similarity and attraction among groups. Could this work help to explain a phenomenon first reported in the 1970s by Masahiro Mori of the Tokyo Institute of Technology when he proposed the concept of the 'uncanny valley' to describe his observation that as robots appear more and more human, they become more appealing until a person's affinity or positive emotional responses are countered by stronger feelings of unease as they descend into his valley? The concept has been found to be equally valid when applied to the creation of characters or avatars in the film, games and virtual reality market sectors.

Contemporary neuro-imaging studies [36] are now employing naturalistic sources such as film, avatars and androids to provide stimuli to accelerate research into a better understanding of how the brain processes complex situations and perception whilst maintaining experimental control. An exaggerated caricature can be accepted as familiar if the viewer expects it to be less human-like but still retains recognisable characteristics. An example is the high-quality characterisations achieved in the two Disney© *Maleficent* films, made even more complex by requiring a seamless transition between conventional performances by well-known actors and their anthropometrically different *fairy* avatars. Similarly, Rémi Brun has employed his extensive special effects and motion capture experience to create amazing and seminal light sculptures, using small numbers of carefully placed and synchronised lights, to create compelling illusions of the motion of bipeds and quadrupeds in a viewer's imagination.

The goal of every animator remains to elicit a desired emotional response from audiences to keep them engaged with their story. Technologies now exist to create realistic human-like characters to a degree where the subtlety of gait, body language and expressions of emotion can be perceived by most current observers to be aligned to that presented by physical actors, with similar behavioural, physiological and neuronal stimulations, in particular empathy. We have all identified weaknesses in the anatomical realisation of humanoid characters that even recently would have been considered innovative and ground-breaking. Producers, creators and manufacturers can continue to expect ever-increasing demands for greater reality within imagery from their viewing public in the years to come.

A recent example of the application of basic clinical research outcomes to directly benefit animators was the work undertaken by Dr Jean-Charles Bricola of Vicon Motion Systems Ltd on modelling the human spine to further improve motion capture body tracking and solving. He chose to describe the spine as seven 'hardy-spicer joints' from the vertebra prominens to an estimation of the location of the sacrum between the two palpable posterior superior iliac spines. He added slider joints between each bone to account for the actions of the spinal muscles, in particular the erector spinae, upon the vertebral column. Jean-Charles faced the challenges to correlate the poses of the underlying vertebral bodies to track spinal motions using a sparse set of surface markers and to then verify that his model responded as expected (Figure 11.4).

The solutions came from research into quantifying changes in back surface shape among patients diagnosed with Adolescent Idiopathic Scoliosis, a nasty disease where a spinal deformity can progress over time by simultaneously curving towards the arms and rotating. In some cases, it can force the ribs outwards causing a characteristic hump. The recent discovery of the skeleton of Richard III confirmed he had suffered from scoliosis, described unkindly by William Shakespeare that he was 'rudely stamp'd', 'deformed, unfinish'd'. Suspicion continues to linger widely that the playwright, with a mastery of perception, intentionally depicted the last Plantagenet king as a 'hunchbacked spider' to curry favour with his Tudor masters and audiences.

Jean-Charles' first challenge was addressed by drawing on published research undertaken by Dr Alan Turner-Smith of the Oxford Orthopaedic Engineering Centre in 1988 [37] where he measured a selection of cadaver spines to create a mathematical relationship in order to calculate the offset in the sagittal plane between each spinous process and locus of the vertebral body acquired from typically developed subjects and later patients.

A small number of scoliosis patients require surgery to stabilise the progression of the disease with a growing emphasis within the clinical community to include an assessment of an individual's quality of life, interests and personal goals when planning a treatment. The intervention can introduce some physical impairment as the procedures rigidly fix vertebral bodies that are normally capable of

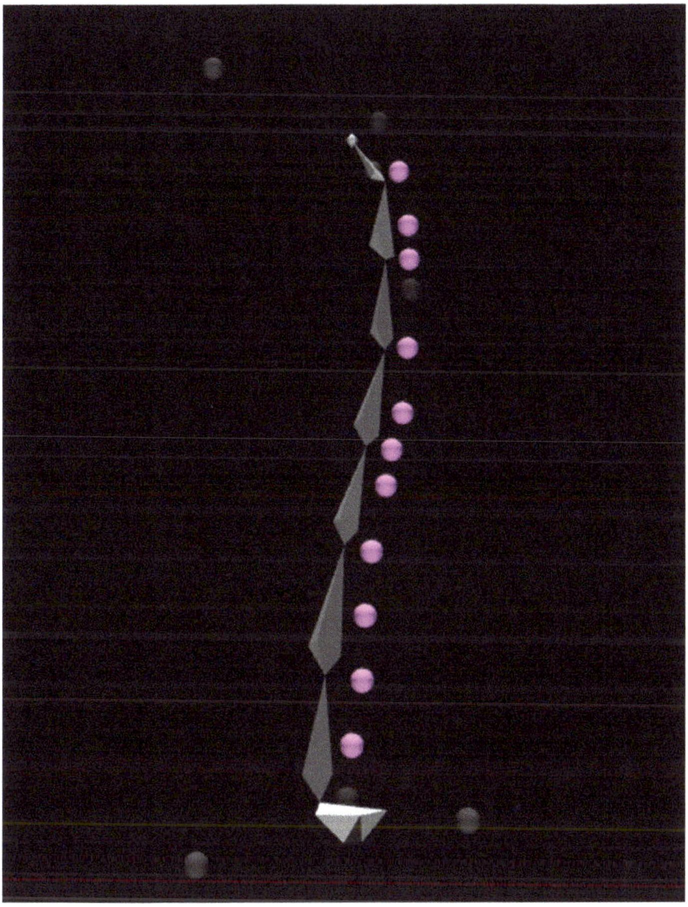

Figure 11.4 Spinal motion modelling (Vicon Motion Systems Ltd)

inter-segmental motion and affecting spinal global ranges of motion. A number of researcher centres have presented papers where they compared pre- and post-operative forward and lateral trunk flexion and transverse rotation, concluding that post-operative range of motion was reduced in both fused and unfused regions above and below the surgery when compared against that acquired from un-affected subjects. Jean-Charles was given access to anonymised motion capture trials acquired from 30 typically developed adults [38]. All subjects followed a common exercise protocol which he was able to use to successfully compare so to verify the performance of his model against the published normative clinical results.

The experience gained through this and other studies towards improved body tracking and solving is now incorporated into the latest versions of the Shōgun software (Vicon Motions Systems Ltd, Oxford, UK) (Figure 11.5). This is just one

Figure 11.5 Vicon Shogun (Vicon Motion Systems Ltd)

example of where multi-faceted and lateral reviews of fundamental research can have a positive impact on the development of computer-generated imagery as it continues to evolve.

All major manufacturers have responded to the challenge by offering ever more sophisticated software to optimise visual effects production pipelines to deliver robust and high-quality skeletal data solutions to meet industry time-lines.

Animators have exploited technological improvements to convert the movement of an actor's face into tri-dimensional data using conventional motion capture cameras, head-mounted cameras or laser scanners. The spatial and temporal data can be readily ingested into computer graphics software to create animations for movies, games or real-time avatars. Computer-generated characters derived from the motion of real people are always more realistic and nuanced compared to those created manually. Facial motion tracking is more challenging than conventional full-body capture due to the higher resolution requirements to identify and track subtle expressions from small movements of a few millimetres from the eyes, lips and musculature. Similarly, ubiquitous hand motion tracking technologies and software are now widely available in the same sectors and virtual reality environments. Multiple commercial offerings including the VICON Shōgun application reconstruction solutions are routinely used to drive body, face and hand motion mapping of third party avatars and characters for the enjoyment of movie-goers and gamers across the world. The results of the efforts in these sectors are also defining improvements in current versions of dedicated and agnostic clinical display and presentation software such as Visual3D (C-Motion Inc., Germantown, MD).

The challenge for animators and manufacturers is to continue to ensure that their computer-generated characters do not approach the edge of Mori's 'valley'. For this reason, the reliable and reproducible acquisition of ever more subtle motions offered by marker-based capture systems will still have a place in the entertainment sector well into the future. Through percolation and symbiosis from significant investment, research effort and advancements in non-clinical motion capture sectors will surely continue to directly benefit biomechanics, research, generic life sciences and clinical gait analysis activities.

11.5 Data fusion

The definition of a gait analysis system within the global medical device nomenclature (35757) is as

> an assembly of mains electricity (AC-powered) devices and instruments intended to be used to study walking or running patterns. The device uses a platform that measures ground reaction forces and/or cameras to determine knee, ankle, and hip joint movements and measure forces and torques. It can be used to diagnose problems related to walking or running and help plan corrective actions.

The concept that kinematic, kinetic and electromyographic data streams could be temporally fused to provide a comprehensive measure of lower limb motion, ground reaction forces applied, joint moments and powers derived, with the commensurate electrical activity of muscles has been verified and validated as clinically useful since the acceptance of the method into routine practice. The nomenclature does not mention the rising and ubiquitous application of inertial measurement units (IMU) containing nine-axis, accelerometers, gyroscopes and magnetometers into the field (Figure 11.6). In 2009, Favre *et al.* [39] proposed using two synchronised IMU devices placed proximally and distally to a typically developed knee joint to determine if the technique could offer predictable tri-dimensional knee joint angle measures outside the strictures of the gait laboratory. Their goal was to establish if the method could be of benefit during routine clinical examinations and therapeutic comparisons that required a new joint coordinate system calibration procedure based only on quaternion data streams from each device. They reported moderate accuracy of three-axis knee joint accuracy of between 4.0° and 8.1° when compared against data acquired from a conventional magnetic marker motion capture system (Liberty (R), Polhemus, USA). In 2014, Palermo *et al.* [40] proposed a lower limb functional calibration methodology to correlate seven sensors (MTws Xsens Technologies, Enschede, The Netherlands) and body coordinate systems. They measured the vertical axis of each sensor during two static sessions and validated their approach from data acquired from ten typically developed adults using a reference motion capture system (MX System, Vicon Motions Systems Ltd, Oxford). They reported accuracy and repeatability values less than 4° for each angle except for the ankle internal–external rotations (9.7°, 7.2°, respectively).

Continuous improvements, in particular, physical size and well-established connection protocols to miniature 'System of a Chip' (SOC) with on-board radio capabilities, and the availability of small batteries with significant storage, have resulted in wide acceptance of the technology in biomedical and biomechanical applications. Iosa *et al.* [41] and Nazarahari and Rouhni [42] reported applications

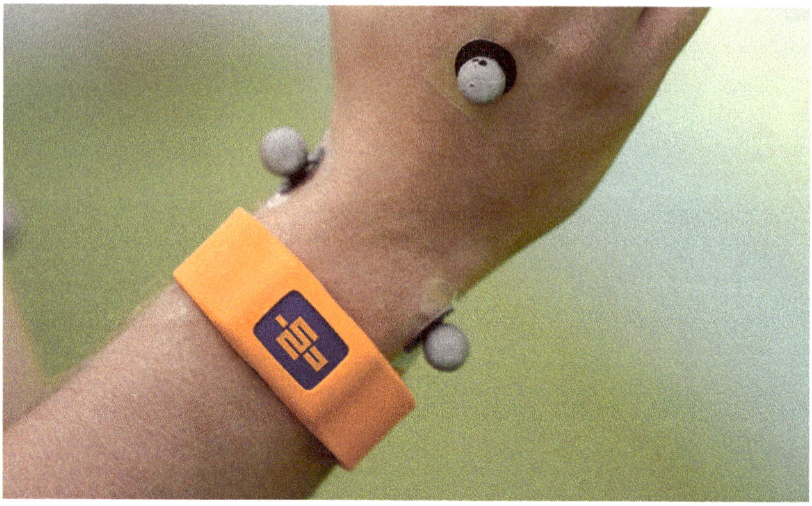

Figure 11.6 Optical and IMU data fusion (Vicon Motion Systems Ltd)

such as gait analysis, fall detection and physical activity analysis in 2016 and 2018, respectively.

It has long been known that passive marker-based motion capture technology can provide accurate and precise temporal and spatial data streams of palpated anatomical landmark trajectories but are weak when measuring subtle rotations unless markers are placed on short rods to function as a moment arm. Conversely, IMUs are poor in identifying their tri-dimensional location over time. This is due to drift-related integration of gyroscopic data and being prone to disturbances induced by any presence of ferromagnetic materials in the environment. They do provide the capability to provide subtle angular measures, expressed in terms of quaternions when dedicated algorithms are applied to the nine-axis data streams. Recent work by Nazarahari *et al.* [43] expands on the requirement to establish a reliable alignment between the inertial measurement device and the corresponding anatomical frames to identify meaningful joint and segment kinematics from the acquired data streams.

With further advances in inertial measurement miniaturisation, the potential exists to fuse marker-based motion, markerless motion and quaternion data streams to draw on the benefits of each technology to further improve the sensitivity and specificity of clinical gait analysis and generic motion capture applications.

11.6 Application of computer vision

In the years prior to the end of the last century [44], researchers were identifying the potential to apply computer vision algorithms to identify, isolate and interpret features in kinematic gait data. By 2006, Moeslund *et al.* [45] reported that human motion capture and analysis continued to be an active research activity with over 350 publications in the first 6 years of the twenty-first century. They identified advances through applying novel computer vision methodologies to initialisation, motion tracking, pose estimation and movement recognition. The researchers grouped potential reviewed research into three topics: surveillance, control interfaces and analysis. Surveillance applications identified focussed on security concerns including abnormal action recognition, crowd or individual behaviours and person identification. Control applications included improvements in the human-computer interfaces through gesture recognition and motion capture for entertainment. Moeslund *et al.* identified research proposing orthopaedic automated diagnostics, analysis and optimisation of athletic performance.

Research efforts since the Moeslund *et al.* paper have developed exponentially along two inter-related paths; the application of artificial intelligence/machine learning algorithms to existing kinematic image databases and computer vision feature extraction directly from video streams to reduce the dependence on marker-based motion capture systems. Major opportunities exist within these emerging capabilities to extract tri-dimensional meaning from two-dimensional data streams that may be drawn from outdoor activities, including sporting events.

In 2006, Corazza *et al.* [46] published a methodology to correlate two-dimensional images acquired from surroundings and multiple synchronised video cameras to derive a three-dimensional visual hull reconstruction and 'a priori' model of a typically developed subject in view. They developed a simple aligned skeleton model to the visual hull. The researchers assessed tracking capability and evaluated a quantitative validation of the method in a virtual environment for a complete gait cycle. They reported achieving mean errors for knee and hip flexion, respectively, 1.5° ($\pm 3.9^\circ$) and 2.0° ($\pm 3.0^\circ$), and for knee and hip adduction, 2.0° ($\pm 2.3^\circ$) and 1.1° ($\pm 1.7^\circ$). The use of multiple video cameras to derive visual hulls remains an accepted method to acquire markerless motion capture data in a number of contemporary applications including commercial applications such as the offerings from The Captury (Saarbrücken, Germany).

11.7 Application of cameras with a depth measurement capability

In 2008, Freedman *et al.* [47] published a patent application for a new apparatus designed to project fixed patterns of spots in the infra-red spectra into a measurement volume. A co-located camera within the device captured an image of the pattern projected into the scene that was then processed to reconstruct a three-dimensional map of any observed objects where the depth information can be estimated using a triangulation method if the baseline between the camera and projector is known. The idea was commercialised into the form of many low-cost devices with the most prevalent being the Kinect ® (Microsoft Corporation, Seattle, USA) supplied as an optional extra to enhance their console video gaming experiences.

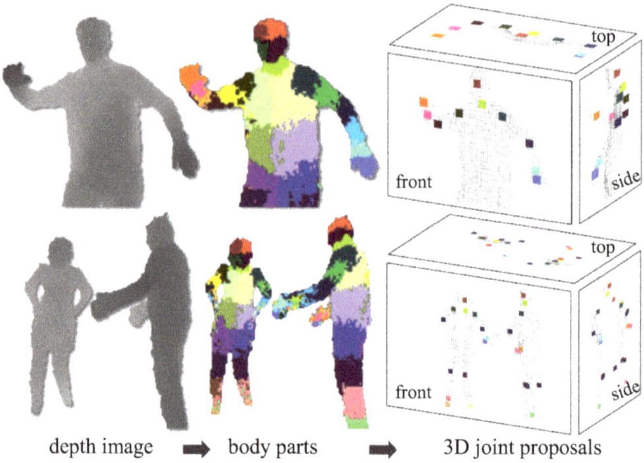

Figure 11.7 Shotton et al. *(CVPR2011)*

By 2013, Shotton *et al.* [48] had proposed a new method to estimate the tri-dimensional position of 20 body joints using the Kinect camera (Figure 11.7). The researchers synthesised a very large dataset of ground truth imagery to train a classifier to reliably predict body segment labels, invariant of pose, body shape and clothing to estimate and segment body parts in a scene. For the tracking solution, Shotton *et al.* proposed a method to predict the tri-dimensional positions of body joints from a single depth image without '*a* priori' knowledge. The researchers took an object recognition approach by designing an intermediate body parts representation that mapped the complex pose estimation problem into a per-pixel classification problem. The classification confidence-scored proposals of several body joints were then re-projected to find local optimised solutions using a randomised decision tree approach. This meant that they could embed the algorithm into a bespoke micro-chip device within the camera, not requiring access to any external computational power, but still achieving the desired tracking reconstruction speed. The practical implementation was not robust for general application as the training data was focussed towards the anterior–posterior view and no kinematic constraints were applied to restrain bone segment motion. The manufacturer launched an enhanced version, the Kinect One, in 2014 with significant hardware improvement and 25 joint skeleton tracking using a 'random forest' computer vision protocol. Time-of-flight was employed as the depth estimation method, in which the distance to an object was determined by the time taken for emitted light to illuminate an object and return to a co-located sensing element. Microsoft stopped production of the commercial device but in 2019 launched the Azure Kinect designed for industrial and research applications that can be integrated with a cloud-based library of artificial intelligence applications including image analysis. The new device employs deep learning with key points acquired from two-dimensional convolutional neural networks (CNN) and a depth image feed to derive 31 three-dimensional joint locations, orientations and temporal labels from a supporting kinematic model at around 30 frames s^{-1}. The devices can be used alone or in synchronised groups but are not standalone and do require access to specific graphical processing units (GPU) to carry out the image processing.

Albert *et al.* [49] evaluated the relative tracking performance of the Kinect One, Azure Kinect and a conventional 10 camera passive optical motion capture system (Vicon Bonita10, 39 marker Plug-in-Gait model, Vicon Motion Systems Ltd, Oxford, United Kingdom) acquiring the gait of five typically developed subjects walking on a treadmill at three different velocities (3.0 km h^{-1}, 3.9 km h^{-1} and 4.7 km h^{-1}) (Figure 11.8). They applied a bespoke spatial calibration method to align the skeleton data from each device to derive spatio-temporal gait parameters including step length, time, width and stride time. The researchers observed that the tracking accuracy of foot marker trajectories and for spatial parameters demonstrated by Azure Kinect was significantly improved over the previous offering but not for the commensurate temporal parameters, both findings were independent of treadmill velocity (Figure 11.9). They concluded that the equipment may have a place in the continuous monitoring of gait rehabilitation progression in older people without reliance on access to a clinical gait laboratory.

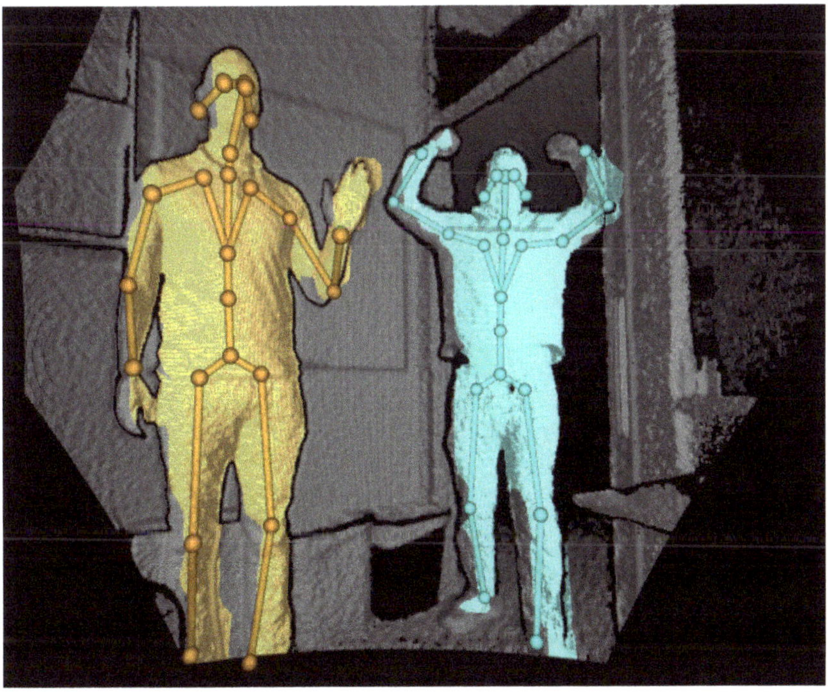

Figure 11.8 Azure kinect body tracking joint location estimations (Microsoft Corporation)

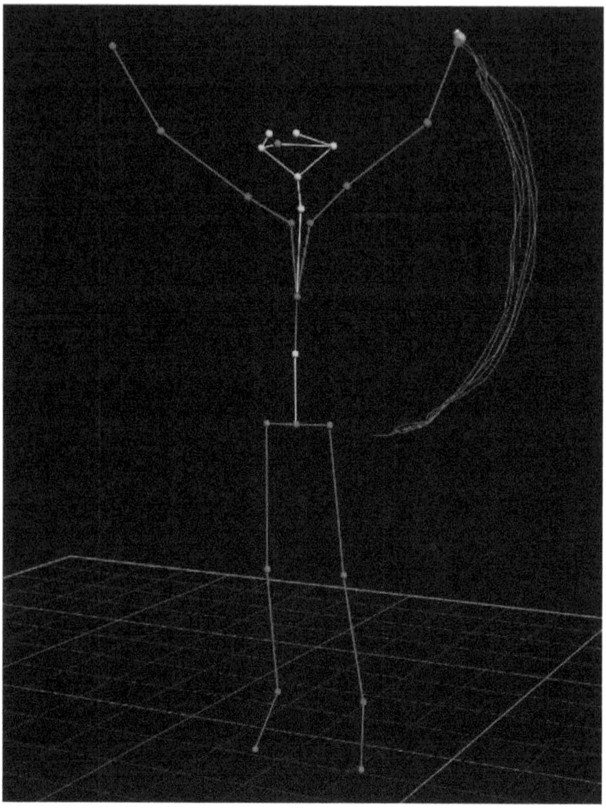

Figure 11.9 Azure kinect body tracking joints (Vicon Motion Systems Ltd)

Research continues in many centres to establish if there is a role for contemporary depth cameras within clinical gait analysis of presenting pathologies but to date, they are assessed as not suitable [50,51] for this application over conventional marker-based optical gait analysis but may well offer a simple means to measure patient performance in routine physical therapy, rehabilitation and general exercise regimes.

11.8 Deep learning and neural networks

Deep learning and neural networks in their various forms (CNN, RNN, DNN, etc.) are being rapidly applied to many different machine learning problems due to the co-incident arrival of powerful and affordable computational platforms, the ready availability of huge quantities of training data and the continuous development of new algorithms, e.g. very deep convolutional networks [52]. The technology already powers many aspects of modern society [53] from web searches to content filtering, object and scene identification, facial recognition and speech transcription to have become a ubiquitous presence in consumer products such as automobiles, cameras, smart speakers and smartphones.

Interest is forming in the academic and commercial communities to apply the technology to the problem of human pose estimation. One 2015 example from Stanford University was the work of Bearman and Dong [54] who used 180,000 training and 30,000 validation images in the MPII pose dataset (410 activities within 20 categories) plus 16,500 training and 1,000 validation images in the Leeds sports activity dataset (Figure 11.10). They achieved an 80.51% correct category classification (20 classes) but only a validation accuracy of 31% for the activity classification problem (410 classes). Other human pose estimation and action recognition research [55–60] have also been presented at well-respected computer vision conferences implying a growing interest in the topics and applications.

The problem of creating, synthesising and validating human pose datasets was addressed by Rogez and Schmid [61] who identified the significant challenge of a lack of availability of training data for their interest in planar images of humans annotated with tri-dimensional pose labels. For their study, they used the Carnegie Mellon University (CMU) Motion Capture dataset consisting of 2,500 sequences with a total of 140,000 tri-dimensional poses derived from movement analysis data [62] and the Human3.6M dataset [63] using a subset of 190,000 tri-dimensional poses also from this source. Using annotated planar pose databases, they also achieved 80,000 images for testing. Rogez and Schmid proved that they could artificially augment a tri-dimensional annotated dataset of real images with synthetic images. Research using CNN remains focussed towards pose estimation from planar imagery including by those with an interest in markerless movement analysis [64]. A particular focus of the research is towards outdoor applications such as in the measurement of sporting performance in large volumes where the use of conventional marker-based systems would not be appropriate.

Figure 11.10 *Bearman* et al. *'Human Pose Estimation and Activity Classification Using Convolutional Neural Networks' (2015)*

11.9 Markerless gait analysis

Simi Motion® (Simi Reality Motion Systems GmbH, Unterschleissheim, Germany) provides commercial offerings based on single and multiple, synchronised industrial cameras, force plates and electromyography devices to achieve two- and three-dimensional markerless gait analysis. The company uses different computer vision algorithms such as image segmentation, silhouette tracking and calibrated tri-dimensional metrology from multiple images to provide products into the gait analysis, sports biomechanics, industrial and animation market sectors.

The Theia3D (Theia Markerless Inc., Kingston, Ontario) system [65–67] employs a deep learning algorithm protocol with an array of synchronised video cameras, with known global locations and orientations, surrounding a measurement volume, to achieve a three-dimensional pose estimation of human subjects (Figure 11.11). Video data are captured and processed using bespoke software containing deep convolutional neural networks (DNN) to undertake recognition in each camera two-dimensional views so as to identify and locate specific features on a frame-by-frame basis. Extrinsic camera calibration photogrammetric parameters are used to estimate the three-dimensional location of anatomical landmarks based on the acquired features. An articulated 17 segment inverse-kinematic model is

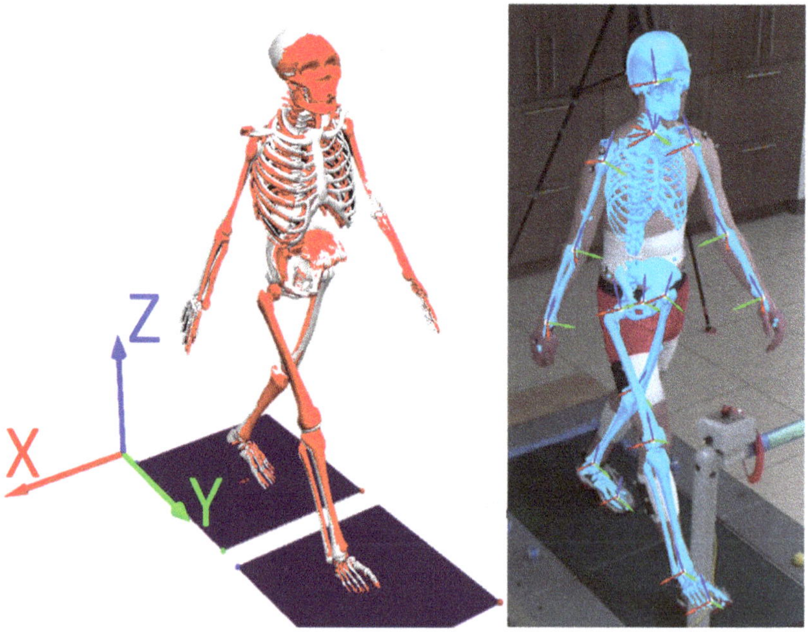

Figure 11.11 Theia Markerless Kanko et al. *'Concurrent assessment of gait kinematics using marker-based and markerless motion capture' (2021)*

applied to the landmarks to estimate the temporal and spatial pose of the subject throughout the analysis of subject motion.

In 2020, Kanko *et al.* [65] compared the metrological performance of the Theia3D system and a seven-camera Qualysis 3+, marker-based motion capture system (Qualysis AB, Gothenburg, Sweden) in the acquisition of spatiotemporal parameters used to characterise human gait patterns. The researchers synchronously acquired nine kinematics-based gait events from 30 typically developed adults performing treadmill walking at self-selected speeds. The derived parameters from each modality were compared and they reported, in using the Bland–Altman method, no systematic biases were observed. Pearson correlation coefficients between systems were found to be $r \geq 0.96$ for gait speed, cadence, step time, step length, stride length, stride width and stance time, with $r \geq 0.87$ for double-limb support and swing time. *T*-tests did identify differences in stance, swing and double-limb support parameters between the capture methodologies. Intraclass correlation coefficients (ICC) of the form (C-1) and (A-1) were calculated to establish the consistency between systems and agreement within subjects, the lowest observed being $ICC_{(A\text{-}1)}$ of 0.84 for both double-limb support and swing time. In the same year, Kanko *et al.* also published their inter-session repeatability of gait kinematics findings when using the equipment and software. The researchers used eight synchronised and calibrated cameras to acquire the kinematic patterns of multiple gait cycles captured from seven healthy participants over three sessions. They reported average inter-trial errors across all joints of 2.61° with average inter-session errors at 2.75°. These results offer encouragement to the potential feasibility of using markerless offerings firstly in the gait analysis of healthy subjects but with more peer-reviewed research and development, in the measurement of presenting pathologies.

11.10 Application of artificial intelligence and machine learning

Horst *et al.* [68] reported the application of DNN to identify the uniqueness of kinematic and kinetic features from the gait patterns of 57 typically developed adults, not exhibiting any applicable pathologies or lower extremity injuries. They employed a Layer-Wise Relevance Propagation (LRP) method to attribute portions of a model prediction back to the input ground reaction forces and joint angles. Horst *et al.*'s reported promising results using non-linear machine learning such as DNN for the analysis of the human gait. They presented the caveat that the interpretation and an understanding of any algorithmic predictions remain essential to avoid the current disadvantage of an over-reliance on the methodology in any clinical decision-making. Horst *et al.*'s approach provides a sound framework for future studies in biomechanical analysis with potential clinical applications, once the spatiotemporal parameters, kinematic and kinetic patterns of presenting gait pathologies have been fully considered. Khera and Kumar [69] reported on the results of a literature survey of research between 1980 and 2019 on the progress of the integration of machine learning algorithms to assist in the analysis and interpretation within the gait analysis domain. They identified 943 studies, concluding that 43 met their inclusion criteria. The statistical outcomes revealed that support vector machines (SVM) achieved the best classification (mean-score $= 0.87 \pm 0.07$). Kehra and Kumar's research gave insight into the potential of the application of machine learning techniques to the quantification of gait disorders that may offer value in a clinical setting.

11.11 Gait analysis in the twenty-first century

Clinical gait laboratories providing detailed kinematic, kinetic and electromyographic analyses will continue to play a central role but there will also be a supportive place for wearable devices such as IMUs, depth cameras and artificial intelligence capable applications initially for home and outdoor use but potentially in future clinical environments. The continually expanding experience and competencies in the robust measurement of human motion combined with the rapid development and improvements in models, existing and emerging technologies will ensure that the future of gait analysis remains positive. The most likely future will draw upon a fusion of data acquired from available technologies that will further enhance treatment decision-making and so benefit patients and subjects alike.

References

[1] Sutherland D.H. (2002). The evolution of clinical gait analysis. Part II: kinematics, *Gait & Posture* 16: pp. 159–179.

[2] Kadaba M.P., Ramakrishnan H.K., Wootten M.E., Gainey J., Gorton G. and Cochran G.V. (1989). Repeatability of kinematic, kinetic, and electromyographic data in normal adult gait. *Journal Orthopaedic Research* 7: pp. 849–860.

[3] Stagni R., Leardini A., Cappozzo A., Benedetti M.G. and Cappello A. (2000). Effects of hip joint centre mis-location on gait analysis results. *Journal of Biomechanics* 33: pp. 1479–1487.

[4] Sangeux M., Peters A. and Baker R. (2011). Hip joint centre localization: evaluation on normal subjects in the context of gait analysis. *Gait & Posture* 34: pp. 324–328.

[5] Kadaba M.P., Ramakrishnan H.K. and Wootten M.E. (1990). Measurement of lower extremity kinematics during level walking. *Journal of Orthopaedic Research* 8(3): pp. 383–392.

[6] Davis R.B., Õunpuu S., Tyburski D. and Gage J.R. (1991). A gait analysis data collection and reduction technique. *Human Movement Science* 10(5): pp. 575–587.

[7] Harrington M.E., Zavatsky A.B., Lawson S.E.M. and Yuan Z. (2007). Prediction of the hip joint centre in adults, children, and patients with

cerebral palsy based on magnetic resonance imaging. *Journal of Biomechanics* 40(3): pp. 595–602.

[8] Camomilla V., Cereatti A., Vannozzi G. and Cappozzo A. (2006). An optimized protocol for hip joint centre determination using the functional method. *Journal of Biomechanics* 39(39): pp. 1096–1106.

[9] Charlton I.W., Tate P., Smyth P. and Roren L. (2004). Repeatability of an optimised lower body model. *Gait & Posture* 20: pp. 213–221. doi: 10.1016/j.gaitpost.2003.09.004.

[10] Sangeux M., Pillet H. and Skalli W. (2014). Which method of hip joint centre localisation should be used in gait analysis? *Gait & Posture* 40: pp. 20–25.

[11] Wu G., Siegler S., Allard P., *et al.* (2002). ISB recommendation on definitions of joint coordinate system of various joints for the reporting of human joint motion—Part I: ankle, hip, and spine. International Society of Biomechanics. *Journal of Biomechanics* 35: pp. 543–548.

[12] Schache A.G., Baker R. and Lamoreux L.W. (2006). Defining the knee joint flexion-extension axis for purposes of quantitative gait analysis: an evaluation of methods. *Gait & Posture* 24: pp. 100–109.

[13] Sauret C., Pillet H., Skalli W. and Sangeux M. (2016). On the use of knee functional calibration to determine the medio-lateral axis of the femur in gait analysis: comparison with EOS biplanar radiographs as reference. *Gait & Posture* 50: pp. 180–184.

[14] Davis R. and DeLuca P. (1996). Clinical gait analysis: current methods and future directions. In: *Human Motion Analysis: Current Applications and Future Directions. Piscataway, NJ: IEEE Press;* pp. 17–42.

[15] Ehrig R.M., Taylor W.R., Duda G.N. and Heller M.O. (2007). A survey of formal methods for determining functional joint axes. *Journal of Biomechanics* 40: pp. 2150–2157.

[16] Chang L.Y. and Pollard N.S. (2007). Robust estimation of dominant axis of rotation. *Journal of Biomechanics* 40: pp. 2707–2715.

[17] Baker R., Finney L. and Orr J. (1999). A new approach to determine the hip rotation profile from clinical gait analysis data. *Human Movement Science* 18: pp. 655–667.

[18] Sangeux M., Barré A. and Aminian K. (2017). Evaluation of knee functional calibration with and without the effect of soft tissue artefact. *Journal of Biomechanics* 62: pp. 53–59.

[19] Firooz S., Wolf S., Simeon J., *et al.* (2020). Estimation of ankle joint parameters in typically developed adults using functional calibration methods, *Gait & Posture* 77: pp. 95–99.

[20] Leboeuf F., Baker R., Barré A., Reay J., Jones R. and Sangeux M. (2019). The conventional gait model, an open-source implementation that reproduces the past but prepares for the future. *Gait & Posture* 69: pp. 126–129.

[21] Cappozzo A., Catani F., Leardini A., Benedeti M.G. and Della C.U. (1995). Position and orientation in space of bones during movement: anatomical frame definition and determination. *Clinical Biomechanics* 10: pp. 171–178.

[22] Barre A., Thiran J.P., Jolles B.M., Theumann N. and Aminian K. (2013). Soft tissue artifact assessment during treadmill walking in subjects with total knee arthroplasty. *IEEE Transactions on Biomedical Engineering* 60 (11): pp. 3131–3140. doi:10.1109/TBME.2013.2268938.

[23] Leardini A. Sawacha Z., Paolini G., Ingrosso S., Nativo R. and Benedetti M.G. (2007). A new anatomically based protocol for gait analysis in children. *Gait & Posture* 26(4): pp. 560–571.

[24] Manca M., Leardini A., Cavazza, S., *et al.* (2010). Repeatability of a new protocol for gait analysis in adult subjects. *Gait & Posture* 32(2): pp. 282–284.

[25] Seth A., Sherman M., Reinbolt J.A. and Delp S.L. (2011). OpenSim: a musculoskeletal modelling and simulation framework for in silico investigations and exchange. *Proceedings of International Union for Theoretical and Applied Mechanics* 2 (0): pp. 212–232. https://doi:10.1016/j.piutam.2011.04.021.

[26] Leardini A. and Caravaggi P. (2016). Kinematic foot models for instrumented gait analysis. In: Müller B. *et al.* (eds), *Handbook of Human Motion*. Cham: Springer. https://doi.org/10.1007/978-3-319-30808-1_28-1

[27] Krzak J.J., Corcos D.M., Damiano D.L., *et al.* (2015). Kinematic foot types in youth with equinovarus secondary to hemiplegia. *Gait & Posture* 41(2): pp. 402–408.

[28] Stebbins J., Harrington M., Thompson N., Zavatsky A. and Theologis T. (2010). Gait compensations caused by foot deformity in cerebral palsy. *Gait & Posture* 32(2): pp. 226–230.

[29] Dubbeldam R., Nester C., Nene A., Hermens H. and Burke J. (2013). Kinematic coupling relationships exist between non-adjacent segments of the foot and ankle of healthy subjects. *Gait & Posture* 37: pp. 159–164.

[30] Rao S., Baumhauer J.F., Tome J. and Nawoczenski D.A. (2009). Comparison of in vivo segmental foot motion during walking and step descent in patients with midfoot arthritis and matched asymptomatic control subjects. *Journal of Biomechanics* 42(8): pp. 1054–1060.

[31] Chang R., Rodrigues P.A., Van Emmerik R.E. and Hamill J. (2014). Multi-segment foot kinematics and ground reaction forces during gait of individuals with plantar fasciitis. *Journal of Biomechanics* 47(11): pp. 2571–2257.

[32] Portinaro N., Leardini A., Panou A., Monzani V. and Caravaggi P. (2014). Modifying the Rizzoli foot model to improve the diagnosis of pes-planus: application to kinematics of feet in teenagers. *Journal of Foot and Ankle Research* 7(1): pp. 754.

[33] Baker R., Leboeuf F., Reay J. and Sangeux M. (2018). The conventional gait model – success and limitations. In: Müller B., Wolf S. (eds), *Handbook of Human Motion*. Cham: Springer.

[34] Scally G. and Donaldson L. (1998). Clinical governance and the drive for quality improvement in the new NHS in England. *British Medical Journal* 317: pp. 61–65.

[35] Pinzone O., Schwartz M.H., Thomason P. and Baker R. (2014). The comparison of normative reference data from different gait analysis services. *Gait & Posture* 40 (2): pp. 286–290.

[36] Mori M., MacDorman K.F. and Kageki N., (2012). The Uncanny Valley [from the field]. *IEEE Robotics & Automation Magazine* 19(2): pp. 98–100.
[37] Turner-Smith A.R., Harris J., Houghton G.R. and Jefferson R.J. (1988). A method for analysis of back shape in scoliosis. *Journal of Biomechanics* 21 (6): pp. 497–509.
[38] Shannon T.M.L. (2010). Dynamic surface topography and its application to the evaluation of adolescent idiopathic scoliosis. In: C. Aubin, I.A.F. Stokes, H. Labelle, A. Moreau (eds.), *Research into Spinal Deformities 7.* Amsterdam: IOS Press.
[39] Favre J., Aissaoui R., Jolles B.M, de Guise J.A. and Aminian K. (2009). Functional calibration procedure for 3D knee joint angle description using inertial sensors, *Journal of Biomechanics* 42(14): pp. 2330–2335.
[40] Palermo E., Rossi S., Marini F., Patanè F. and Cappa P. (2014). Experimental evaluation of accuracy and repeatability of a novel body-to-sensor calibration procedure for inertial sensor-based gait analysis. *Measurement* 52: pp. 145–155.
[41] Iosa M., Picerno P., Paolucci S. and Morone G. (2016). Wearable inertial sensors for human movement analysis. *Expert Review of Medical Devices* 13: pp. 641–659.
[42] Nazarahari M. and Rouhani H. (2018). Detection of daily postures and walking modalities using a single chest-mounted tri-axial accelerometer. *Medical Engineering and Physics* 57: pp. 75–81.
[43] Nazarahari M., Noamani A., Ahmadian N. and Rouhani H. (2019). Sensor-to-body calibration procedure for clinical motion analysis of lower limb using magnetic and inertial measurement units, *Journal of Biomechanics* 85: pp. 224–229.
[44] Moeslund T.B.E. and Granum E. (2001). A survey of computer vision-based human motion capture. *Computer Vision and Image Understanding* 81(3): pp. 231–268.
[45] Moeslund T.B., Hilton A. and Krüger V. (2006). A survey of advances in vision-based human motion capture and analysis. *Computer Vision and Image Understanding* 104: pp. 90–126.
[46] Corazza S., Mündermann L., Chaudhari A., Demattio T., Cobelli C. and Andriacchi T. (2006). A markerless motion capture system to study musculoskeletal biomechanics: visual hull and simulated annealing approach. *Annals of Biomedical Engineering* 34: pp. 1019–1029.
[47] Freedman B., Shpunt A., Machline M. and Arieli Y. (2008). Depth mapping using projected patterns. Patent Application, 10 2008. WO2008/120217A2.
[48] Shotton J., Sharp T., Kipman A., *et al.* (2013). Real-time human pose recognition in parts from single depth images. *Communications of the ACM* 56(1): pp. 116–124.
[49] Albert J.A., Owolabi V., Gebel A., Brahms C.M., Granacher U. and Arnrich B. (2020). Evaluation of the pose tracking performance of the azure kinect and kinect v2 for gait analysis in comparison with a gold standard: a pilot study. *Sensors* 20(18): p. 5104.

[50] Springer S. and Yogev S.G. (2016). Validity of the kinect for gait assessment: a focused review. *Sensors* (Basel, Switzerland) 16(2): p. 194.
[51] Ma Y., Mithraratne K., Wilson N.C., Wang X., Ma Y. and Zhang, Y. (2019). The validity and reliability of a kinect v2-based gait analysis system for children with cerebral palsy. *Sensors* (Basel, Switzerland) 19(7): p. 1660.
[52] Simonyan K. and Zisserman A. (2015). Very deep convolutional networks for large-scale image recognition. In: *The International Conference on Learning Representations.*
[53] LeCun Y., Bengio Y. and Hinton G. (2015). Deep learning, *Nature* 521: pp. 436–444.
[54] Bearman A. and Dong C. (2015). Human Pose Estimation and Activity Classification Using Convolutional Neural Networks, CS231n Course Project Reports.
[55] Tompson J.J., Jain A., LeCun Y. and C. Bregler. (2014). Joint training of a convolutional network and a graphical model for human pose estimation. In *Neural Information Processing Systems* (NIPS).
[56] Toshev A. and Szegedy C. (2014). Deep pose: human pose estimation via deep neural networks. In: *Conference on Computer Vision and Pattern Recognition*, 2014.
[57] Wang C., Wang Y., Lin Z., Yuille A. and Gao W. (2014). Robust estimation of 3D human poses from a single image. In: *Conference on Computer Vision and Pattern Recognition*, 2014.
[58] Wei S.-E., Ramakrishna V., Kanade T. and Sheikh Y. (2016). Convolutional pose machines. In: *Conference on Computer Vision and Pattern Recognition*, 2016.
[59] Yang W., Ouyang W., Li H. and Wang X. (2016). End-to-end learning of deformable mixture of parts and deep convolutional neural networks for human pose estimation. In: *Conference on Computer Vision and Pattern Recognition*, 2016.
[60] Yasin H. Iqbal U., Krüger B, Weber A. and Gall J. (2016). A dual-source approach for 3D pose estimation from a single image. In: *Conference on Computer Vision and Pattern Recognition*, 2016.
[61] Rogez G. and Schmid C. (2016). MoCap-guided data augmentation for 3D pose estimation in the wild. In: *Neural Information Processing Systems* (NIPS) 2016.
[62] http://mocap.cs.cmu.edu
[63] Ionescu C., Papava D., Olaru V. and Sminchisescu C. (2013). Human3.6m: large scale datasets and predictive methods for 3D human sensing in natural environments. *Primary Angioplasty in Myocardial Infarction* 36(7): pp. 1325–1339.
[64] Mehta D., Rhodin H., Casass D., Sotnychenko O., Xu W. and Theobalt C. (2017). Monocular 3D Human Pose Estimation Using Transfer Learning and Improved CNN Supervision.
[65] Kanko R., Strutzenberger G., Brown M., Selbie S. and Deluzio, K. (2020). Assessment of spatiotemporal gait parameters using a deep learning

algorithm-based markerless motion capture system. *bioRxiv*. 10.31224/osf.io/j4rbg.

[66] Kanko R., Laende E., Selbie S. and Deluzio K. (2020). Inter-session repeatability of Theia3D markerless motion capture gait kinematics. *bioRxiv*. 10.1101/2020.06.23.155358.

[67] Strutzenberger G., Kanko R., Selbie S. and Deluzio K. (2020). Assessment of kinematic cmj data using a deep learning algorithm-based markerless motion capture system. In: *ISBS Proceedings Archive*, 38(1): Article 218.

[68] Horst F., Lapuschkin S., Samek W. *et al.* (2019). Explaining the unique nature of individual gait patterns with deep learning. *Nature.com, Science Report* 9, 2391. https://doi.org/10.1038/s41598-019-38748-8.

[69] Khera P. and Kumar N. (2020). Role of machine learning in gait analysis: a review, *Journal of Medical Engineering & Technology* 44(8): pp. 441–467.

Index

www.ingramcontent.com/pod-product-compliance
Lightning Source LLC
LaVergne TN
LVHW051049210926
848486LV00001B/51
9781839531316